Geology of the
Coyote Mountains, Southern California

2015 SDAG FIELD TRIP PARTICIPANTS

On outcrop: Greg Cranham, Brian White, Jean Moran, Ann Bykerk-Kauffman, Germar Bernhard, Annie Hoppe, JR Morgan, Diane Murbach, Monte Murbach, Matthew Bula, Dave Bloom, Sean Lucas

3rd row: Scott Snyder, Brian Olson, George Morgan, Monte Marshall, Diana Piolo, Sean Alperson, Karen Young, Don Cooke

2nd row: Jen Morton, Bob Stroh, Sam Eisenberg, Christine Eisenberg, Rupert Adams, Carole Ziegler, Mike Jensen, Thomas Lopez, Lowell Lindsay, Norrie Robbins, Jerry Treiman, Robert Thiele, Steve Jacobs

Front row: Brad Eisenberg, James Morton, Izzy Tihanyi, Isabelle Wirths, Todd Wirths, Joe Corones, Chris Buchen, Randy Wagner

Not shown: Benji Morton, Phil Farquharson, Donna Gooley, Jeff Hull, Chris Lynch, Tek Olson, Mary Walke

Photo: Jennifer Morton

Participants on Andrade Canyon hike, January 9, 2016. See Field Trip D, Andrade Canyon 6 Mile Geological Hike, page 55. Photo: Josselyn Molina

GEOLOGY OF THE COYOTE MOUNTAINS

SOUTHERN CALIFORNIA

Published in conjunction with the

2015 Field Trip of the

San Diego Association of Geologists

November 6-8, 2015

RANDALL WAGNER, *Editor*

SAN DIEGO ASSOCIATION OF GEOLOGISTS

Distributed by Sunbelt Publications
www.sunbeltbooks.com

Geology of the Coyote Mountains, Southern California

Copyright © 2015 San Diego Geological Society, Inc.

Individual papers, photos, and artwork copyright by authors and used with permission. All rights reserved.

First edition 2015, fifth printing 2022.

Edited by Randall Wagner
Cover and Book Design by Michael Schrauzer
Project Manager: David M. Bloom
Printed in United States of America

San Diego Association of Geologists
3130 North Evergreen Street
San Diego, California 92110
www.sandiegogeologists.org

Sunbelt Publications
664 Marsat Court, #A
Chula Vista, California 91911
(619) 258-4911, fax: (619) 258-4916
www.sunbeltpublications.com

25 24 23 22   8 7 6 5

Library of Congress Cataloging-in-Publication Data

Names: Wagner, Randall (Geologist), editor. | San Diego Association of Geologists.
Title: Geology of the Coyote Mountains, southern California / editor, Randall Wagner.
Description: First edition. | San Diego, California : San Diego Association of Geologists : Sunbelt Publications, 2015. | "Published in conjunction with the 2015 field trip of the San Diego Association of Geologists, November 6-8, 2015."
Identifiers: LCCN 2015041942 | ISBN 9781941384213 (softcover : alk. paper)
Subjects: LCSH: Coyote Mountains (Calif.) | Coyote Mountains (Calif.)--Guidebooks. | Geology--California--Coyote Mountains.
Classification: LCC QE90.S116 G457 2015 | DDC 557.94/99--dc23 LC record available at http://lccn.loc.gov/2015041942

San Diego Geological Society, Inc., gratefully acknowledges the generous contributions of our sponsors to the publication of this book.

Front Cover Photograph: From eastern flank of the Coyote Mountains above Painted Gorge looking north to the Fish Creek Mountains. Yellow east-dipping Pliocene mudstones of the upper Imperial Group dominate the Carrizo Badlands between the two ranges. Photograph by Dave Bloom, 10-7-2015.

Inside Front Cover: Geologic map of the Coyote Mountains, excerpted from two Thomas W. Dibblee, Jr., maps. Eastern portion from Geologic map of the Plaster City and Brawley 15 minute quadrangles, Imperial County, California, by Thomas W. Dibblee, Jr., 2008, (field work, 1943), Dibblee Geology Center Map #DF-406: First printing, September 2008, edited by John A. Minch (George C. Copenhaver Jr. Honorary Map). Western portion from Geologic map of the Mount Laguna and Carrizo Mountain 15 minute quadrangle, San Diego and Imperial Counties, California, by Thomas W. Dibblee, Jr., 2008 (field work, 1943-4), edited by John A. Minch, Dibblee Geology Center Map #Df-407: First printing, September 2008 (Thomas K. Rockwell Honorary Map). Explanation of mapped units on fold out, last page.

Inside Back Cover: Map of field trip routes and stops by Dave Bloom.

Back Cover: Geologic map of the Coyote Mountains, excerpted from "Geologic map of Imperial County showing mines and mineral deposits," compiled by Paul K. Morton (1966), Scale 1:250,000, Plate 1 of Morton, P. K., 1977, Geology and mineral resources of Imperial County, California: California Division of Mining and Geology, County Report No. 7. Explanation of mapped units on fold out, last page.

PLEASE READ: This book describes driving and hiking situations and other potentially hazardous activities that are the sole responsibility of the reader. Exploring geological features in mixed terrain from remote desert wilderness to urban environments is not without risk. All who engage should be aware of their own capabilities, be knowledgeable, prepared, and alert. When exploring such environments, please abide by laws of the road, respect private property, and don't forget about common sense. The conditions, access, and features described in this book can and do change with time.

Table of Contents

Introduction and Acknowledgments

THIS GUIDEBOOK ACCOMPANIES THE SAN DIEGO Association of Geologists' 2015 field trip and covers the geology of the Coyote Mountains and general vicinity of the Western Salton Trough area of Imperial County, California. This book is useful for anyone interested in the unique and complex geology of the Coyote Mountains which appears strikingly different from what is typically seen in the geology just a few miles away. In just a few square miles so much geology is exposed: Paleozoic metamorphic rocks of a passive continental margin, Mesozoic tonalite and other igneous rocks from a magmatic arc, and Cenozoic rift continental red beds and volcanic rocks overlain by marine, fluvial, and pluvial sequences. All these capped by fanglomerates and cut by faults, some of which are active. I am not aware of any location in the region where such diverse lithologies and so much geologic time is so richly represented in such a small area.

The region discussed in this publication and covered by the multi-day field trip consists of the area along the southeastern flank of the Coyote Mountains that is bounded by the Elsinore fault zone; the eastern flank of the mountain range and the badlands to the east that has spectacular volcanic features and left-lateral faults transferring slip from the Elsinore-Pescadores fault zone to the San Jacinto fault zone; and the top of the Coyote Mountains with impressive views of the central and western portions of the Coyotes and the surrounding mountain ranges and basins of the Western Salton Tough region.

This book has three main sections: road logs, new papers, and reprints of associated geologic papers. The road log follows the route of both trips and provides instructions and useful information for self-guided driving and hiking tours through the Coyotes. The new papers section includes an updated new paper by George Morgan and JR Morgan based on the extensive work in the Coyotes over the last dozen or so years, and excellent new papers by other local geologists and scientists. The reprinted papers section represents a collection of applicable previously published geology-related papers that cover the Coyote Mountains region.

The 2015 SDAG field trip will begin at the new Ocotillo Community Park located on Imperial Highway/S2 just outside Ocotillo to the west. The Ocotillo Community Park will also be our camp site and the starting point each day. This year's field trip includes a day on a rugged jeep trail up to the top of the mountain range that requires a true four wheel drive vehicle with high-ground clearance. Refer to the road log for additional details and field trip stops.

In planning the field trip, I would like to thank the 2015 field trip committee and specifically the following people: George and JR Morgan for their extensive geologic mapping and knowledge of the Coyotes and for providing a significant amount of the data used in this guidebook and field trip stops; Dr. Ann Bykerk-Kauffman of California State University, Chico for making available her and her students' geologic maps and field interpretations; Dr. Tom Rockwell of SDSU for his notable expertise; David Bloom for his considerable support and seemingly infinite efforts during our numerous planning trips out to the Coyotes (even during the summer months) and during the editorial process in completing this guidebook; Greg Cranham for his valuable knowledge of field trip logistics; Lowell Lindsay for his superb editorial and publication support and passion for the written word; to my cohorts on the SDAG Committee; anyone else coerced at the last minute; and finally to my wife, Diane, and daughter, Liesl, for allowing me the time on this tremendous project.

As Walter Mendenhall of the USGS stated in 1910 and still holds true today "it is to be said that but the barest outlines of the history of this fascinating region are yet known. Its rocks contain a rich upper Miocene fauna, probably in large part new, and its exposures and structures are so clear that its geology will be an open book to the fortunate student to whom falls the pleasant task of deciphering it in detail."

Randall Wagner, CEG
LGC Valley Geotechnical
SDAG 2015 *Vice President and Field Trip Chair*

In Memoriam: **David Huntley**

DR. DAVID HUNTLEY, PROFESSOR EMERITUS, SDSU Dept. of Geological Sciences, passed away Saturday, July 4, from breathing complications related to ALS (Lou Gehrig's Disease). He is survived by his wife, Linda (Lina) Clark, daughter, Sabrina Parker, and grandson, Jake Parker. For the past 30 years Dr. Huntley has been a principal technical expert for San Diego County's groundwater regulations and various technical issues. In 2014 the Groundwater Resources Association of California presented him with a Lifetime Achievement Award in recognition of his contributions to environmental science and his impact on a generation of hydrogeologists, and was also honored by being named the 2013 David Keith Todd Distinguished Lecturer. The San Diego County Board of Supervisors proclaimed March 11, 2014 "Dr. David Huntley Day."

Dr. Huntley earned his PhD from the Colorado School of Mines in 1976 and subsequently taught at the University of Connecticut then San Diego State University from 1978–2005. During his tenure he supervised over 60 Masters' theses and taught groundwater courses to more than 1,000 undergraduates. Over the years he received numerous job offers from industry, but his response was always the same: "It would be exponentially better for them [industry] if I continued my research while teaching students to be excellent groundwater hydrologists." He was known as a demanding professor who approached hydrogeology in a mathematically rigorous and challenging manner, but his students often became his friends for life and frequently give him credit for their successful careers.

In addition to being a highly respected instructor, his groundbreaking work on the movement of petroleum hydrocarbons changed how practitioners viewed the effectiveness of certain remediation practices. He provided quantitative measures of potential recovery rates via simple tests, methods that were adopted by industry and continue to inspire research. Dave also served as associate editor of the professional journals *Groundwater* and *Groundwater Monitoring and Remediation*, and as a private consultant for both industry and regulatory agencies throughout the United States.

Dave's life was not defined by his career however. Growing up in San Mateo, California, he became a lifeguard, scuba diver, and wrestler, and earned a black belt in judo. Later, his passions expanded to surfing, downhill skiing, sailing, racing sailboats and cars, photography, and traveling all over the world. Many San Diegans know him for his involvement with triathlon and the Challenged Athletes Foundation (CAF), for which he raised tens of thousands of dollars to help wounded warriors and challenged athletes of all ages to lead healthy and rewarding lives. In 2011 he added to his long list of achievements, completing both the Boston Marathon and Ironman New Zealand. He completed an Ironman 70.3 in St. George, Utah, in 2013, even after he suspected he had ALS. He was officially diagnosed in June 2013, but nevertheless completed a one mile ocean swim, continuing his support of the Challenged Athletes Foundation four months later.

In addition to fundraising for CAF, Dave donated generously to numerous charities throughout his life. The consummate coach, he was a mentor to many whether instructing them to ski, swim, sail, or compete in their first triathlon. Dave was extraordinarily capable, generous, a genuine friend, the voice of reason, and comforted many with his words of wisdom. His many friends and family will miss his passion, his compassion, his biting wit, his intellect, and his knowledge.

Dave is a founding member of the global grassroots campaign to urge the FDA to use Accelerated Approvals (AAP) for promising drugs to treat terminal illnesses like ALS (www.change.org/ALSMedication). Up until his death he remained active as a science advisor to the organization he and his wife helped create, Hope NOW for ALS. His wife continues to be a leader in this organization, which is meeting with the FDA to foster pathways that will accelerate the pace of drug discovery and access to those affected by ALS, who have an average life expectancy of just 2–5 years.

Last but not least, we gratefully acknowledge Dr. Huntley's service as president of the San Diego Association of Geologists for 1982.

—SDAG 2015 Field Trip Committee:
Randall Wagner, Dave Bloom, Greg Cranham,
Monte & Diane Murbach, George Morgan,
JR Morgan, and Lowell Lindsay.

Sponsors

DIAMOND SPONSORS ($1000)

H & P Mobile Geochemistry, Inc.
(800) 834-9888
www.handpmg.com

Hargis + Associates, Inc.
www.hargis.com

RUBY SPONSORS ($500)

David Bloom/Tetra Tech

Crux Subsurface, Inc.

Geocon, Inc.

EMERALD SPONSORS ($100)

Dr. Pat Abbott, *SDSU Prof. of Geology, Emeritus*
James R. Ashby, *Mission Geoscience, Inc.*
Marty and Sherry Bloom
Curtis Burdett
Joe Corones, *City of San Diego*
Greg Cranham
Eco Rental Solutions
Dr. Margaret Eggers, CHG, *Eggers Environmental, Inc.*
Phil Farquharson, *CG-Squared Productions*
Carolyn Glockhoff, *Caro-Lion Enterprises*
Cari Gomes
Leighton Group
Lowell A. Lindsay and Diana Lindsay, *Sunbelt Publications/AESE*
Dr. Monte Marshall, *SDSU Prof. of Geology and Geophysics, Emeritus*
John Minch
JR Morgan
Monte and Diane Murbach, *Murbach Geotech*
Property Matrix — Property Management Software
Les Reed, *Geotechnical Exploration Inc.*
Scott Snyder, *Snyder Geologic*
Gerald Shiller
Southwest Geophysics, Inc.
David and Jan Steller
Dr. Anne Sturz
Mary Walke
Carole L. Ziegler

The generosity of these sponsors allows our organization to keep membership dues low, reduce rates for students, produce monthly newsletters, purchase supplies, host the annual field trip, produce our geologic guidebooks, and provide scholarships to local geology students.
To contact any of the above sponsors, please visit www.sandiegogeologists.org.

Officers

SAN DIEGO ASSOCIATION OF GEOLOGISTS

PRESIDENT
Jennifer Morton

VICE PRESIDENT
Randy Wagner

SECRETARY
Rupert Adams

TREASURER
Chris Livesey

PUBLICATIONS
Lowell Lindsay

WEBMASTER
Carolyn Glockhoff

SAN DIEGO GEOLOGICAL SOCIETY, INC.

PRESIDENT
Dave Bloom

VICE PRESIDENT
Mike Hart

SECRETARY
Diane Murbach

TREASURER
Greg Cranham

DIRECTOR
Lowell Lindsay

DIRECTOR
Monte Murbach

DIRECTOR (*ex officio*)
Jennifer Morton

San Diego Association of Geologists is a program of San Diego Geological Society, Inc., a California 501(c)(3) nonprofit corporation.

San Diego Association of Geologists Publications

1973 *Studies on the Geology and Geologic Hazards of the Greater San Diego Area, California*, Arnold Ross and Robert J. Dowlen, eds.
1974 *Recent Geological and Hydrologic Studies, Eastern San Diego County and Adjacent Areas: A Guidebook*, Michael W. Hart and Robert J. Dowlen, eds.
1975 *Studies on the Geology of Camp Pendleton, and Western San Diego County, California*, Arnold Ross and Robert J. Dowlen, eds.
1977 *Geology of Southwestern San Diego County and Northwestern Baja California*, Gregory T. Farrand, ed.
1978 *Natural History of the Coronado Islands, Baja California, Mexico*, Herman T. Kuper, ed.
1979 *Earthquakes and other Perils, San Diego Region*, Patrick L. Abbott and William J. Elliott, eds. *(GSA field trip guide)*
1981 *Geologic Investigations of the Coastal Plain, San Diego County, California*, Patrick L. Abbott and Shannon O'Dunn, eds.
1982 *Geologic Studies in San Diego*, Patrick L. Abbott, ed.
1985 *On the manner of Deposition of Eocene Strata in Northern San Diego County*, Patrick L. Abbott, ed.
1988 *Landslides in Crystalline Basement Terrain*, James R. Evans, ed.
1989 *Field Trip Road Log: Fault Features between La Jolla, U.S.A. and Ensenada, Mexico*, by Diane Murbach: *in* The Seismic Risk in the San Diego Region.
1990 *Geotechnical Engineering Case Histories in San Diego County*, John H. Hoobs, ed.
1991 *Environmental Perils, San Diego Region,* Patrick L. Abbott and William J. Elliott, eds. *(for a GSA annual meeting).*
1992 *Natural History of the Coronado Islands, Baja California, Mexico* (revisited 1992), Lyne Perry, ed.
1993 *Colorado Desert and Salton Trough Geology*, Joe Corones, ed.
1994 *Geology and Natural History of Camp Pendleton, U.S.M.C. Base, San Diego County, California*, Phillip S. Rosenberg, ed. *(*2nd Edition, 2010*).*
1995 *Paleontology and Geology of the Western Salton Trough Detachment, Anza-Borrego Desert State Park, California*, Paul Remeika and Anne Sturz, eds.
1996 *Geology and Natural Resources of Coastal San Diego County, California*, Tissa Munasinghe and Phillip Rosenberg, eds.
1997 *Santa Cruz Island: Geology Field Trip Guide*, James R. Boles and Werner Landry, eds.
1998 *Geology and Geothermal Resources of the Imperial and Mexicali Valleys*, Lowell E. Lindsay and William Hample, eds.
1999 *Water for Southern California: Water Resources Development at the Close of the Century*, Gregory T. Cranham, ed.
2000 *Geology and Enology of the Temecula Valley, Riverside County, California*, Barbara B. Birnbaum and Kerry D. Cato, eds.
2001 *Coastal Processes and Engineering Geology of San Diego, California*, Robert C. Stroh, ed.
2003 *Geology of the Elsinore Fault Zone, San Diego Region*, Monte L. Murbach and Michael W. Hart, eds. *(joint publication with SCGS)*
2004 *Mining History and Geology of Joshua Tree National Park*, Margaret R. Eggers, ed.
2005/06 *Geology and History of Southeastern San Diego County, California*, David M. Bloom, Phillip T. Farquharson, and Carole L. Ziegler, eds.
2006 *Mines and Geology of the Randsburg Area, An Historical Gem of the Mojave Desert*, by D.D. Trent.
2008/09 *Palms to Pines: Geological and Historical Excursions through the Palm Springs Region*, Jeff Miller, Bryan Miller-Hicks, and Scott Snyder, eds. *(joint publication with SCGS).*
2010 *Geology and Lore of the Northern Anza-Borrego Desert Region*, Monte L. Murbach and Charles C. Houser, eds. *(joint publication with SCGS).*
2011 *Picacho and the Cargo Muchachos*, Todd Wirths, ed.
2012 *Roadside Geology along Sunrise Highway*, by Michael J. Walawender
2012 *Waiting for Tsunami, Coastal Hazards of Northern San Diego County*, Cari Gomes, ed.
2013 *San Luis Rey on Display, Geoscience in Northern San Diego County, California*, Brian J. Olson, ed.
2014 *Coast to Cactus: Geology and Tectonics, San Diego to Salton Trough, California*, Jennifer Morton, ed.

For more information, visit www.SanDiegoGeologists.org.

Geologic Time Scale

EON	ERA	PERIOD (SYSTEM)		EPOCH (SERIES)	Age estimates of boundaries (in mega-annum [Ma] unless otherwise noted)
Phanerozoic	Cenozoic	Quaternary (Q)		Holocene	11,700 ± 99 yr
				Pleistocene	2.588
		Tertiary (T)	Neogene	Pliocene	5.332 ± 0.005
				Miocene	23.03 ± 0.05
			Paleogene	Oligocene	33.9 ± 0.1
				Eocene	55.8 ± 0.2
				Paleocene	65.5 ± 0.3
	Mesozoic	Cretaceous (K)		Late	99.6 ± 0.9
				Early	145.5 ± 4.0
		Jurassic (J)		Late	161.2 ± 4.0
				Middle	175.6 ± 2.0
				Early	199.6 ± 0.6
		Triassic (TR)		Late	228.7 ± 2.0
				Middle	245.0 ± 1.5
				Early	251.0 ± 0.4
	Paleozoic	Permian		Lopingian / Guadalupian / Cisuralian	299.0 ± 0.8
		Carboniferous	Pennsylvanian	Late / Middle / Early	318.1 ± 1.3
			Mississippian	Late / Middle / Early	359.2 ± 2.5
		Devonian		Late / Middle / Early	416.0 ± 2.8
		Silurian		Pridoli / Ludlow / Wenlock / Llandovery	443.7 ± 1.5
		Ordovician		Late / Middle / Early	488.3 ± 1.7
		Cambrian		Late / Middle / Early	542.0 ± 1.0
Precambrian					~4600

Reference: U.S. Geological Survey Geologic Names Committee, 2010, Divisions of geologic time—major chronostratigraphic and geochronologic units: U.S. Geological Survey Fact Sheet 2010–3059, 2 p.

I. ROAD LOGS

2015 SDAG FIELD TRIP
GEOLOGY OF THE SOUTHERN COYOTE MOUNTAINS

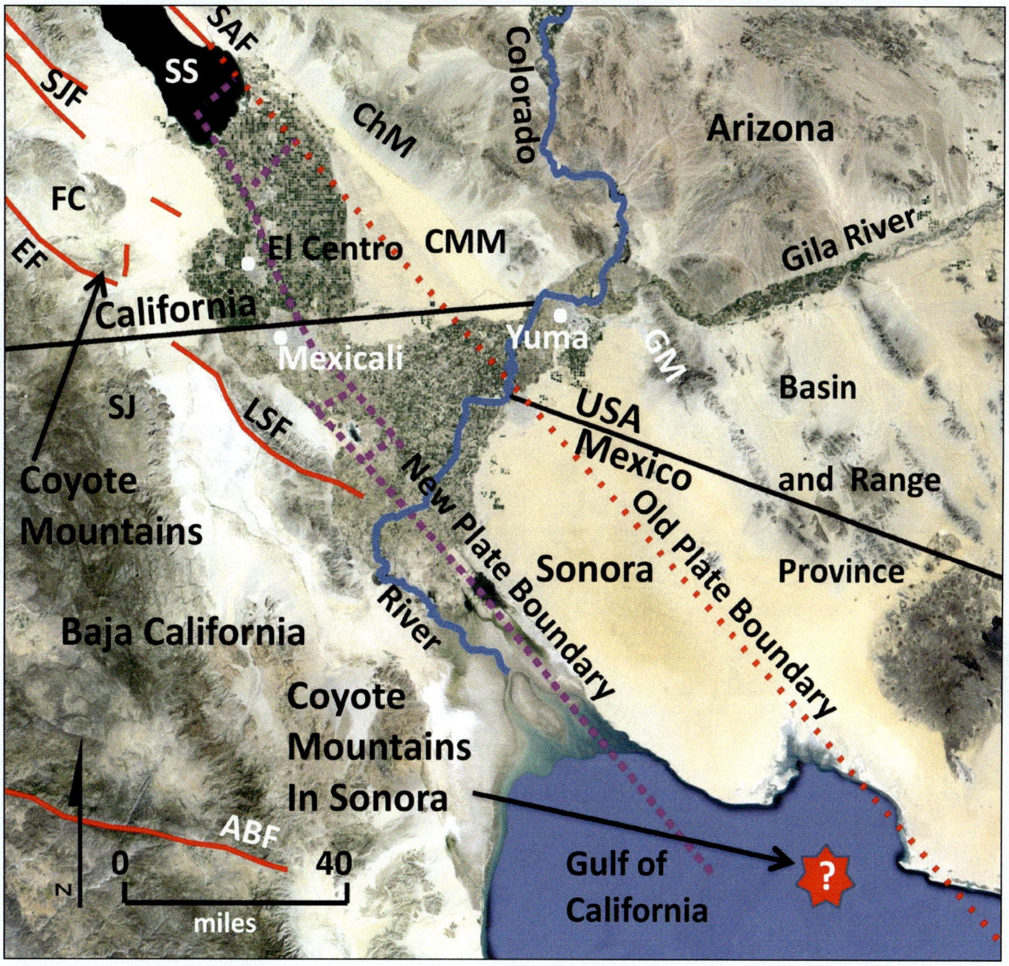

ABF-Agua Blanca Fault,**CMM**-Cargo Muchacho Mountains,**ChM**-Chocolate Mountains, **EF**-Elsinore Fault, **FC**- Fish Creek Mountains, **GM**-Gila Mountains, **LS**-Laguna Salada, **LSF**-Laguna Salada Fault, **SAF**-San Andreas Fault, **SJF** San Jacinto Fault, **SJ**- Sierra Juarez, **SS**-Salton Sea. **Queried star** is the paleogeographic location of the Coyote Mountains 5.5 Ma (Winker and Kidwell, 1996).

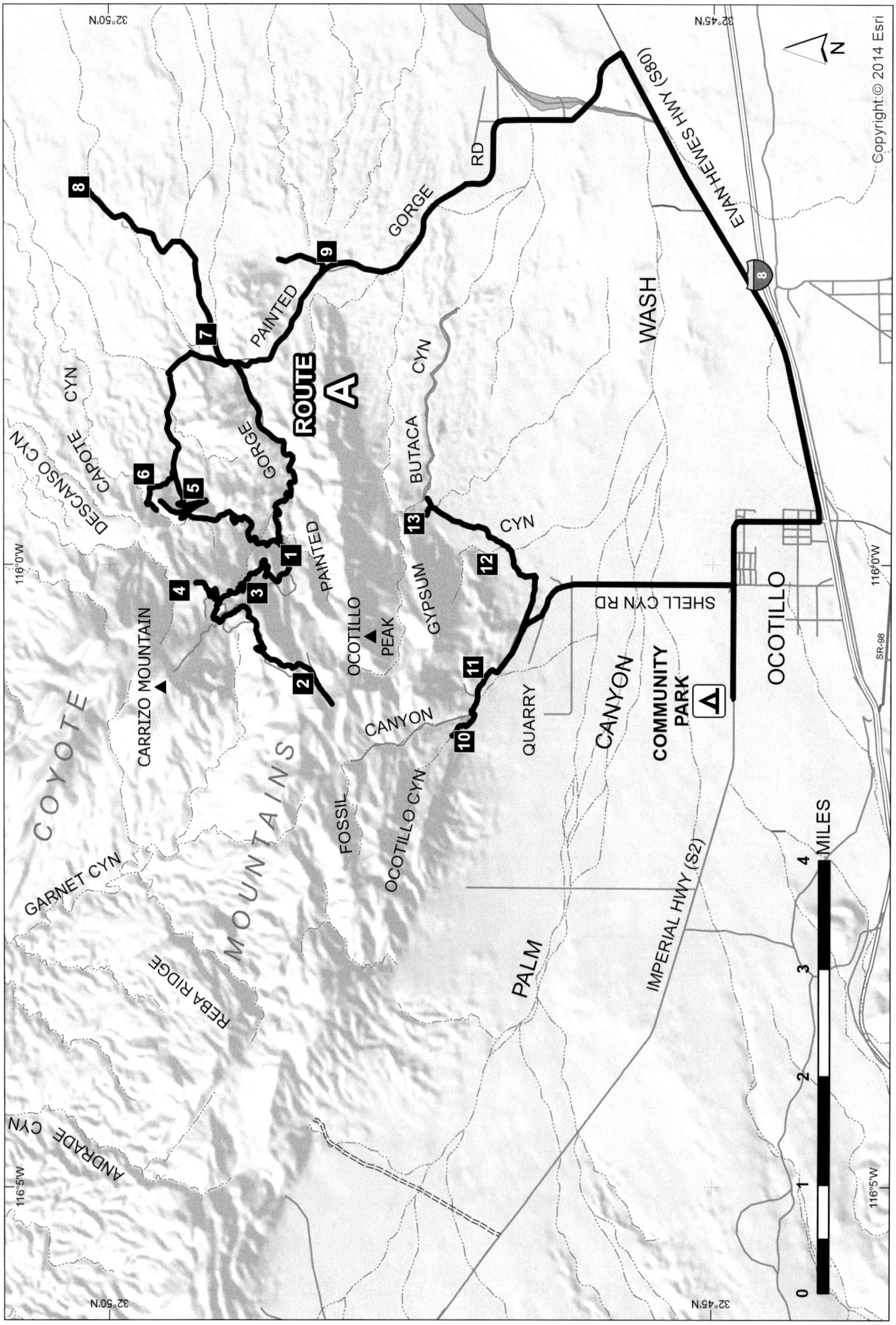

Index Map: Route A

FIELD TRIP A:
Painted Gorge Loop and Alverson and Butaca Canyons Loop

GEORGE MORGAN, ***Field Trip Leader***
JR MORGAN, ***Co-Field Trip Leader***

THE COYOTE MOUNTAINS OF SOUTHWESTERN Imperial County, California, provide evidence for the tectonic evolution of the Gulf of California and Salton Trough for the last roughly 6 million years. The range also shows evidence for previous continental extension that started in the early Miocene.

This road log covers the southern Coyote Mountains from Painted Gorge up to the crest of the range near the summit of Carrizo Mountain, and back to the lower flanks to Fossil Canyon and Butaca Canyon (Figure 1, overleaf). The stratigraphy and structural relations are well exposed in this arid and nearly unvegetated range. The Coyote Mountains are a young range, uplifted by tectonic forces perhaps only in the last million years. Erosion has cut deeply into the range revealing formations down to crystalline basement, but enough geologic strata remain to adequately interpret its history.

Upon initial inspection of the range, it might appear that the latest period of faulting and deformation has rendered the stratigraphic and structural relations so complex as to be beyond rational interpretation. However, the authors show that by carefully mapping the area at large scale, such relationships can be successfully worked out.

The purpose of this road log is to interpret an area of structural complexity into a coherent framework, make observations that have not been previously reported or have been underreported, and present findings that differ from those that have been previously published.

The core of the range is underlain by pre-Cenozoic metamorphic and igneous crystalline basement. Coarse fanglomerate sequences and a complex of basaltic, andesitic, and rhyolitic volcanic rocks of Miocene age form the lower Split Mountain Group. The overlying Miocene to Pliocene Imperial Group consists of sequences of fresh water fanglomerates intermingling with marine sandstones and mudstones. Miocene and Pliocene marine fossils can be found in abundance in Imperial strata. The Imperial is overlain by nonmarine sedimentary units of the Palm Spring Group.

A note to those attempting to follow this road log on their own. There are two gates, one in Painted Gorge and one north of Painted Gorge. The Bureau of Land Management closes and locks the gates from January 1 to June 30 for lambing season. The roads beyond the gates are challenging and should be only attempted by those experienced drivers with high clearance, 4WD vehicles.

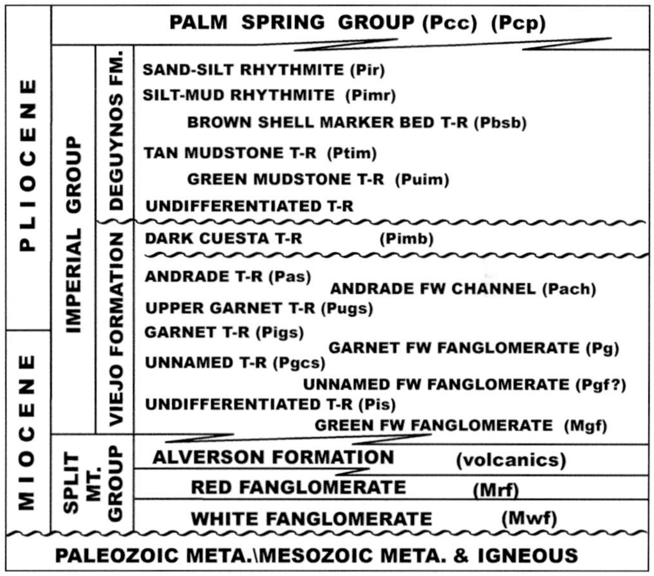

Figure 2. Our working simplified columnar section for the southern Coyote Mountains. Wavy lines are unconformities. All transgression-regression (T-R) sequences are marine, and **FW** is freshwater. Viejo Formation is a newly proposed unit for the lower Imperial Group in the Coyote Mountains (Morgan and Morgan this volume). Letters in parentheses are map designations.

The authors are mapping the southern Coyote Mountains using a working stratigraphic section that has evolved since their first field season in 2001. Figure 2 illustrates a simplified version of this section. Previous workers who produced maps in the area assigned some units to different intervals, grouped them differently, or used different unit names, leading to confusion. Figure 3 provides a simplified comparison of the major units found in the Coyote Mountains by Dibblee (1954 and 2003) who first mapped in 1943, Christianson (1957), Winker and Kidwell (1996), and Morgan and Morgan (this volume).

One of the units that remains difficult to interpret is the Alverson Formation, which consists largely of a volcanic complex that includes flows, tuffs, cinder beds, plugs, and other features. Attempts to carry individual beds from the Alverson (Fossil) Canyon area, to the Butaca and Gypsum Canyon area, or on to the Painted Gorge area have been, so far, unsuccessful (Figure 4).

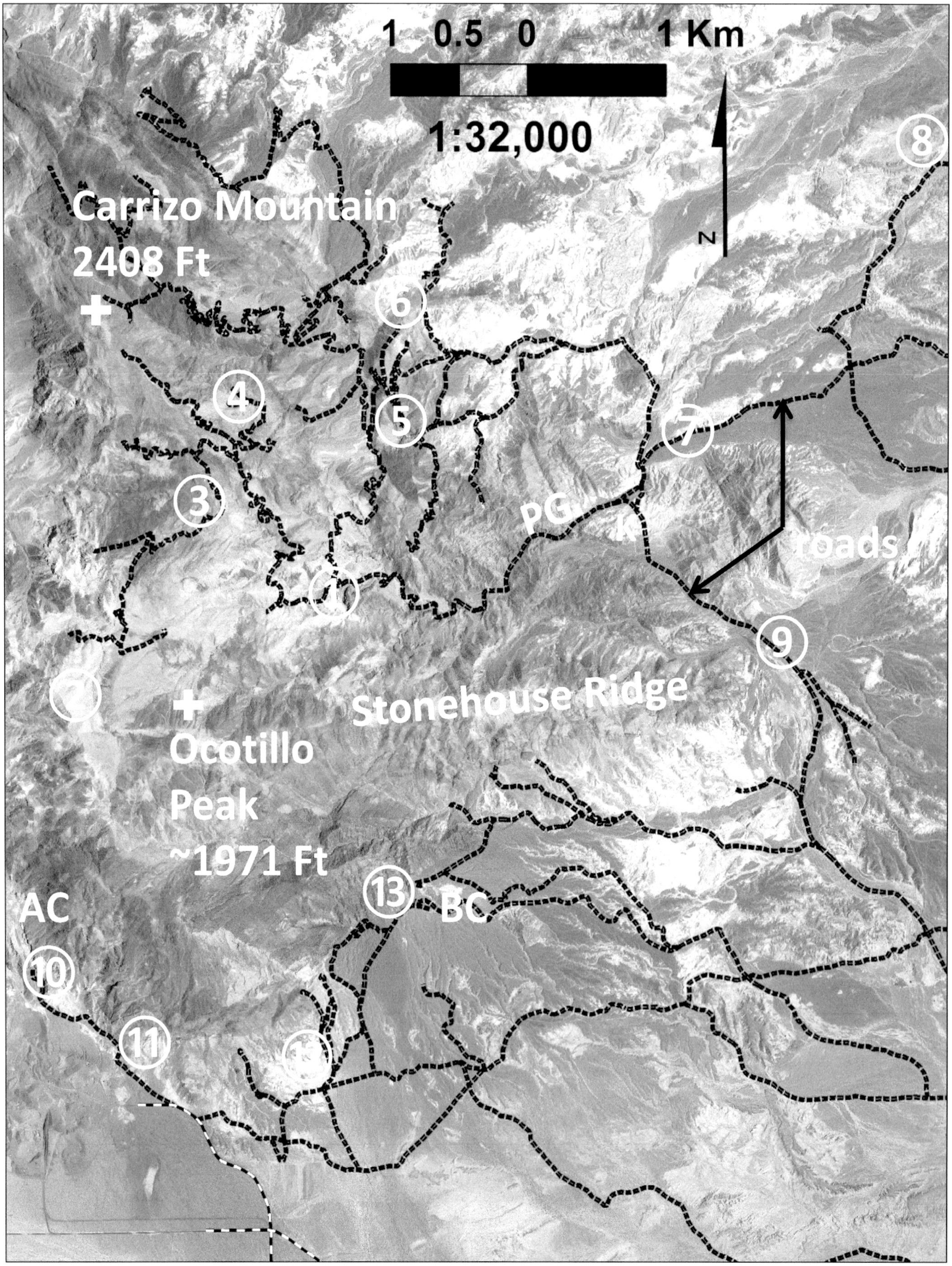

Figure 1. Numbers are Field Trip Stops. **AC**-Alverson Canyon, **BC**-Butaca Canyon, **K**-kiosk, **PG-**Painted Gorge.

Dibblee, 1954	Christensen, 1957	Dibblee, 2003	Winker and Kidwell, 1996	Morgan and Morgan, 2015
Palm Spring Formation	Garnet Fm.	Palm Spring Formation	Palm Spring Group	Palm Spring Group
	Canebrake Fm.			
Imperial Formation	Palm Spring Formation	Imperial Formation (Garnet)	Imperial Group: Deguynos Formation	Imperial Group: Deguynos Formation
Split Mountain Cgl. (Garnet)	Imperial Formation		Imperial Group: Latrania Formation	Imperial Group: Viejo Formation (Garnet)
			Split Mountain Group: Garnet Formation	Split Mountain Group: Alverson Formation
Alverson Volcanics	Alverson Volcanics	Alverson Volcanics	Split Mountain Group: Alverson Formation	Split Mountain Group: Red Fanglomerate
Red Rock Beds	Split Mountain Formation	Anzo Red Bed	Split Mountain Group: Red Rock Formation	Split Mountain Group: White Fanglomerate
basement	basement	basement	basement	basement

Figure 3. Edited columnar sections for units used in the Coyote Mountains from Dibblee (1954, 2003), Christensen (1957), and Winker and Kidwell (1996). Morgan and Morgan (2015) shows our working columnar section for the Coyote Mountains. **(Garnet)** designates units that contain Garnet Formation rocks. Only relative age of units shown. **Fm**-formation, **Cgl**-conglomerate.

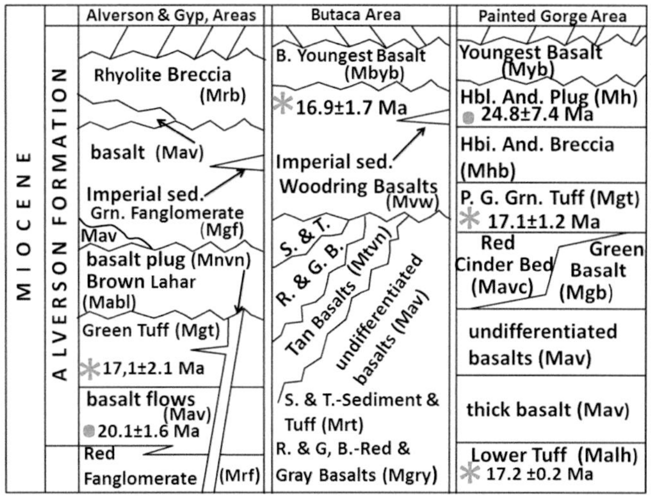

Figure 4. Our working simplified columnar section for the Alverson Formation. We have been unable to find a marker bed between any of the three areas. Letters in parentheses are map designations. Grey asterisk indicates U-Pb date from zircons. Grey asterisk in the Butaca Area Woodring basalt (Morgan and Morgan) is a detrital zircon date. Grey dots indicate K-Ar dates from Ruissard (1979). Hbl.-hornblende, And.- andesite.

DAY ONE

0.0 MILE The kiosk in Painted Gorge drainage is the starting point for this road log (Figure 5). All figures show simplified geology. Set your odometers now. The Painted Gorge drainage runs from west to east and out through the narrows to Painted Gorge Wash. On the skyline due west are two hills with a saddle between them. The hill to the south is Winker and Kidwell (1996) Marble Mountain. The northern hill has an outcrop of light tan Andrade marine calcarenite (Pas), and with binoculars you can see a road on the Andrade outcrop. Stop 3, Figure 1, is on the Andrade outcrop.

Closer in and to the east, the high hills are the Eastern Hornblende Andesite plug (Mh) and to the northwest the high hills to the northwest comprise the Western Hornblende Andesite plug. These two plugs are part of the Alverson Formation and have similar composition and internal structures. To date it is unknown whether these are two independent plugs, represent a bifurcated plug, or one plug that has been faulted right-laterally or for that matter left-laterally.

The two andesite plugs and the other five-plus basaltic plugs in the Painted Gorge area were likely fed by a single magma chamber. The two andesite plugs are younger than most of the Alverson basalts in the Painted Gorge area (Figure 3). There are several basaltic flows (Myb) younger than the andesite plugs. These younger flows are the youngest volcanics in the Painted Gorge area. We interpret this sequence as indicating that the magma chamber started mafic rich and become silica-enriched through eruptions until the emplacement of the andesite plugs. The chamber was then replenished with new mafic-rich magma to produce the youngest basalt.

Terrace 3 (Q3), Figures 5 and 6, is the terrace on which Painted Gorge Road traverses to the south of the kiosk. The terrace can be followed to the south on the western side of the Eastern Hornblende

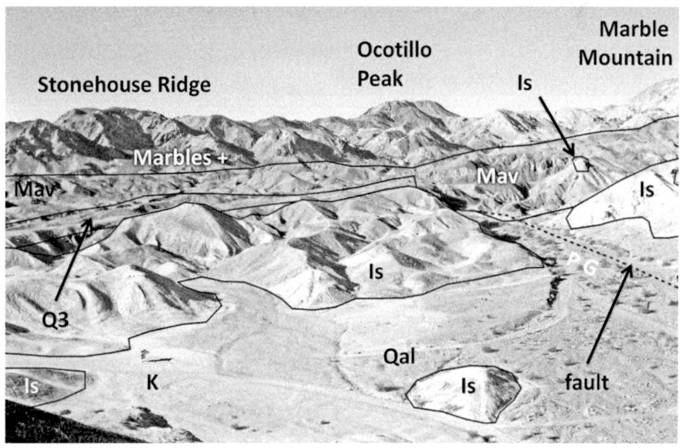

Figure 5. Looking southwest at the kiosk area. **K**-kiosk, **Qal**-alluvium, **Is**-Imperial Group sediments, **Q3**-Terrace 3, **Mav**-Alverson Formation, **Marbles+** = schist + quartzites + amphibolites.

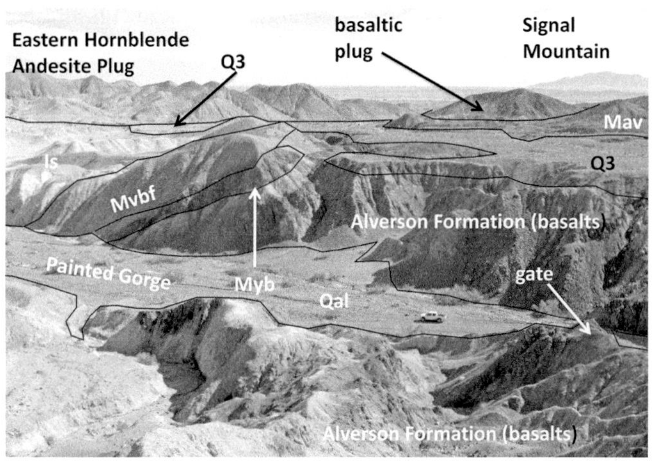

Figure 6. Looking southeast across Painted Gorge. **Q3**-Terrace 3, **Qal**-alluvium, **Is**-Imperial Group sediments, **Mvbf**-hornblende andesite breccia fanglomerate, **Mav**-Alverson Formation, **Myb**-youngest basalt flow in the Painted Gorge area. The gate is closed from Jan. 1–June 30.

Andesite plug. To the west the terrace forms a prominent bench above the Painted Gorge drainage. This terrace represents an older path that the Painted Gorge drainage once used. A stream draining to the east through the narrows captured the old Painted Gorge drainage. To the east beyond the narrows, Terrace 2 represents the first sediments deposited by the newly diverted Painted Gorge drainage. The pirating of Painted Gorge drainage happened after the formation of Terrace 3 (Q3) and before the formation of Terrace 2 (Q2).

On the skyline to the south is Stonehouse Ridge (Figure 5). The spine of the ridge, an east-west horst, is made up of basement rock consisting of marbles (mb), schist (as, s, sch, msch), quartzites (qz), amphibolies (am), Jurassic gneiss (Jwc), and intrusive Jurassic igneous rocks (Jt, Jgd). The overall structure and foliation of the ridge is east-west. Unconformably resting on the basement rocks of the ridge are volcanic rocks of the Alverson Formation and sediments belonging to the Imperial Group.

As we drive west, note the light yellow-brown (tan) mudstone and resistant shell beds of Imperial Group sediments north and south of the road. On the north side of the road the Imperial Group rests on the Western Hornblende Andesite plug (Mh) and dips, for the most part, to the east. A fault runs east-west under the alluvium on the north side of the canyon. West along the road, on the south side of the canyon, Terrace 2 (Q2) is on (in descending order) Imperial mudstone, sandstone, a gray-tan hornblende andesite breccia fanglomerate (Mabf), all dipping east, and flows and cinder beds of the Alverson. The top basaltic flow of the Alverson is the youngest basalt (Myb) in the eastern and northeastern part of the Coyote Mountains (Figure 6).

0.4 MILE Before the road passes through the Painted Gorge gate and turns west the east-west fault that was covered by alluvium (Qal) is exposed in the north wall of the Gorge. The fault crosses the Gorge and is exposed on the south wall of the canyon.

As the road continues west, Terrace 5 (Q5) is high on the north side if the canyon. Terrace 5 is higher than Terrace 3 which is on the south side of the canyon and both rest uncomfortably on Alverson flows and basaltic tuffs.

The Gorge makes a turn to the south. The west side of the gorge is capped by the Green Basalt flow (Mbg) (Figure 4). The Green Basalt is on the eastern side of the Gorge but cannot be seen from the road. Under the Green Basalt is a marker unit, a purple basaltic tuff or cinder bed. Three- quarters up the canyon on the west side is a paleomagnetic site (Mace, 1981) that shows 130 degrees of counterclockwise rotation.

An east-west high-angle north-side-down fault separates the Alverson volcanic rocks in Painted Gorge from the basement rocks of Stonehouse Ridge. This fault is referred to in this guide as Fault A (Figures 7, 8, 11, 12, 13 and 15).

The Gorge turns west again. A tuff (Mavt) is on both sides of the road but is more prominent on the north side (Figure 9). We have a zircon age-date of 17.2 ± 0.48 Ma. In this same area, Ruisaard (1979) dated a basalt at 14.9 ± 0.05 Ma using K-Ar from plagioclase.

The road turns south for a short way. Just before it turns west again the fault we noted at Painted Gorge gate can be seen on the east side of the gorge (hard to see going up the gorge). Terrace 4 (Q4) is on the east side of the gorge just south of the fault.

1.1 MILES The road approaches the second narrows of the canyon where large red blocks of a cinder unit partially block the canyon. A wider trail in the Gorge just south of the red blocks can be taken if you cannot squeeze through the narrows. As we drive through the second narrows the road turns south. Mace (1981) has a paleomagnetic sample location on the west side of the gorge that has 43 degrees of clockwise rotation.

After the narrows, the road turns west again. An old road, now a trail (Figure 8), comes into the canyon from the north. The road goes west and then turns north. As the road turns north, look north at the top of Alverson basalts, where light colored gravels of Terrace 3 (Q3) overlie darker older gravels of Terrace 4 (Q4).

As the road turns west and then again to the north, marbles covered by the Alverson Formation and Terraces 3 and 4 can be seen on the eastern side of canyon. This Terrace 4 (Q4) is represented by a weathering surface and darker sediment that is commonly found under Terrace 3 (Q3). Terrace 4 has only been seen in Painted Gorge. Another Terrace 4 is seen in Alverson Canyon and is also covered in part by a Terrace 3. The two terraces, 4 and 3, represent the only evidence for subsidence we have found in the rapid, episodic Quaternary uplift of the Coyote Mountains. The basement rocks under the terrace deposits represent part of a north-south graben that is buried to the south by the Alverson Formation (Figure 7).

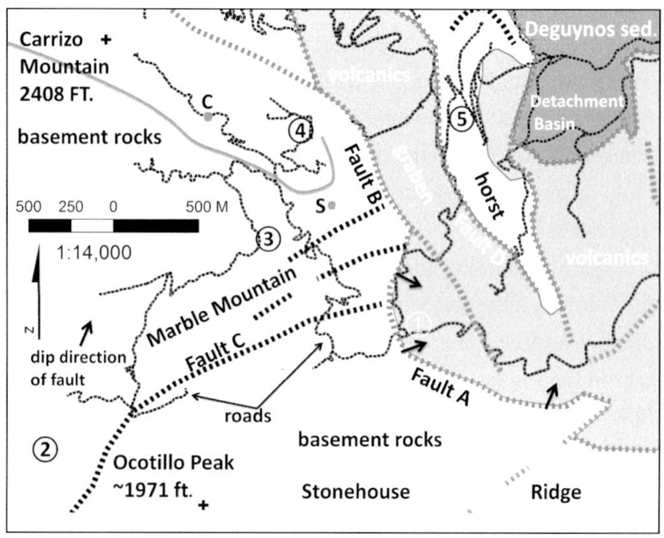

Figure 7. Geological map centered on Marble Mountain. Numbers represent field trip stops. Dashed lines are faults. The area labeled "horst" is made of basement rocks. Solid grey line is the approximate trace of Yellow Band, a dolomite marker unit in the marbles. **Sed.**-sediments. Grey dot labeled "C" is the conodont location of Miller and Dockum (1983) and the grey dot labeled "S" is our scolecodont location.

As the road turns back to the west directly ahead is a north-south fault that offsets the Alverson Formation but not the overlying terraces (Q3 and Q4) (Figure 10). This is Fault B, west side up. Fault B forms the west side of a graben (Figure 7, 11, 15, 25, 26, 27, 36). This graben is one of several horsts and grabens that can be found on the east side of the Coyote Mountains.

Fault B is one of several north-south faults (Figure 7). To the south, these faults do not cut Fault A, or they die out before reaching Fault A (Figure 8). Where the north-south faults cut the metamorphics, the fault zone is wider than where the faults cut the Alverson. These north-south faults may be older faults that have been reactivated after the Alverson Formation was deposited. To the north Fault A is truncated by Fault B (Figure 7).

Turning north at Fault B the road follows the fault up the Gorge toward outcrops of muscovite (retrograde after andalusite) schist (as) that are on both sides of the road. Terrace deposits of Q3 and Q4 are on the west side of the canyon. On the east side a tuffaceous unit under a basaltic flow yielded zircons dated at 17.2 ± 1.2 Ma. On the west side of the road the muscovite schist (as) is faulted (Fault B) against Alverson basalts (Mav).

As the road turns to the west it crosses Fault B. In this area Mace (1981) has another paleomagnetic site with 130 degrees of counterclockwise rotation. Looking north, the west side of Fault B is composed of basalts of the Alverson Formation (Mav) and the east side is composed of white marble (mb).

1.8 MILES Leaving Painted Gorge drainage the road makes the turn up the hill. The view to the north reveals Fault B and north-south Dog Tooth Ridge (Figures 11 and 15). Also on the north side of the road are depositional contacts between basalts of the Alverson Formation, the overlying Terrace 4 fanglomerates, and the overlying Terrace 3 fanglomerates. As the road reaches the top of Terrace 3, the road crosses back onto the basalts of the Alverson. Just to the north of the road is the marine calcarenite of the Andrade Transgression-Regression sequence. A thin marine Andrade basal conglomerate is at the bottom of the marine Andrade calcarenite. The gravels on top of the Andrade calcarenite belong to Terrace 5 (Q5).

STOP 1

Painted Gorge Amphitheater. Relationship between three Viejo Transgression-Regression Sequences and Fault A.

2.0 MILES To map the Coyote Mountains we have used the newly named Viejo Formation (Morgan and Morgan, this volume).The Viejo Formation is found only in the Coyote Mountains. We have divided the Viejo into marine Transgressional and Regressional (T-R) sequences (Figures 2 and 14) and fresh water sediments. A T-R sequence is the result of rising sea level with deposition followed by a falling sea level, erosion, and an unconformity. A typical, ideal T-R sequence produces sediments with a recognizable, predictable order. The first and lowest is a

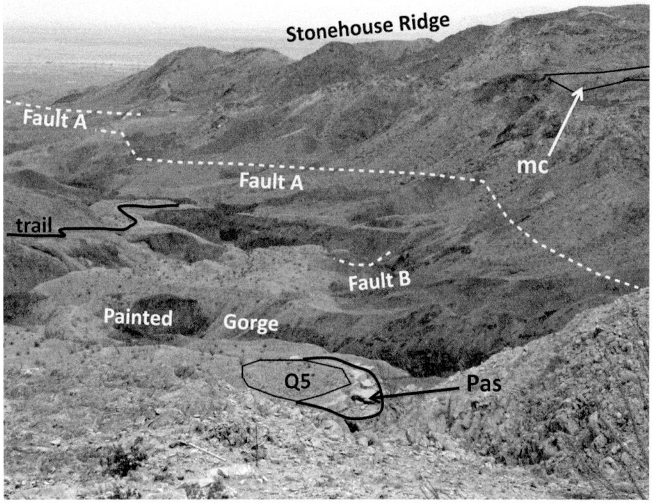

Figure 8. Looking southeast at the geology on the north side on Stonehouse Ridge. **mc**- marble lag gravel, **Q5**- Terrace 5. **Pas**- Viejo sandstone of the Andrade T-R just northeast of the amphitheater.

Figure 9. Looking north at an Alverson Formation tuff on the north side of the road.

Figure 10. Looking at Fault B. **Qal**-Alluvium, **Qoa**- Older Alluvium.

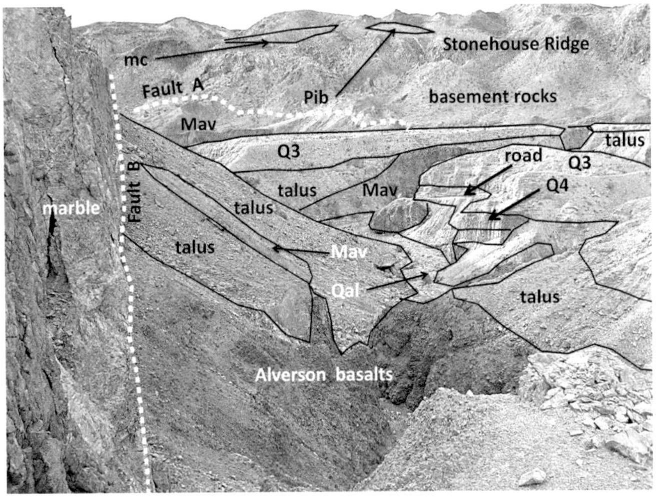

Figure 11. Looking south along Fault B towards Fault A on the north side of Stonehouse Ridge. "Road" is the route as it climbs out of Painted Gorge. **Pib**- Viejo marine basal conglomerate undifferentiated, **mc**-marble lag gravel (Garnet?), **Mav**-Alverson Formation, **Q3**-Terrace 3, **Q4**-Terrace 4 **Qal**-alluvium

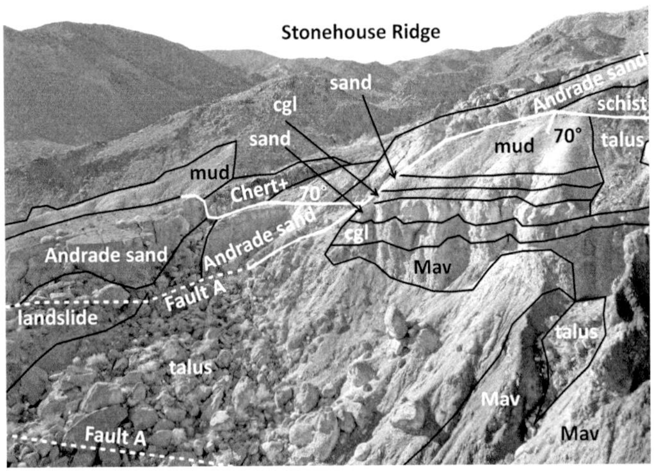

Figure 12. Looking south and down into the amphitheater of Painted Gorge. There are three predicable and recognizable Viejo T-R sequences exposed in the amphitheater. **cgl**-basal conglomerate, **sand**-sandstone, and **mud**-mudstone belong to Viejo T-R C and D Sequences. The **mud** that belongs to T-R D sequence is missing. Note that the T-R B and C Sequences are in a sliver of Fault A (see Figure 7). The Andrade sandstone is a calcarenite. The mudstone unit (**Pam**) above the Andrade belongs to the Viejo Andrade T-R Sequence. **Mav**-Alverson basalts **Chert+**-chert+ marble. Numbers are the dips on the faults.

marine conglomerate (high energy) facies formed at sea-level by waves cutting into the underlying rocks. As sea-level rises the second sedimentary facies is a marine sandstone (lower energy) deposited below wave base. With a continuing rise in sea-level the third sedimentary facies deposited is a marine mudstone representing a low-energy environment. At times sea level continues to rise and a marine limestone facies is deposited. This succession of sedimentary facies represents the transgression of a T-R sequence. If sea-level starts to fall a reverse order may be deposited starting with limestone, then mudstone followed by sandstone, and finally a conglomerate capped by an unconformity. This later succession of sediments represents the regression part of a T-R sequence. There are instances when the erosional part of the regression is great enough to erode the units that were deposited leaving only part of the transgression.

The amphitheater is one of the areas where the stratigraphy of Dibblee (1954, 2003), Christensen (1956), and Winker and Kidwell (1996) could not be used for detailed mapping. These authors only recognized a single marine transgression-regression sequence in the Coyote Mountains. Detailed mapping of the outcrops in this amphitheater indicates more than one Imperial sandstone unit and thus more than one transgression-regression sequence (Figures 2, 12, 13, and 14). Our mapping in the Coyote Mountains indicates there are several (8+) Viejo Formation marine Transgressional-Regressional (T-R) sequences (Morgan and Morgan, 2014).

In the Viejo Formation (found only in the Coyote Mountains) the T-R sequences are predictable: high energy marine conglomerates, followed by lower energy marine sandstones and silts and ending in low energy marine mudstone capped with an angular unconformity (Figure 13 and 14). The amphitheater area is one of the few areas in the Coyote Mountains where the predicable Viejo T-R sequences are easily recognized. Interfingering with the Viejo T-R sequences are non-marine high-energy fanglomerates, sandstones and a few low-energy limestones, but they have not been deposited in this area of the Coyote Mountains. A more extensive description of the Viejo can be found in Morgan and Morgan (this volume).

This is the first good look we have of Fault A, a high angle normal fault that dips east at the amphitheater. The fault dips to the southeast north of the amphitheater. On the northern side of Stonehouse Ridge, the fault dips to the north (Figure 7). We suspect this fault may have been a detachment. Two strands of Fault A form a fault sliver crossing the amphitheater of Painted Gorge. This fault sliver contains two Viejo marine T-R Sequences, C and D (Figure 12 and 13). Another strand of Fault A offsets the Viejo Andrade calcarenite by only a few meters. The Andrade calcarenite with its thin schist-rich basal conglomerate does not correlate with the sandstone and volcanic-rich basal conglomerate of the T-R C and D Sequences. The Andrade T-R sequence on the south and west sides of Fault A is deposited basement rocks. The Andrade north of the amphitheater and north of the road is east of Fault A and is deposited on Alverson. The T-R C and D Sequences in the fault sliver are deposited on Alverson volcanics and the next outcrops of T-R C and D sequences are found up slope near Marble Mountain. Because of the greater offset of the fault sliver T-R C and D sequences compared to the Andrade offset, we believe that the sliver T-R sequences are older than the Andrade (Figures 12, 13, and 14). These three T-R sequences are found on the southern and western sides of Marble Mountain. In the Marble Mountain area the Andrade T-R Sequence is also stratigraphically above the two older T-R sequences.

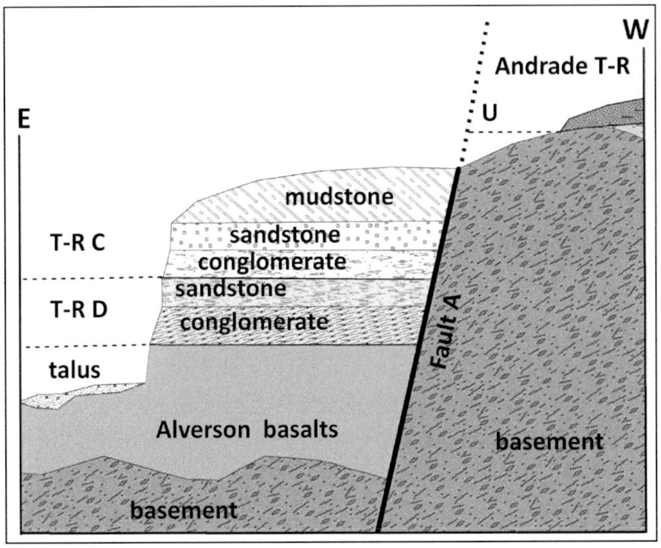

Figure 13. Simplified geological cross section at the amphitheater of Painted Gorge. Not to scale. These predicable sequences are typical for T-R sequences in the Coyote Mountains. All T-R sequences are marine.

On the southeast side of the amphitheater, under the thick basalts is a tuff that we have dated with zircons at 17.2±0.4 Ma (Figure 3).

Farther north Fault A (Figure 15) is finally truncated by a north-south Fault B. One branch of Fault B is on the west and another branch is on the east side of Dog Tooth Ridge. The Dog Tooth of Dog Tooth Ridge is a marble landslide faulted right-laterally from the marble block to the north that contains a landslide.

Fault C (Figure 15) is southernmost of two faults that are on the west side of Fault A north of the amphitheater. These faults strike to the southwest and we encounter Fault C at several places during the field trip.

Turn east. The trail (Figure 8) drops into Painted Gorge from the north at mileage 1.1 on page 6. The trail starts at the end of a road that comes in from the east. The trail and road are in a graben even though the rocks appear elevated. This is the graben labeled in Figure 7.

From the fork in the road at Stop 1 we take the road to the west. By the time we see schist on the south side of the road we will have crossed Fault A. Turn south on the marbles. More Andrade outcrops are to the west up the canyon. The red areas on the north side of the canyon below the marbles are muscovite schist (sandstone). Note the eastern outcrop of brecciated marble, high on the ridge; this is a landslide. The mountain at the western end of the canyon is Marble Mountain (Winker and Kidwell, 1996).

As the road continues south, look south at the top of Stonehouse Ridge. There is a marble lag gravel that may be Garnet fanglomerate (mc). Before the road turns west we drive up on top of the Viejo marine Andrade calcarenite. Looking south across a canyon the Andrade appears to thin on top of metamorphic rocks and a 14 meter-wide dike of resistant White Dike Pegmatite (wd). The Andrade forms a buttress unconformity with the marble hill to the southwest.

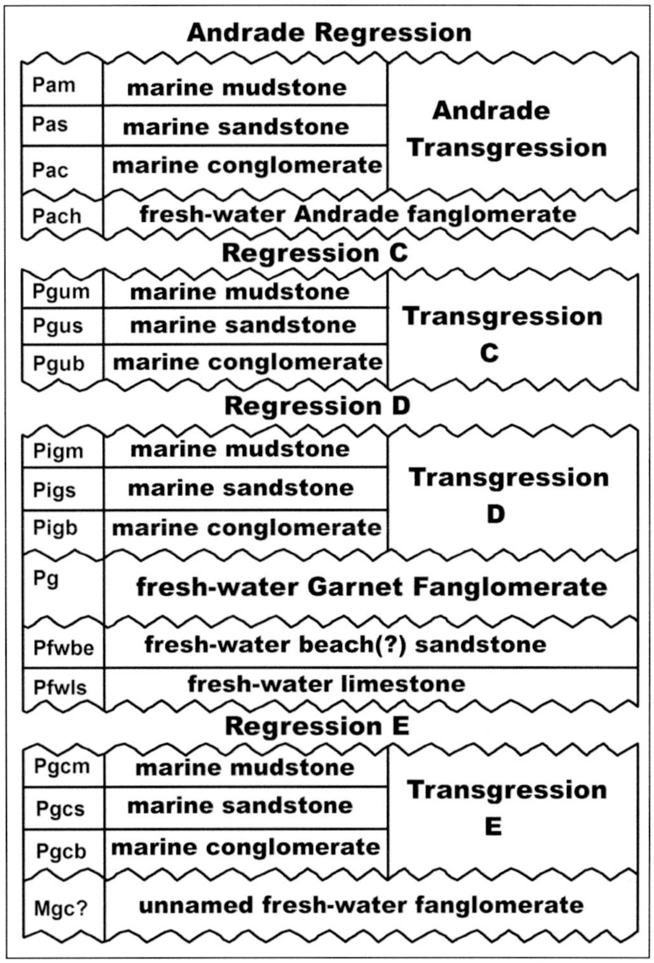

Figure 14. Stratigraphic section of the Viejo Formation sedimentary rocks in the Painted Gorge amphitheater and central section of the southern Coyote Mountains. Wavy lines are unconformities. In the T-R sequences note the repeating cycles of conglomerate, followed by sandstone, and ending with mudstone, which form recognizable and predictable, typical T-R sequences.

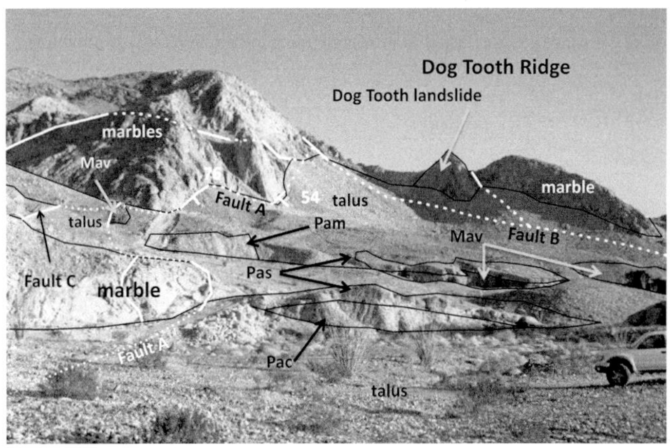

Figure 15. Looking northeast at the geology at Dog Tooth Ridge. **Mav**-Alverson basalts, **Pas**- Andrade calcarenite, **Pac**- Andrade marine basal conglomerate, **Pan**- Viejo marine Andrade Mudstone. The three Andrade units form a typical, predictable outcrop pattern for T-R sequences that we find in the Coyote Mountains. Faults shown by white lines, dotted where concealed. Dips on Fault A shown in white numbers. Fault B truncates Fault A here, but Fault B dies out (?) to the south and does not cut Fault A (Figure 7), which forms the northern side of Stonehouse Ridge.

The road continues west and crosses a small fault, a contact between Andrade calcarenite and Andrade basal conglomerate. Note the pholad holes, what we call "bullet holes," in the large marble clasts just south of the road. Pholads are small bivalves that drill holes for themselves in marbles and sometimes even basalts. The holes are about 2.5 cm wide and do not last long without a sedimentary protective cover. If you find bullet holes start looking for Viejo Formation sediments. Pholad holes by themselves are an indication that the Viejo Formation was present even if eroded away. The pholads are filter feeders and like high energy environments. The buttress unconformity between marbles and volcanics, and the Viejo Formation marine conglomerates and sandstones is a good place to find the "bullet holes."

Continuing west the road climbs and crosses into Andrade mudstone that is capped by a marble conglomerate (mc) (Garnet??). Looking west is a good view of Marble Mountain and the Andrade calcarenite that has been deposited on the southeast side of the mountain (Figure 16).

At the fork we take the road that turns north. The other road goes southwest to a marble prospect. Our road crosses onto marbles and schists at the end of the ridge with Andrade calcarenite on top. The road turns west and cuts beds of the Miocene Red Fanglomerate (Mrf) unconformably overlain by Andrade calcarenite. This Red Fanglomerate has been correlated by Dibblee, Winker and Kidwell, Ann Bykerk-Kaufman (personal

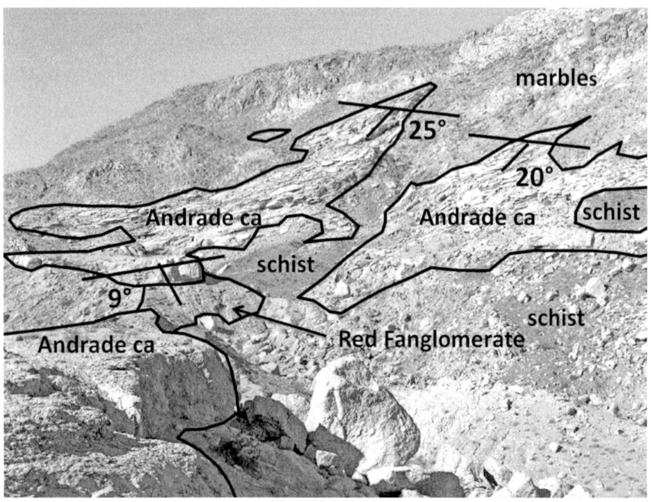

Figure 16. Looking northwest at the southern side of Marble Mountain. The dips of the Andrade calcarenite on the south slope on Marble Mountain are not depositional but are the result of uplift of Marble Mountain. The highest Viejo Andrade calcarenite on the southeast side of Marble Mountain has a dip of 25 degrees. The ridge of Andrade that is above the road is very close to horizontal. It is unrealistic for the Andrade to have been stably deposited on a 25 degree slope. Rather, the Andrade was deposited at a very shallow dip and later differentially uplifted with Marble Mountain. Similar deformed Andrade is on the west side of Marble Mountain. Bullet holes, which are near the top of Marble Mountain, indicate that Marble Mountain at one time was completely covered by the Andrade. Winker and Kidwell (1996) considered Marble Mountain to be an island during the deposition of their Andrade.

communication) with the Red Rock Formation that is found in the Fish Creek Mountains. Morgan and Morgan (2014, and in this volume) provide evidence showing difficulty correlating the Red Fanglomerate with the Red Rock Formation.

The road turns north into Ship Rock Basin and crosses into the drainage we have been following since leaving Stop 1.

2.5 MILES Just to the west of the road is the Red Fanglomerate (Mrf) outcrop that forms the namesake of this small basin (Figure 17). As we pass the "ship" we also cross north-side-down Fault C (Figures 7, 15, and 17), offsetting Red Fanglomerate (Mrf) on the north side and marble on the south side. This is the same Fault C that we noted at Stop 1. We again encounter Fault C at Stop 2.

Within this small basin the Red Fanglomerate is represented by sand dune deposits covered by fanglomerates containing Alverson clasts. The Red Fanglomerate is resting on basement. No White Fanglomerate was found under the Red Fanglomerate in this basin. Sand dune deposits are also reported in the Red Rock Formation in the Red Rock Canyon area, Fish Creek Mountains.

White and Red Fanglomerate, and similar fanglomerates found in the Fish Creek Mountains, Volcanic Hills, Davies Valley area, and Sierra Cucupa in Mexico, are usually found under Miocene volcanics.

We believe that these fanglomerates represent Miocene sedimentation in extensional basins during Miocene Basin-and-Range extension in Sonora, Mexico (Nourse et al. 1994). In the extensional Miocene basins in Sonora, McDowell et al. (1997) found sediments similar to the White and Red Fanglomerates with volcanics similar to the Alverson Formation. Each of the fanglomerates found in the Salton Trough are derived from local rocks. In the Salton Trough area some of these fanglomerates, including the White Fanglomerate, contain reworked exotic clasts of silicified meta-volcanics from Eocene gravels sourced in Sonora, Mexico (Link, and Abbott, 1991) filling fossil river channels (Los Palmas and Poway) that crossed the region onto the Peninsular Range (Minch, 1972).

The road heads north and crosses out of the drainage. Looking west and east you can see the light-colored sand dune deposits under the fanglomerates. At the first switchback the road turn on to muscovite schist and climbs up and crosses onto Red Fanglomerate (Mrf) outcrops of both sand dune and the fanglomerates. At the next switchback, the Mrf is faulted (Fault C, Figure 15) against the marble to the south. As the road curves around to the north, the Red Fanglomerate (Mrf) forms a buttress unconformity (Figure 17) against the marble just below the road above.

As the road curves around the next switchback to the east, look back at the area just driven (Figure 8). Most of Painted Gorge, the area around the kiosk, and the east and west hornblende andesite plugs can be seen. Look down to the east at Dog Tooth Ridge. The extent of Terrace 3 and the turn the terrace makes to the south through the area just west of the Eastern Hornblende Andesite Plug can be seen. To the east is a view of the Superstition Hills,

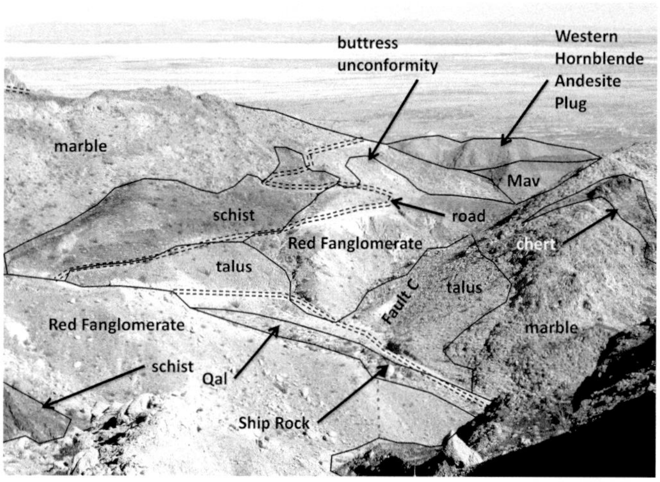

Figure 17. Looking down and easterly at Ship Rock Basin. **Mav**-Alverson basalts, **Qal**-alluvium

Superstition Mountains, and beyond to the Chocolate Mountains that are east of the Salton Trough. To the south is Signal Mountain just south of the border.

Continuing on, the road crosses a marble pod and onto schist.

3.0 MILES At this intersection, stay straight ahead (west branch). The road east goes to a drill hole location. Less than 0.1 mile from last intersection is another fork in the road. The east branch goes to another drill hole, we will take the northern branch. We leave the schist at a saddle and are on Red Fanglomerates (Mrf). The large boulder of Viejo Andrade calcarenite on the east side of the road is not in place. Looking west is a good view of the Viejo Andrade calcarenite on the sloping side of Marble Mountain. Watkins (1990) studied the Imperial in these outcrops of Andrade calcarenite. The road climbs up and onto marble and curves around the hill. Looking east and down is another marble lag gravel (mc) (Garnet?). Most of the marble-rich talus in this area came from this mc and the mc that is higher on the hill. Further up the road we leave the marble, cross a fault and move onto a lahar (?) (Malh) or a red sediment that is very rich in basaltic fragments. We consider Malh to be the part of the Ship Rock Basin and a member of the Alverson Formation.

3.3 MILES As the road continues to climb up the mountain it switches back and forth on the covered red lahar and finally onto the upper marble lag gravel (mc) (Garnet?). Christensen (1957) considered this lag gravel and the lower gravel to be part of the Garnet Formation. The clast count Christensen took for the Garnet was from a "prospect trench" west of the road about 30 m.

Looking east, the Yellow Band (a dolomite marker unit), can be seen on the west and north sides of the marble promontory. This Yellow Band outlines a synform in the metamorphics (Figure 7). The Yellow Band is on the south side of Carrizo Mountain.

While we are on top of the marble lag gravels, look north and west at thick outcrops of Andrade calcarenite. Both Andrade outcrops rest on metamorphic basement rocks. As the road rounds the outcrop of Andrade and heads north, we cross onto marbles, schists, and quartzite pods. The road then crosses onto another pod of the Viejo Andrade calcarenite.

3.6 MILES At this intersection take the northwest fork. The road stays on the Andrade. Where the road turns west, it crosses back onto marbles and muscovite schist. As the road crosses the marbles, bullet holes can be seen.

3.9 MILES At this fork, turn south. The road to the west ends in Viejo marine sandstones a little farther down the slope.

Continuing south the road crosses onto a flat pod of Viejo Andrade calcarenite with a view to the south and east across the Salton Trough. This is the outcrop we pointed out when we were at the kiosk. We stop here on the way back (Stop 3).

Continuing south the road leaves the Andrade and turns west onto marble (look for bullet holes). Note Andrade calcarenite across the canyon forming a buttress unconformity against marble. Turning south we cross a saddle onto the north side of Marble Mountain. Schist pods on Marble Mountain are covered somewhat by talus. The road turns south and crosses onto another pod of Andrade calcarenite. The road climbs up and crosses the Andrade and onto schist and then marbles. Get your cameras ready for the view to the northwest as the road descends.

The Deguyños Formation (Figure 2, 3 and 18) is not found in the Coyote Mountains but is found in the flats around the Coyote Mountains and throughout the Salton Trough (Winker and Kidwell, 1996). The Deguyños Formation is characterized by olive green

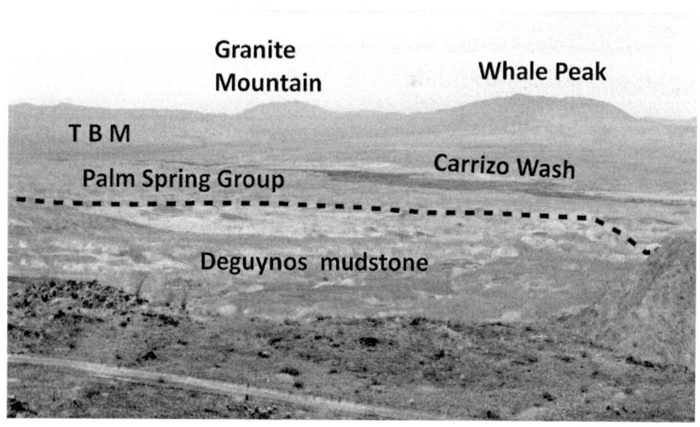

Figure 18. View to the north. Prominent Whale Peak is the foot wall antiform of the West Salton detachment fault (Axen and Fletcher, 1998). The dark area in Carrizo Wash and Creek (which flows from the Bow Willow area) is vegetation. Granite Mountain is northwest of Whale Peak. The valley that starts at the Bow Willow area and goes north is Carrizo Valley. At the northern end of this valley is Agua Caliente. In between the Bow Willow area and Agua Caliente is the Tierra Blanca Mountains (**TBM**). The Elsinore fault, on the western side of this valley along Tierra Blanca, can be followed north as it passes on the west side of Granite Mountain (Magistrale and Rockwell, 1996). In the flats between us and the green area of Carrizo Creek are the tan sediments of the Deguyños Formation marine mudstones and, farther in the distance, the pink sediments of the younger freshwater Palm Spring Group. Look north at the basement rocks of Carrizo Mountain including the distinctive Yellow Band dolomite marker unit. The road switches back and down onto schist and quartzite sequences on the west side of Marble Mountain.

[weathering to light yellow brown (tan)] mudstone, light yellow brown (tan) siltstones and brown sandstones with wave base shell beds. The Deguyños is considered to be marine with Colorado River deltaic influences (Winker and Kidwell, 1996). No conglomerates or fanglomerates have been seen in the Deguyños. Angular unconformities are found in the Deguyños and may represent T-R sequences. The Deguyños Formation lacks the high energy marine conglomerates and fresh water fanglomerates that are found in the Viejo Formation. A more detailed description of the Deguyños is found in Morgan and Morgan (this volume).

Palm Spring Group (Figure 2) is characterized by freshwater pink mudstones, tan to white friable silt beds, dark brown sandstones, and concretion beds in the flats around the Coyote Mountains (Winker and Kidwell, 1996). Some units contain petrified wood. A more detailed description of the Palm Spring is found in Morgan and Morgan (this volume).

4.7 MILES At this intersection go south. The west road goes out to the ridge of marble. Walk starting from the end of this road to views of the northern Coyote Mountains and the badlands to the north.

Going south along the western side of Marble Mountains the road crosses pods of Viejo marine Andrade calcarenite and Andrade basal conglomerate that rest on marbles. Look for bullet holes. These Andrade pods dip as steeply as 29 degrees to the west.

5.0 MILES The Andrade outcrops to the west of the road form anticlines and synclines resulting from soft sediment deformation as the Andrade slipped downhill during the differential uplift of Marble Mountain.

The road continues south (straight) and down between two marble outcrops. As we pass through the narrow gap, a small flat ridge of Viejo marine T-R C basal conglomerate (Pgub) is just west of the road. This basal conglomerate was deposited on the Viejo freshwater Garnet. The prominent peak to the south is Ocotillo Peak.

5.2 MILES Stop and park at the triangle for the walk over to Stop 2. We recommend a good pair of boots and a walking stick. The road going south and east ends in a marble prospect. The road that goes to the west ends just beyond the hill to the west.

Walk to the southwest 0.4 mile, toward the top of the north end of the ridge of Viejo freshwater Garnet fanglomerates (Mg). This ridge and the one south of the triangle are made up of Viejo Garnet fanglomerates dipping gently to the east. The Garnet is largely composed of clasts of granitic rock, quartzite, schist, volcanic rock, and marble. The Garnet surface is rich in marbles because of differential weathering discussed below. This part of the Garnet has few volcanic clasts. The best outcrops of the Garnet are exposed in the bottom of stream channels.

The main north-south drainage we cross is Garnet Wash. We believe Christenson (1957) used this wash to name the Garnet Formation and thus this may be considered the type section. Christensen did not indicate a type section in his thesis.

STOP 2

View of the Upper Alverson Canyon (Fossil Canyon) area and the Garnet Wash Area. Discuss the relationship between the freshwater Garnet and marine T-R sequences of the Viejo Formation.

The Garnet was named by Christensen (1957) for the fanglomerates in the surrounding area and he believed the unit to be Quaternary in age (Figure 3). Depending on who was mapping and when they mapped the Garnet influenced where the unit was put in a columnar section. Dibblee (1954) mapped the fanglomerates of the Garnet as Split Mountain Conglomerate. In later mapping, Dibblee (2003) combined the Garnet and other similar fanglomerates with the Imperial. Winker and Kidwell (1996) expanded the Garnet to include all fanglomerates in the Coyote Mountains stratigraphically above the Alverson Formation and below the Imperial Group. They then put the Garnet in the Split Mountain Group, but indicated it had marine tongues (?). Our mapping indicates that the Garnet, as Christiansen originally described, and other similar fanglomerates interfinger with the marine T-R sequences of the Viejo Formation.

The Garnet is composed of light brown to red-brown weathering gray, matrix-supported, fanglomerates with light brown coarse sandstones. Clast composition includes granitic, quartzite, schist, volcanic, and marble rocks. The Garnet is marble poor, but the surface is marble rich. The marble cobbles are able to "heal" themselves, while the clasts of the other rock types cannot heal themselves after they have been weathered. Water chemically attacks the cobbles of the Garnet. When the water evaporates in this arid environment, the calcium carbonate precipitates from the water and heals the marble. Different facies of the Garnet are poor in volcanic clasts. Some of the marble lag gravels(?)(mc) we see on the field trip may be remnants of the Garnet.

We believe the freshwater Garnet is a part of the Viejo Formation because our mapping indicates that the Garnet is between the marine T-R D and E sequences of the Viejo Formation (Figure 14, 19, 20, 21, and 22). We have demoted the Garnet Formation to Garnet member of the Viejo Formation.

Upon reaching the top of the ridge we have a view to the west of the upper Alverson Canyon (Fossil Canyon). The Garnet we are standing on has been deposited on schist and to the south on an unnamed darker freshwater fanglomerate of the Viejo. Looking to the west below us, Garnet continues into the distance (Figures 19, 20, and 21) and is capped by Viejo marine T-R D sequence sediments (Pigs). Downslope and between us and the western outcrops of the Garnet is a detachment fault (Figure 21, 49 and 50) that has rotated the hanging wall and the Garnet down and away from us. The detachment fault is expressed in the basement rocks with up to a 3 m wide fault zone plus a wider zone of brecciated wall rock. This western detachment fault (Figure 49) has been reactivated as a right-lateral, strike-slip fault (Figure 50). In Alverson (Fossil)

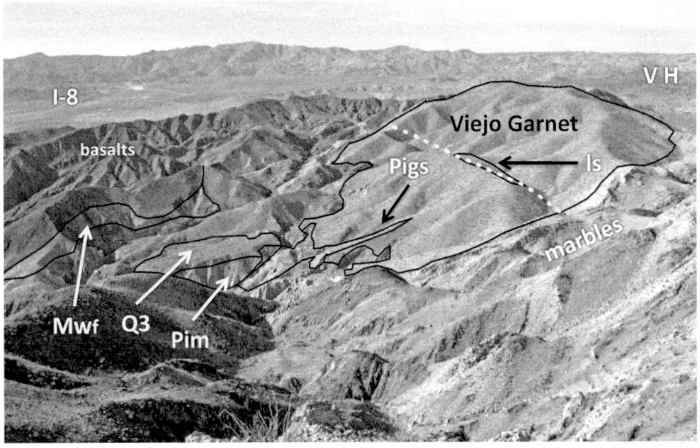

Figure 19. Looking southwest at the upper Alverson Canyon (Fossil Canyon). **V H**-Volcanic Hills, **ls**-landslide, **Pim**- Viejo marine mudstone, **Pigs**- sandstone of Viejo marine Transgression-Regression D Sequence, **Q3**-Terrace 3, **Mwf**-White Fanglomerate.

Figure 20. Looking east at Carrizo Mountain. **Pis**- undifferentiated Viejo marine ss, white line fault dipping 70 degress to the north west, **Pis** east of the fault is dipping east 6° to the southeast. This photo is at the western end of the cross section in Figure 21.

Canyon we see a continuation of this reactivated fault.

From this vantage point we can see that the large western Garnet outcrop and Garnet outcrops further to the northwest rest on the basement of the Coyote Mountains. Small outcrops of the Garnet also rest on basalts of the Alverson along the ridge south of the large Garnet outcrop. On the western side of the large Garnet outcrop the Garnet not only rests on the basement but is also resting on undifferentiated Viejo T-R marine sediments (Figure 21 and 22). An even older T-R sequence is downstream of the large Garnet outcrop

Near the western undifferentiated Viejo T-R marine sediments directly under the Garnet is an unusually large, ~1.6 m isolated boulder from the Garnet fanglomerates. Further to the north in another outcrop of the Garnet is another large, ~1.6 m, isolated boulder. In Arizona fanglomerates associated with detachments often have large, 3 m, isolated boulders.

There are no Viejo marine sediments directly under the Garnet that we are standing on but there are Viejo marine sediments (Viejo T-R E Sequence) deposited on the already noted older

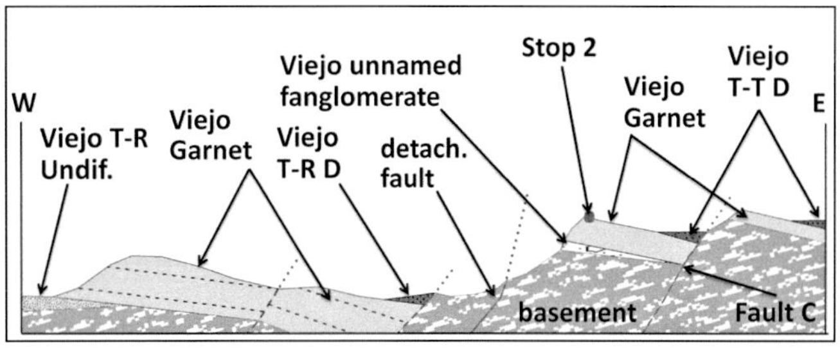

Figure 21. Simplified cross section at Stop 2. Not to scale. **T-R D**- Viejo Transgression-Regression D Sequence, **detach**. fault- detachment fault; **T-R Und.-** Viejo Transgression- Regression Undifferentiated. Faults are in grey. Dotted black lines are bedding planes.

Figure 22. Looking east, from Stop 2, at the eastern end of cross section in Figure 21. The road south of the triangle is just off the northern side of the photo. **Pigs** is the Viejo marine sandstone of T-R D Sequence, **Q1** is Terrace 1. Dip of fault is in black. Dips of the sediments are in white.

unnamed freshwater fanglomerate of the Viejo below the Garnet to the south (Figure 14). This relationship can be found about 800 m to the south of us in the next drainage.

Looking east at the next ridge is a repeated section of the Viejo freshwater Garnet, deposited on schist. The fault that repeats the section can be found along the contact between the Garnet we are on and the schist. This is Fault C, west side down, that we saw in Ship Rock Basin (Figure 7, 15, 17, 21, 22, and 50). On top of the Garnet we are standing on are outcrops of Viejo marine T-R D sediments (Pigs) (Figure 21 and 22) that are along the western side of Fault C. Fault C continues south before it turns west and is truncated by the reactivated western detachment fault in Figure 50.

At the southeastern end of the Garnet on the eastern ridge and under the Garnet are the small outcrops of a freshwater limestone. This is not the only freshwater limestone in the section (Morgan and Morgan, 2013). At one time we thought that this Viejo freshwater limestone under the Garnet was possibly related to the Bouse Formation that is found along the lower Colorado River (Spencer et al., 2013). It now appears that this Viejo freshwater limestone is too old to be related to the Bouse.

On the way back we cross Viejo marine sediments.

Once back at the car retrace our route.

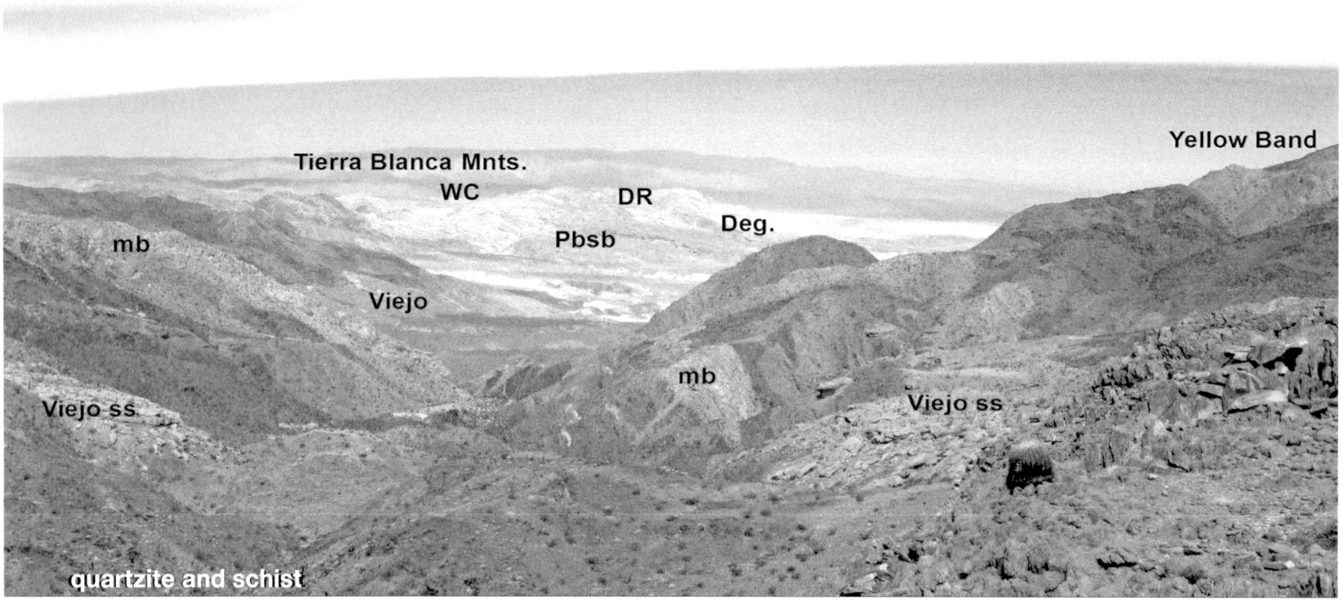

Figure 23. Looking west at the northwest (Dolomite Ridge) Coyote Mountains. **WC**- Wind Cave area, **Pbsb**- Marker bed in the Deguyños Formation, **Deg.**- Deguyños Formation, **DR**-Dolomite Ridge, **Viejo**- marine Viejo Formation, Viejo ss- Viejo marine sandstone undifferentiated, **mb**-marble, **Yellow Band**- a dolomite marker unit (Figure 7). To the east of Dolomite Ridge is Deguyños marine tan mudstone covered by pink muds of the freshwater Palm Spring Group. The northern end of Dolomite Ridge is the Canyon Sin Nombre area; a little further north is Carrizo Wash. The Elsinore fault is on the east side of the Tierra Blanca Mountains.

STOP 3

Viejo sandstone (limestone?) of the Andrade T-R Sequence, Pbsb marker bed and a regional geological view.

6.4 MILES The Viejo sandstone we are parked on is the calcarenite of the Viejo marine Andrade T-R Sequence. Note that most of the Andrade calcarenite, including this outcrop, is flat or gently dipping. The differences in the elevation between these outcrops of Andrade calcarenite can be explained by the Quaternary differential uplift of the Coyote Mountains. Winker and Kidwell (1996) considered the Andrade to be a limestone and here the unit does look like a limestone. We have chosen to call this unit a calcarenite because of the high-energy depositional environment the unit represents.

The view, starting from the eastern end of Ocotillo Peak to the south, is the Sierra Juarez escarpment (western edge of the Salton Trough) and at night, headlights on the grade up to La Rumorosa. The low hills just above Stonehouse Ridge and in front of the Sierra Juarez are in the Davies Valley area. The smooth area on the crest of Stonehouse Ridge is lag gravel (Garnet?). Laguna Salada is the flat area between the Sierra Juarez and the next mountains to the east, starting at the border to south: Signal Mountain, the Sierra Cucapas, the Sierra Major, and on a really clear day the Sierra Pinta. The view east includes the Salton Trough, Plaster City, the new plate boundary, and the old plate boundary (Figure 1). East of the Salton Trough (the San Andreas fault and the old plate boundary) are the Cargo Muchachos Mountains and Pilot Knob that are west of Yuma, Arizona. On a clear day you can see the Gila Mountains south and east of Yuma. Going north from the Cargo Muchachos are the buttes of Picacho Peak. Further north are the Chocolate Mountains.

In the Coyote Mountains, are the two Alverson hornblende andesite plugs and the Kiosk. Note Terrace 3 turning and going south along the western side of the Eastern Hornblende Andesite plug. Painted Gorge makes the dark canyon north of Stonehouse Ridge. The Alverson (red-brown) is in fault contact with the basement rocks (dark gray) along the north side of Stonehouse Ridge. This is Fault A (Figures 7, 8, 11, 12, 13, and 15). As the Gorge disappears behind a marble ridge, note the dark marble that is half way up the north side of Stonehouse Ridge. The flat area back from the dark marbles is covered by lag gravels (Figures 8 and 11). From north to south, the first is a marble lag gravel (Garnet?) and then a Viejo marine basal conglomerate.

Look back to the road we came up to see Ship Rock Basin and Fault C (Figure 17). To the southwest of Ship Rock Basin, the Viejo marine Andrade calcarenite laps onto the marbles.

Walk north about 200 feet along the road and then west to a viewpoint where the marble ridge is low. Look at the northwestern ridge of the Coyote Mountains (Dolomite Ridge) and the sediments to the east of this ridge (Figure 23). Pbsb is a marker bed in the Deguyños Formation southeast of the Wind Cave (WC) area. We have mapped this marker bed on the east side of the Coyote Mountains.

A little more to the north is the Yellow Band, a dolomite marker unit on the southern side of Carrizo Mountain (Figure 7). A little further east is the road to the base of Carrizo Mountain. On the skyline the road goes around what we call Wollastonite Hill and comes out on the saddle just west of the hill. The hill is named for the large wollastonite crystals that are found there. The area around saddle is a conodont location (Figure 7) documented by Miller and Dockum (1983). We have not been able to reestablish their conodont location.

Return to vehicles and retrace some more of our route.

6.9 MILES We came up the road from the south earlier. Take the north fork of the intersection. Note the small inselberg of marble surrounded by Viejo Andrade calcarenite on the west side of the road and the bullet holes on the east. In this area we cross the Yellow Band marker bed.

7.0 MILES At this fork, take the road to the east. The northern road goes higher up the mountain toward Carrizo Mountain. This is the road to the conodont location of Miller and Dockum (1983).

Just west and a little southwest of the end of the northern road is an outcrop of Viejo marine sediments at an elevation of ~2100 feet. This is only 300 feet short of the top of the Carrizo Mountain (elevation of 2408 feet). This and other indirect evidence leads us to believe that at one time the Coyote Mountains were eroded flat and completely covered by Viejo marine sediments.

7.1 MILES Continuing east across the marbles the road forks again. Take the northern route. The eastern route goes up and ends in a small alluvial valley. The small alluvial valley is a small back basin related to the landslide we are now on. Look north and down at Alverson basalt outcrops on the landslide. This is one of the indications that the Alverson Formation at one time completely covered the area what was to become the Coyote Mountains.

STOP 4

Lunch stop. Reactivated landslide. Deguyños Formation Pbsb marker bed, Painted Gorge Wash Fault and regional geology to the east.

7.3 MILES Note brecciated marbles. From the edge of the landslide the eastward view (Figure 24) is across the Salton Trough to the Chocolate Mountains, Picacho Peak, Cargo Muchacho Mountains, Pilot Knob, and possibly the Gila Mountains east of Yuma. At times, in the farm areas, steam can be seen venting from the geothermal plants. The Fish Creek Mountains and the Salton Sea are to the north. South of the Fish Creek Mountains and low in the badlands is Carrizo Wash that drains to the Salton Sea.

In the flats to the east are tan Deguyños marine sediments in contact with younger freshwater pink mud facies of the Palm Spring Group. Some of the contacts are depositional while others are faults. Some of the Deguyños and Palm Spring sediments have a Colorado River signature (Winker and Kidwell, 1996).

The Deguyños section is made up of marine mudstones, siltstones, rhythmite beds, and dark shell beds that represented a higher energy environment. The rhythmites can be divided into sand–silt and silt-mud map units. There are several angular unconformities within this section that may represent T-R sequences. Several of these unconformities we see later. Much of the eastern side of Coyote Mountains from the north to eastern side of the Western Hornblende Andesite Plug above Painted Gorge can be characterized as a high-angle, mostly west-dipping fault between Deguyños mudstone and basement rocks (Figure 34).

Closer in and directly below us are East Ridge (horst) and Inner Valley (graben). At the northern end of Inner Valley are a red

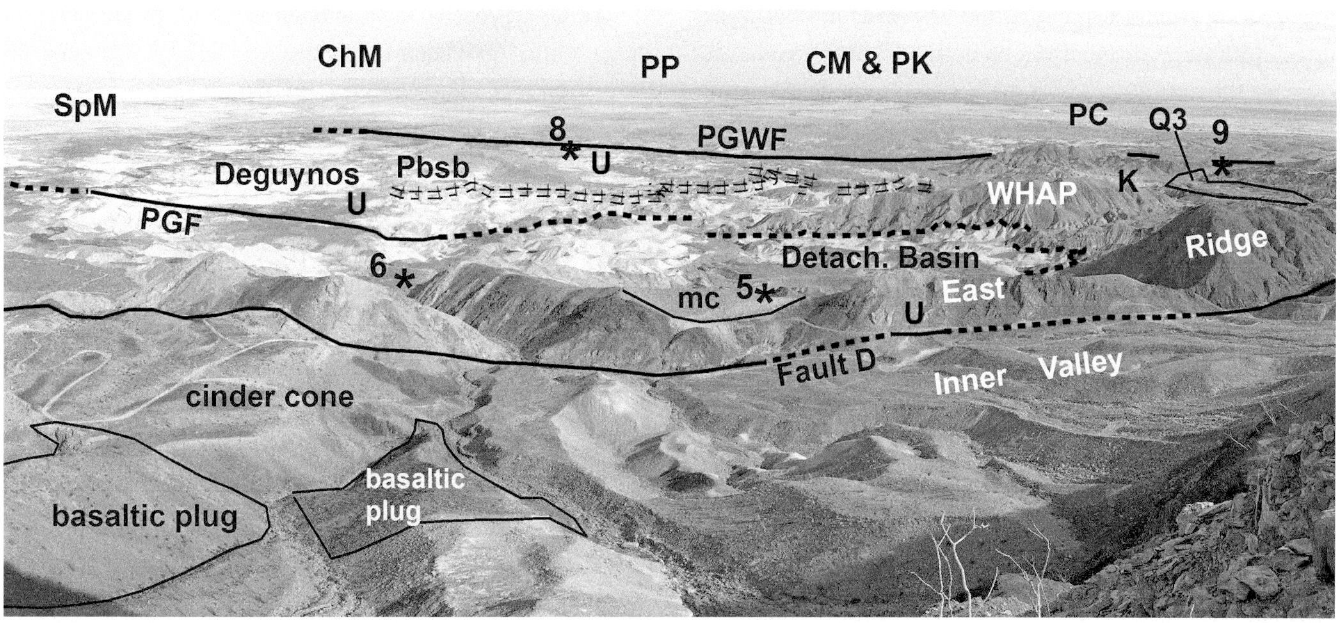

Figure 24. Looking east at the Salton Trough. **SpM**- Superstition Mountains, **ChM**-Chocolate Mountains, **PP**-Picacho Peak, **CM & PK**-Cargo Muchacho Mountains and Pilot Knob, **PC**-Plaster City, **Q3**-Terrace 3 (old Painted Gorge drainage), **K**-kiosk, **PGWF**-Painted Gorge Wash fault, **WHAP**-Western Hornblende Andesite Plug, **PGF**-Painted Gorge fault (Christiansen 1957), **Pbsb**-Deguyños Formation marker bed, **Deguyños**-mudstones, **mc**-marble channel fanglomerates. **Detach.**-detachment fault. **U**-up side of fault. Faults are in black. Asterisks and numbers are field trip stops.

volcanic cinder cone and its dark basaltic plug. The road that climbs up and over East Ridge is on an east-west channel filled with marble fanglomerates. Stop 5 is located at the top on this channel on the road (Figure 24). To the northeast of Stop 6 is Christensen's (1957) Painted Gorge fault offsetting Terrace 2 with a 0.6 m west-facing scarp (Figure 24). Detachment Basin is the flat area between the Western Hornblende Andesite plug and East Ridge. The light tan sediments in the basin are Deguyños mudstone and are fault bounded. Look south to the eastern extension of Stonehouse Ridge.

The landslide of Stop 5 at one time closed Separation Canyon. Separation Canyon is the canyon between Switchback Mountain and Carrizo Mountain north of the landslide. The drainage of Separation Canyon was backed up until it topped the landslide dam and deposited gravels on the slopes of Switchback Mountain (Figure 27 and 28). The drainage also deposited gravels in an east-west channel (mc) on East Ridge. The dam was soon breached and has left gravels on the side of Switchback Mountain. The recent brecciated marble and the small back basins indicate that the landslide has been reactivated.

Backtrack down the mountain to the intersection at Stop 1.

9.3 MILES At the intersection at Stop 1 take the north road. Just after turning on to the north road there is a good example of Viejo marine Andrade basal conglomerate (Pac) under Viejo Andrade calcarenite (Pas) (Figure 15). Remember that the Andrade outcrops along the road north form a predictable and recognizable T-R sequence. Continuing north the road approaches and turns at a marble pod. At the turn are basalts of the Alverson Formation faulted against the marble. Note the brecciated nature of the marble. This is Fault A (Figure 7, 8, 11, 12, 13, 15, and 25). Proceed downhill; Andrade mudstone is exposed on the north side of the road making a complete transgression. The road turns north goes up hill with Andrade mudstone on both sides of the road. After cresting, the road descends; at the bottom, Andrade calcarenite is on both sides of the road. The Andrade unconformity rests on Alverson Formation. Directly north of the road on the marble cliff, Fault A is well exposed dipping 54° south (Figure 25). Note large slickensides (mullion structure) on the marble outcrop.

The road swings away from the marbles and Fault A and then swings back. During this swing, the road passes Andrade basal conglomerate and calcarenite, and outcrops of Alverson. As we make the final swing away from the marble and across the talus slope, look up to the head of the talus slope. This is where Fault A is truncated by north-south Fault B as noted in Figure 7. Somewhere under talus slope on the west side of Dog Tooth Ridge is the western branch of Fault B. Before we swing around Dog Tooth Ridge we cross this western branch of Fault B. Looking south, a resistant brecciated marble marks the fault zone of Fault B. As we swing around Dog Tooth Ridge we cross an eastern branch of Fault B before we reach basalts of the Alverson Formation. This fault marks the western edge of a graben (Figure 7).

Figure 25. Looking north. **MS**-mullion structure along Fault A. The white dot labeled "**S**" is our scolecodont location (Figure 7). Scolecodont are conodont-like teeth from annelid worms.

Continuing up the slope, outcrops of schist and marble are on either side of the road. Near the top the slope is an outcrop of Alverson basalt.

10.0 MILES The road turns north and crosses onto a dark chert. Even though the rocks are elevated in comparison to the valley to the north and east, structurally, we are in a graben. Looking north the road passes below an outcrop of Tertiary marble gravels (Tmc). These marble conglomerates (Tmc) or lag gravels are part of the Split Mountain Group, found under the Alverson Formation. On the east side of the road are marbles along a fault (Figure 26).

We turn east; cross the fault and onto a Terrace 3 that is unconformity on basalts of the Alverson (Figure 26). Looking north, Fault B forms the west side of Inner Valley. The floor of the valley is made up of Alverson overlain by Imperial sediments. These units are covered by various terraces. Directly ahead is East Ridge, a horst, and Fault D (Figures 7, 27, and 35) that make up the eastern edge of Inner Valley. This part of East Ridge is made

Figure 26. Looking northwest at the western side of Inner Valley. **Pib**-undifferated Viejo basal conglomerate, **Pg**-Garnet Formation, **s**-schist, **Mav**-Alverson basalts, **Q3**-Terrace 3.

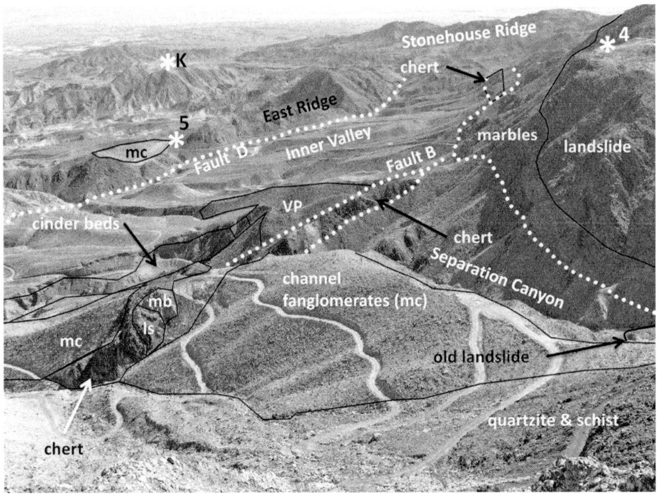

Figure 27. Looking southeast at the geology of Inner Valley from Switchback Mountain. White asterisks are field trip stops, white dotted lines-faults, **K** is kiosk between the two (east, west) andesite plugs, **mc**- marble channel fanglomerates, **ls**-landslide, **mb**-marbles, **VP**-Alverson basalt plug, **old landslide**-the old landslide that blocked Separation Canyon to produce marble channel fanglomerates on Switchback Mountain and East Ridge. The chert at the southern end of Inner Valley is the same chert in Figure 26.

of quartz-rich schist. Looking south Alverson basalts and some marble make up units in the graben. Deposited on the southern end of East Ridge are more Tertiary marble lag gravels (Tmc).

The road switches back and onto the Alverson and then down onto a narrow valley controlled by Fault D. A branch of this fault straight ahead separates Imperial mudstone on the west from basaltic Alverson. Just before reaching the fault an outcrop of Viejo marine basal conglomerate is on the west side of the road. Looking north, another outcrop of Tertiary marble lag gravels (Tmc) is on the top of East Ridge. The road leaves the drainage up a hill to the north passing Terrace 1 and onto Imperial mudstone. Reaching the top, view the graben to the north (Figure 27). The closer part of the graben is somewhat dominated by the terraces that cover the Imperial sediments. The more northern part of the graben is dominated by a dark basaltic plug and a red cinder cone of the Alverson Formation (Figure 27). East Ridge is dominated by marbles.

10.5 MILES At the intersection, take the eastern road. The western road leads to the cinder cone and Switchback Mountain.

As the road heads north it begins to climb onto the west side of East Ridge. Looking west we can see the landslide at Stop 4. Looking north and across the graben is Switchback Mountain and the road that gives the mountain its name. The light colored marbles on the east side of Switchback Mountain are landslides. Separation Canyon is between Carrizo Mountain and Switchback Mountain and is fault controlled.

As the road turns to the east it passes under chert outcrops and onto a dark amphibolite. As the road turns north again it travels along the base of a sedimentary channel (mc) that we noted at Stop 4 (Figure 27). The road turns and climbs through the marble channel fill (mc) to the top of East Ridge and Stop 5.

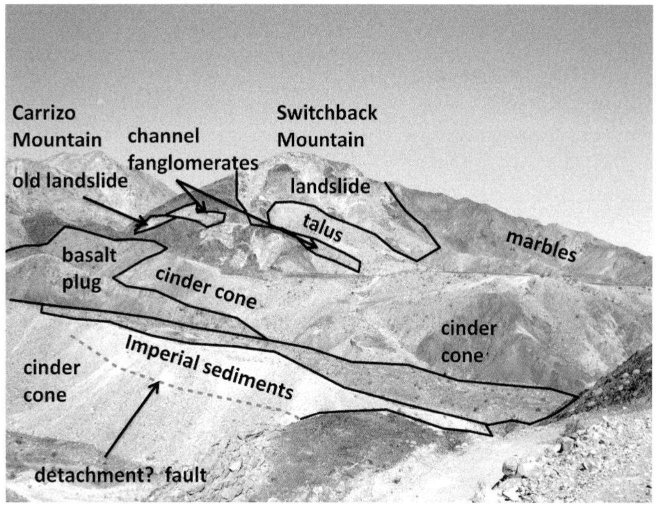

Figure 28. Stop 5 looking northwest at Switchback Mountain. In the foreground a detachment (?) fault may separate Alverson cinder beds from younger Imperial mudstones.

STOP 5

Offset of the marble channel fanglomerates, landslides, cinder cones and plugs, and Detachment Basin.

10.9 MILES Look west to the landslide we visited at Stop 4. Look northwest (Figures 27 and 28) at the red cinder beds of the eroded cinder cone and its dark basaltic plug. Beyond the cinder cone are the landslides on Switchback Mountain.

Outcrops made up of the same fanglomerates that fills the channel we are parked on can be seen on Switchback Mountain (Figures 27 and 28). The small remnant landslide on the east side of Separation Canyon was part of the older landslide at Stop 4. This old landslide dammed Separation Canyon. The gravels on Switchback and this channel represent a drainage that topped the landslide dam and crossed the graben. The channel has a 700 m right-lateral offset across the faults that form the graben.

To the southeast is a view of what we call "Detachment Basin" (Figure 29). In the foreground is another eroded cinder cone with a plug and a klippe (?).

The Deguyños mudstone in the center of Detachment Basin is faulted against the Western Hornblende Andesite plug (Mh). The fault is marked by the light tan mudstones in contact with the darker volcanics (Figures 29, 30, and 31). The mudstone is also faulted against the low Alverson basalts (Mav) on the southern boundary of the basin. The fault then turns north to the edges of the wash just east of the cinder cone below this stop. In this section Deguyños mudstones are faulted against Alverson volcanics (Figure 31 and 35). These faults now dip steeply. Before the uplift of the Coyote Mountains, we think that they were parts of a low-angle fault that was probably dipping to the east. Note cross section in Figure 31.

The Western Hornblende Andesite plug is prominent. With binoculars you can make out the contact in the Western Hornblende Andesite plug between the last volcanic eruptive plug (look for the jointing) that intruded and brecciated an older plug (no jointing).

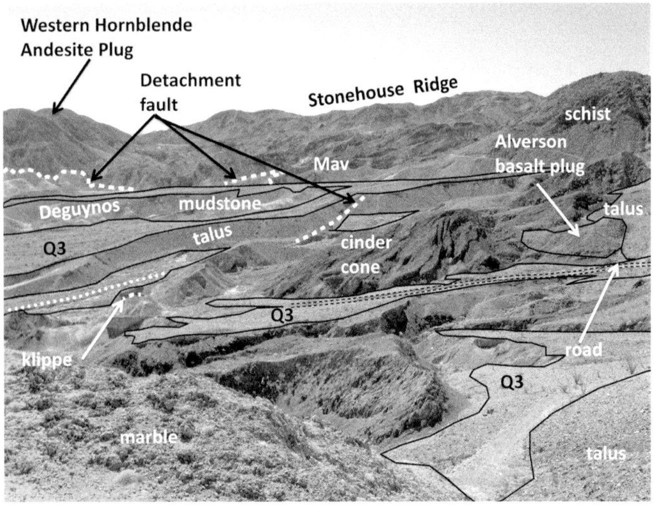

Figure 29. Looking southeast across Detachment Basin. **Mav**-Alverson basalts, **Q3**-Terrace 3. We drive just north of the volcanic plug.

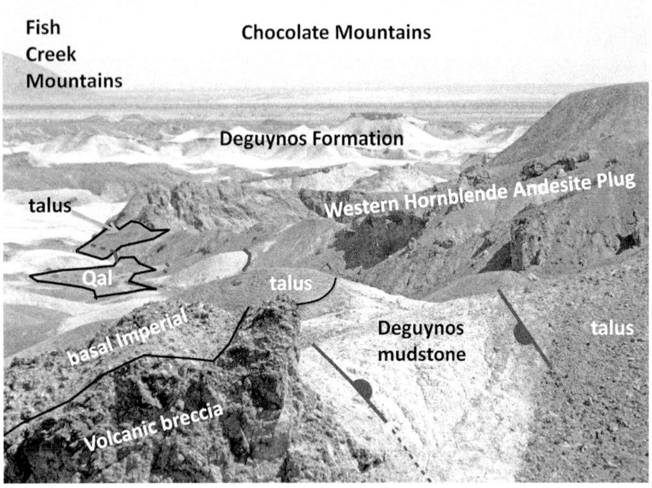

Figure 30. Looking north along the west side of the Western Hornblende Andesite plug and a detachment fault marked with black semicircles on upper plate.

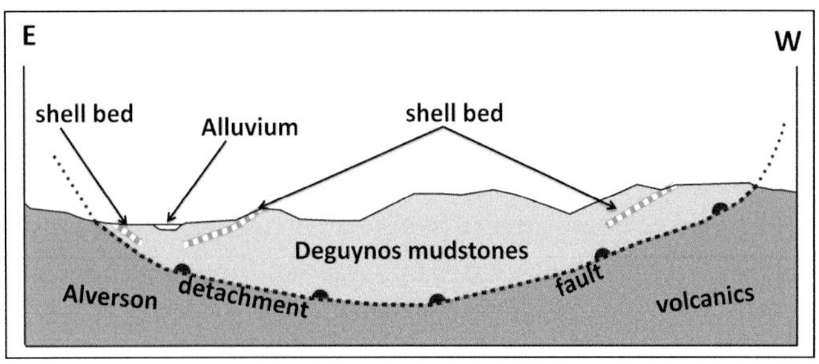

Figure 31. Simplified cross section across Detachment Basin. Not to scale.

Return to vehicles.

Drive down the ridge. Cross back into the amphibolite. From the west side of the horst, cross from marble into cherts and then into amphibolite. This is one of the only metamorphic sequences that may represent an original sedimentary section. The road passes into the talus slope material and stays on the talus except for some exposures of Alverson volcanics.

11.1 MILES At the southern end of the switchback we have a view of a canyon that traverses East Ridge. This canyon is fault controlled with the fault on the south side of the canyon. The fault is marked by the contact between Alverson volcanics (north side) with schist. The fault runs between the western horst boundary fault (Fault D) and the eastern horst boundary fault.

11.4 MILES At the northern switchback the northern fork in the road goes to a marble quarry. We take the south fork. The road shortly turns east onto Terrace 3.

11.5 MILES Look south as the road passes the dark basaltic plug and its red cinder cone. The cinder beds dip as much as 71° to the east. Further down, the road drops off Terrace 3 and onto Alverson brown basaltic tuff (Mb). This basaltic tuff member makes vertical hoodoo outcrops to the south. On the south side of the road is a pedestal of brown basaltic tuff that is capped by hornblende andesite volcanic breccia (Mvb) (klippe) (Figure 32). The contact between these two Alverson units is either a shallow (23 degrees) east-dipping fault (detachment?) or a landslide surface (?).

The road leaves the basaltic tuff and passes through another (younger) basalt plug. The bulbous basaltic outcrops may be caused by the presence of water in the brown basaltic tuff during intrusion of the basalt. From the plug we cross between outcrops of hornblende andesite breccia (Mvb) that cover marbles. On the south side of the road are marble outcrops faulted against the breccia. This fault and other faults that are little to the east represent a high angle fault between the basement rocks and the Deguyños mudstones and siltstones of the flats. This is the eastern fault that makes East Ridge a horst (Figures 7 and 35).

11.7 MILES At the next intersection we make a sharp turn to the north, and onto Deguyños mudstone. The road follows the contact between Tertiary sediments and the basement rocks of East Ridge. This contact is in part depositional and in part a fault. The marble surface to the west is covered with bullet (pholad) holes. The large clasts of marble along the road are also covered with bullet holes. Deposited on the marble are outcrops of Viejo marine basal conglomerate and tan sandstone. The Deguyños mudstone to the east is capped by Terrace 3. A fault is presumed to separate the Viejo sandstones and conglomerate from Deguyños mudstone. The road continues north and passes up onto Terrace 3 where we park for the 0.3 mile walk to Stop 6.

STOP 6

Impressive Mullion Structures, Deguyños Rhythmite, Basal Viejo Formation, and T-R sequences in the Deguyños Formation.

12.0 MILES Rhythmites are thick cycles of uniformly thin silt and mud beds or sands and silt beds. The sands commonly have well-rounded hematite-coated quartz grains derived from

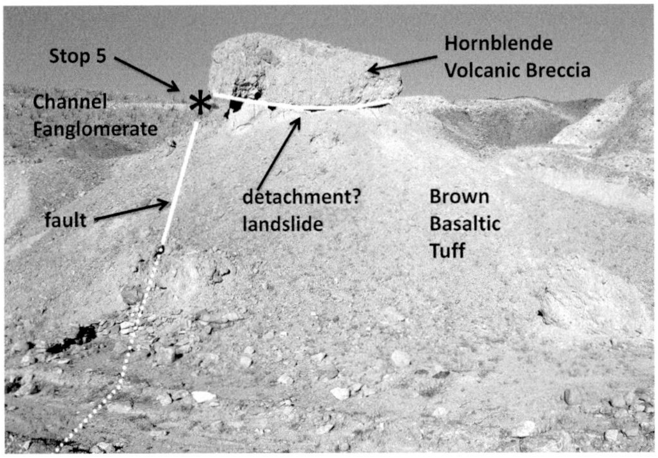

Figure 32. Looking west at the Klippe. The top unit is made of hornblende andesite breccia (**Mvb**). The lower unit is older basalt tuff (Mb) cut by a high angle fault. The contact between the two units is a slip plane dipping 23° east. The hornblende andesite breccia does not appear re-broken as would be expected in a landslide. If this is a low angle fault, it may represent part of a detachment.

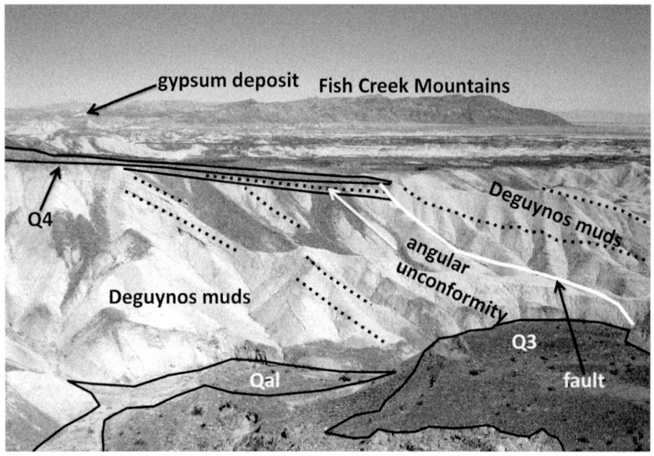

Figure 33. Looking north across Capote Canyon at the T-R sequences in the Deguyños Formation. **Q4**-Terrace 4 **Q3**-Terrace 3. **Qal**-alluvium Dotted lines are bedding planes.

the Colorado River (Winker and Kidwell, 1996). Winker and Kidwell believe the rhythmites are caused by seasonal flooding of the Colorado River. They are a part of the Deguyños Formation.

Walk to the west on Terrace 3. We highly recommend a good pair of high-topped boots and a walking stick for the steep slopes. As we walk, note the next ridge to the north, across Capote Canyon (Dibblee, 1954). The ridge is made up of marine mudstone and siltstones of the Deguyños that are capped by Terrace 4. Just under the western end of the terrace is a shell bed with 15 degree dip (Figure 33). Down the slope, the next shell beds of the Deguyños dip approximately 30°. This angular unconformity is likely in part the result of a change in sea level, i.e. a transgressional-regressional event.

At the edge of Terrace 3 look down and south at the mullion structures (Figure 34). Though unimpressive from here, they are much more impressive the closer you get. The mullion structure in the Coyote Mountains appears to develop only on marble. Note to the south that Terrace 3 extends onto the marble.

In Figure 35, Point B is the mullion structure. Point A is the northern extent of this mullion structure. There appears to be a left-lateral offset of the eastern high-angle fault contact between the basement rocks and the mudstone and siltstone of the Deguyños Formation. This high-angle fault may be part of a detachment fault that has been deformed by uplift of the Coyote Mountains. The detachment (?) at the klippe (K in Figure 35) would be an undeformed part of this detachment. The left-lateral offset may represent a zone of adjustment along the detachment (Figure 35). If this is a detachment fault, there is evidence of recent right-lateral movement.

On the north side of Capote Canyon, across from the mullions, are good outcrops of a Viejo marine basal conglomerate and a Deguyños rhythmite.

Figure 34. The mullion structure. This may be an older detachment that may have been reactivated by a right-lateral strike slip fault.

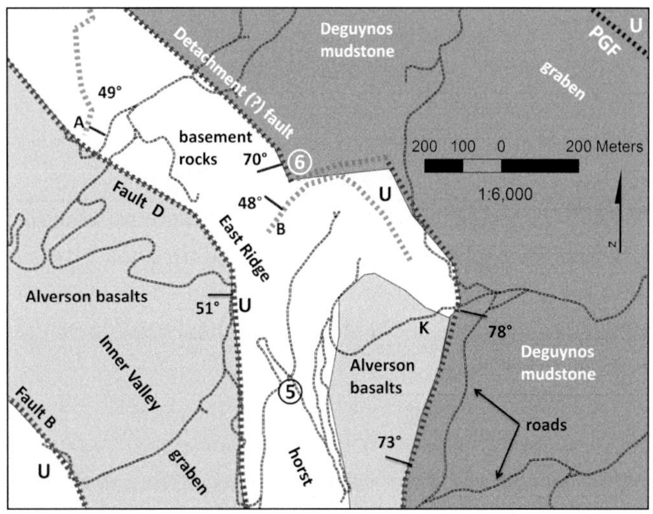

Figure 35. Simplified geological map. **Dotted lines**-faults, **Circled numbers**-field trip stops, **K**-klippe, **U**-up side of a fault, **A** and **B**-are areas of mullion structures. Black short lines on faults indicate dips. Grey fault-zone of adjustment(?).

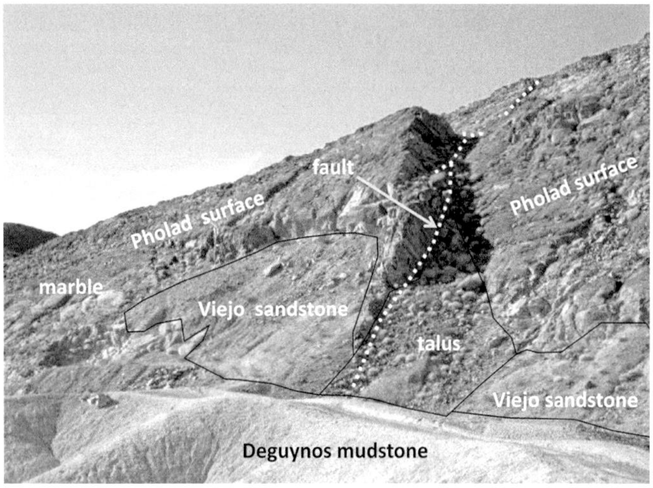

Figure 36. Looking southwest at Pholad or bullet hole surface. White line is a fault.

Return to the vehicles and retrace our route south. The marble surface on the west side of the road is covered with bullet holes (pholad). This surface is offset by some small faults (Figure 36). The scarp surfaces have no bullet holes. The faulting took place after the deposition of the Viejo marine sediments. The pholad bivalves are filter feeders that prefer high-energy environments that form conglomerates and sandstones. Pholads would avoid low-energy environments that form siltstones and mudstones of the Deguyños Formation.

12.3 MILES At the intersection, the road that goes south has views of hoodoos in the Alverson basaltic brown tuff (Mb). The road climbs up on the schist part of the East Ridge. Outcrops of Tertiary marble conglomerates under Alverson basalts occur along the road. The road ends with views of Stonehouse Ridge and Painted Gorge. At one time the road continued all the way to Painted Gorge. The last part is now a trail that was noted after 1.1 miles in this field guide (Figure 8).

The road turns east, with Deguyños mudstone (Pim) on the north and south side of the road.

12.5 MILES Note the shell beds on the north side of the road dipping east 16° just before we pass though the North Gate. This gate is closed between Jan 1 through June 30. The terrace to the south is Terrace 3. To the southeast the Western Hornblende Andesite plug looms large. We pass just north of the andesite plug as it becomes buried by the Deguyños mudstone and siltstone. The contact of the Alverson Hornblende Andesite plug and the Deguyños mudstone on the west side is a moderately west-dipping fault (Figure 36).

13.1 MILES After passing the Hornblende Andesite plug, the first drainage from the south has a road going into Detachment Basin. Continue east. Approach Painted Gorge fault (Christiansen 1957) that cuts Terrace 2 further north as noted at Stop 4. The fault is not traceable south from here. A high bluff just north of the road and just east of Painted Gorge fault is made up of Pbsb (marker bed) Deguyños T-R sequence. The fault on the east side of the Western Hornblende Andesite plug is high angle with a 1.2 m wide fault zone. The contact of the Deguyños mudstones with the andesite plug marks the fault trace.

Continuing east a small valley to the south opens up to a view of some flatirons low in the valley (Figure 37). The flatirons are made of Viejo marine sandstone dipping 41° to the east. On the skyline is a 22° dipping T-R sequence (Pbsb, a marker shell bed) of the Deguyños Formation that is uncomfortably overlying the moderately dipping Viejo Formation.

Further east and to south of the road, Figure 38, the Deguyños marker shell bed (Pbsb) and mudstone Ptim is unconformity deposited on an older Viejo marine basal conglomerate that is in turn unconformity deposited on the youngest Alverson basalt in the eastern Coyote Mountains. Before we dug out the contact, we thought that the contact between the mudstone and the basal conglomerate was a low-angle fault. Ann Bykerk-Kauffman had correctly suggested that this was a depositional contact.

13.1 MILES Just before we turn south, foreset beds within the Deguyños marker shell bed (Pbsb) can be seen on both sides of the drainage. The marker shell bed dips about 15° east. The road south runs along the unconformable contact between the Deguyños (Ptim) mudstone and siltstone, and a Viejo marine basal conglomerate. After topping the saddle the road descends to the south, and at the bottom of the grade, a small drainage enters from the west. Look west up this drainage at the Viejo marine basal conglomerate.

In front of us the shell-bed ridge is a part of the Deguyños marker bed (Pbsb). There is a fault (south side down) along the north side of the ridge. The marker bed dips about 35° to the southeast. To east are dark outcrops of the Viejo marine basal conglomerate and the youngest Alverson basalt on the north side of the fault.

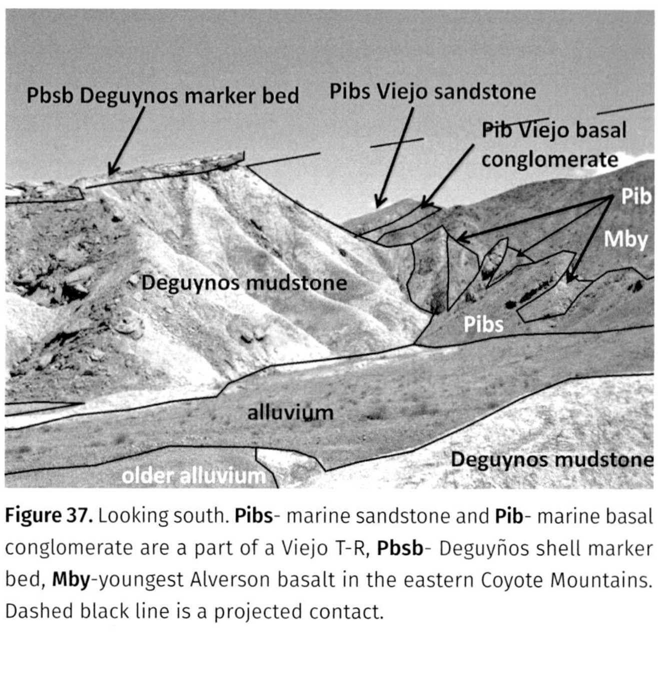

Figure 37. Looking south. **Pibs**- marine sandstone and **Pib**- marine basal conglomerate are a part of a Viejo T-R, **Pbsb**- Deguyños shell marker bed, **Mby**-youngest Alverson basalt in the eastern Coyote Mountains. Dashed black line is a projected contact.

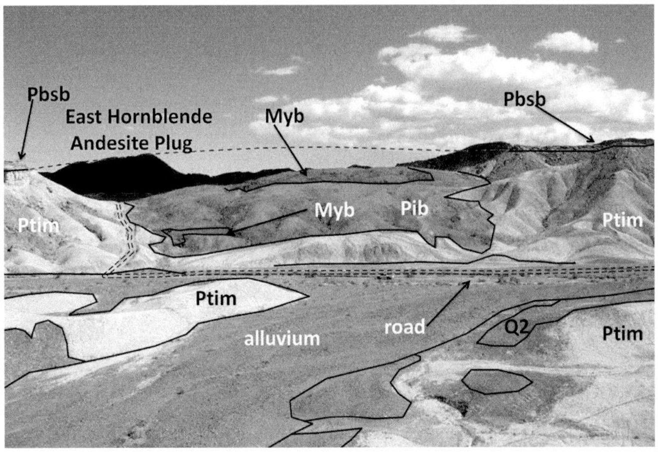

Figure 38. Looking south at erosional window. **Pbsb**-Deguyños marker shell bed. **Myb**-youngest basalt. **Pib**-Viejo marine basal conglomerate undifferentiated. Dotted line-projected contact.

Figure 39. Looking north at a low angle fault on the north side of Painted Gorge Wash in Deguyños mudstone.

Figure 40. Close view of the fault zone without slickensides.

13.4 MILES The road crosses the fault and the ridge, turns southwest, and climbs up on Terrace 2. View this prominent terrace to the south and east. The road drops down into a small drainage. In the valley northwest of the road the youngest basalt (Myb) is deposited on the poorly exposed green tuff (Mgt) beds. Resting on the youngest basalt is marine sandstone of Viejo Formation. These units are unconformably overlain by mudstone (Pitm) and the shell marker bed (Pbsb) of the Deguyños Formation. Looking west Deguyños Formation(?) can be seen faulted against the Western Hornblende Andesite plug.

The road crosses out of the drainage and onto a Deguyños mudstone. Terrace 2 caps the mudstone southeast of the road. This terrace represents the earliest deposits of the diverted Painted Gorge drainage. The road turns south down into the eastern end of the narrows of Painted Gorge (Mile 13.7). Directly across the drainage a Viejo basal conglomerate can be clearly seen on the Eastern Hornblende Andesite plug. We turn east.

We have entered Painted Gorge Wash which widens to the east. Terrace 2 caps the Deguyños mudstone cliffs that bound the northern side of the wash.

STOP 7

Low angle fault-detachment?

13.8 MILES Observe a shallow eastwardly dipping fault (detachment?) (Figures 39 and 40) on the north side of Painted Gorge Wash. The fault dips 15° to the northeast. In the 1.1 m thick fault zone there are no slickensides, only pearly surfaces (Figure 40). This fault was exposed after a small flash flood in 2006(?) and has almost been lost to slope failure.

Across and farther down the wash, Deguyños mudstone is faulted against a part of the Eastern Hornblende Andesite plug.

Return to camp.

DAY TWO

MILE 0 Reset odometer. From the Kiosk go east through the narrows. On the southern wall of the narrows Mace (1981) has a paleomagnetic site that show no rotation, but the site is next to a fault. At the eastern end of the narrows and across from the road to the North Gate, a Viejo basal conglomerate is deposited on the Eastern Hornblende Andesite plug. At 0.3 mile we pass Stop 7 and the road and the drainage stay on the north side of the wash next to the cliffs of Deguyños mudstones and siltstones. The first break in the sedimentary wall is an area where a stream to the north is actively pirating the Painted Gorge Wash drainage.

Look to the east, down the road. Terrace 4 appears to have its upper surface folded into a syncline.

1.3 MILES The hill to the north is capped by Palm Spring Group sedimentary rock that overlies Deguyños mudstone.

Turn north onto a trail that follows the west side up fault scarp of the Painted Gorge Wash fault. The scarp is ~68cm high. The Painted Gorge Wash fault scarp can be followed south across the wash. The trail turns to the northwest but the scarp continues to the north. The cuesta to the northeast is made up of the Palm Spring Group freshwater sequence capped by Terrace 4 with a west facing erosional scarp. The west-side-up fault scarp cuts the Palm Spring Group on the northwest side of the cuesta.

1.5 MILES The trail crosses a wash. We turn to the northeast and follow the wash. The wash turns more northerly as we approach the fault that drops from the south into the wash.

The terrace on the northwest side of the wash is capped by Terrace 2 (Q2) that is unconformity deposited on freshwater Palm Spring Group sediments. The pink mudstones, tan to white friable silt beds, dark brown sandstones, and concretion beds along the walls of the wash are characteristic of the Palm Spring Group. Many beds contain petrified wood. The road follows the wash which is coincident with the trace of the Painted Gorge Wash fault, visible sporadically along the walls of the wash.

2.5 MILES The road and the wash turn east, the fault continues to the northeast. At the north side of the wash stop to observe an exposure of the Painted Gorge Wash fault. The terrace above the bench is capped by Q3.

STOP 8

Painted Gorge Wash left-lateral fault

The Painted Gorge Wash fault is an active left-lateral high-angle fault that is a conjugate to the Elsinore fault (Morgan and Morgan, 2005). It is mapped for 9.5 km, strikes northeast and is east of the Coyote Mountains. This fault is well exposed at this location, providing information on a slip event.

After looking at the fault exposure, retrace route back to the kiosk.

Passing the kiosk we head south up the dugway in the Imperial mudstone. Note the shell bed dipping 58 degrees east on the west side of the road. At the top we cross onto Terrace 3 and follow it south. The volcanic hill straight ahead and west of the road is an Alverson basaltic plug. Based on stratigraphy and weathering profile, this volcanic plug appears to be the oldest one of several plugs in the Painted Gorge area. The road continues south between the basalt plug and the younger Eastern Hornblende Andesite Plug (Figure 3). Dibblee and Christensen put north-south faults between the two plugs; but we found no evidence of faulting. There is brecciation of the western side of the Eastern Hornblende Andesite Plug but this is the result of the intrusion of a younger andesite plug.

Continue south. The drainage is incised in Terrace 3. As we drive down the drainage, Terrace 3 remains at the same level, giving the appearance that the Terrace is raising.

3.5 MILES Note that the wash narrows and basalt is on both sides of the wash. The basalts on the eastern side of the road are faulted against the hornblende andesite plug. The basalts on the western side of the road are capped by Terrace 3.

3.9 MILES The wash widens. To the west, a Viejo marine basal conglomerate has been deposited unconformably on Alverson basalts. Both are capped by fanglomerates of Terrace 3.

To the east the surface of Terrace 3 is cut by a fault. The north-side-up scarp is ~80 cm high. From here Terrace 3 forms an alluvial fan that is being dissected (Figure 43).

Figure 41. Exposure of the Painted Gorge Wash fault, north side of Painted Gorge Wash.

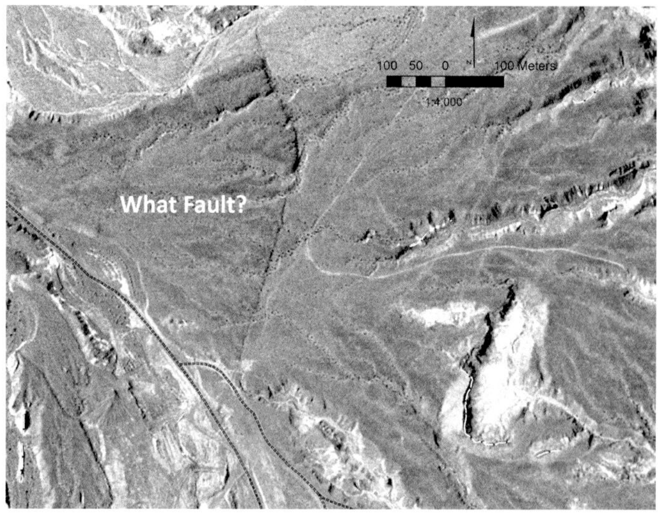

Figure 42. Painted Gorge Wash Fault.

Figure 43. Painted Gorge Wash fault area with geology on aerial orthophoto. The white lines are active faults, dotted where inferred. **Qoa**-older alluvium, **Q2**-Terrace 2, **Q3**-Terrace 3, **Pim**-Deguyños mudstone, **Pio**-top of a Deguyños shell bed, dark thin lines-fault offsets (?).

STOP 9

Scarp of the Painted Gorge Wash Fault and Dissection of the Terrace 3 alluvial fan

4.1 MILES Walk to the scarp of the Painted Gorge Wash fault, northwest side up (Figure 44). The Eastern Hornblende Andesite plug is the prominent hill to the north. Note the widening of Terrace 3 to a fan. The terrace is currently being dissected because of a change in the local base level.

Return to vehicles.

Drive south and look west. On the southern side of Stonehouse Ridge (Figure 45) is a Viejo marine sedimentary channel in the metamorphics. Underlying the Viejo marine sediments are Viejo freshwater fanglomerates. The break in slope on Stonehouse Ridge is a fault between Deguyños mudstones and the basement rocks.

4.8 MILES Road to the west accesses the south side of Stonehouse Ridge, the cuestas and the north side of Butaca Canyon. The road crosses the Painted Gorge Wash fault. Near the Viejo marine channel is the type area of the White Cross Gneiss (Jwc), and is the eastern end of a belt of this gneiss that can be followed west to the Elsinore fault north of Ocotillo Canyon.

Drive down the wash. The sidewalls of the wash consist of Deguyños sediments with shell beds until about mile 4.9, then the freshwater tan silts and pink muds of the Palm Spring Group replace the Deguyños. The Palm Spring sediments dip to the south.

5.5 MILES Intersection with the power line road, a well maintained road that goes west to Alverson (Fossil) Canyon area. The road crosses the southern part of the Painted Gorge Wash fault. A branch of this road goes up Butaca Canyon drainage.

7.3 MILES Cross Coyote Wash. The road begins to climb Palm Spring Group siltstone that dip south (Dibblee, 1954). The overlying Quaternary terrace gravels represent a lake level of one of the many levels of freshwater Lake Cahuilla.

8.0 MILES Turn west on Old Highway 80. To the east is Plaster City. Drive off the Lake Cahuilla terrace. Look south at the railroad and the old cement road. There are no faults cutting the railroad bed or the old cement road.

8.6 MILES A close look at the bridge at the bottom of the slope will reveal that the bridge segments are offset right laterally. After the April 4, 2010, El Major–Cucapah Earthquake, there were new right lateral offsets between the bridge segments.

Note the descending sand dune on the granitic low hills above the Davies Valley area to the southwest of Old Highway 80.

11.2 MILES Turn north on a haulage road and have a good look at the southern Coyote Mountains. Just to the west of the road is the active, east side up, oblique right lateral Ocotillo fault (Treiman, 2014) (Figure 46). This fault showed surface rupture following the April 4, 2010, El Major/Cucapa earthquake. In figure, the dark vegetation side of the fault is on the up-stream side of the fault. The recent small fault scarp and the older, higher, west-facing

Figure 44. Looking north at the scarp, west side up, Painted Gorge Wash fault. White Dotted line-faults, **Q3**-Terrace 3, black lines top of Terrace 3. Unnamed fault cuts and offsets right-laterally the Eastern Hornblende Andesite plug.

Figure 45. Looking west at the Viejo sedimentary channel. Under the Viejo marine sandstones are Viejo freshwater fanglomerates. The sedimentary beds dip about 35° to the south.

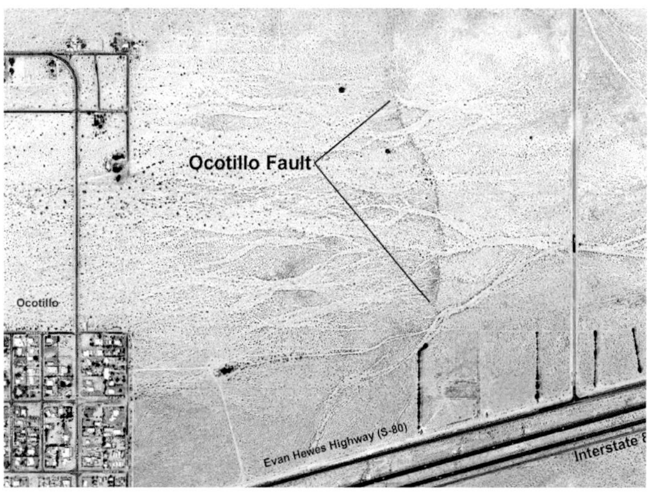

Figure 46. The Ocotillo fault is well-expressed in aerial photos from various years. Its visibility is principally from the vegetation on the upstream (west) side of the fault. Windblown sand and periodic flash flooding have otherwise modified the geomorphic scarp. Image from Google Earth dated 12/30/2005. From Figure 44 of Treiman (2014).

fault scarp (~30 cm) are somewhat restricted to the active alluvial channel of Coyote Wash. The older weakly cemented sediments that make up the banks of Coyote Wash do not have scarps.

Note the dark cuestas on the southern Coyote Mountains. The resistant marker bed that forms the cuestas is part of the Viejo marine Dark Cuesta T-R sequence and is not a volcanic flow. The desert varnish on the marker bed results from the presence of abundant granitic fragments (Morgan and Morgan, this volume).

A synform in basement rock is on the southern side of Ocotillo Peak. The Palm Spring Group makes up the low hills we are driving through (Dibblee 1954).

14.0 MILES Turn north onto Shell Canyon Road. One of the Imperial County landfills is to the northeast. On the northwest side of the dump and east of the road is a terrace composed of the Palm Spring Group (pink muds) capped by Terrace 2 (Q2). On the north side of the terrace is the Elsinore fault, placing Deguyños marine green mudstone (north) against Palm Spring freshwater pink mudstone at the break in slope.

14.6 MILES The power line road is to the east; keep going north.

14.8 MILES Turn northwest. Leave the "paved" road and head toward Alverson (Fossil) Canyon.

14.9 MILES Turn northeast into Alverson (Fossil) Canyon. The terrace just above the road was the old level of the canyon before the sand and gravel operations lowered local base level 25+ years ago.

STOP 10

Mouth of Alverson Canyon. Viejo Dark Cuesta marker bed T-R Sequence, Viejo marine basal conglomerates and sandstone, and Viejo freshwater Green Fanglomerate.

15.5 MILES Patricia Bell (1980) was the first to recognize this marker bed in the Imperial sediments (Figure 47). This quartz-rich marker bed (Pimb) is an example of sandstone and basal conglomerate forming a single bed in a T-R sequence (Figure 47). This thin marker bed is the same unit as the dark cuesta on the southeast side of the Coyote Mountains. We have named this T-R sequence the Dark Cuesta. This marker bed is continuous over 600 m north up Alverson Canyon. Isolated outcrops of this marker bed continue north for almost 3.5 km. This Marker Bed is continuous to the east for over 3 km. This bed ranges in thickness from about 20 cm here at the mouth of Alverson Canyon to as much as 4 m at the cuestas. We consider the sequence to be the youngest Viejo Formation T-R sequences in the Coyote Mountains. This is the first transgressional sequence to cover the Coyote Mountains completely. We will also look at the older Viejo marine T-R sequence under the Marker Bed and its contact with

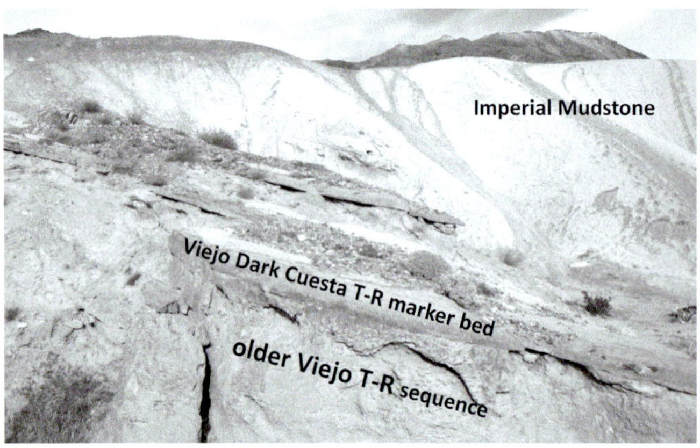

Figure 47. Marker bed at the mouth of Alverson Canyon described by Pat Bell. The marker bed belongs to the Viejo Dark Cuesta T-R Sequence.

Figure 48. Close up of the imbricated rounded pebbles found in the Viejo Dark Cuesta marker bed. The matrix is a coarse, angular, arkose with granitic fragments. Pen for scale.

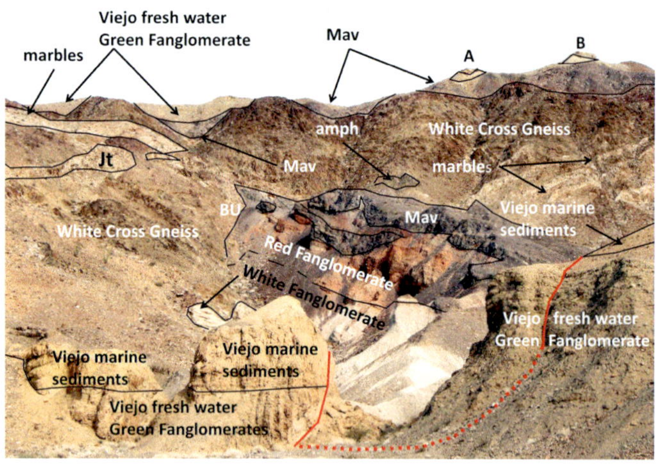

Figure 49. Looking east at an eastern wash of Alverson Canyon (Mile 15.5+). This is the wash where Ann Bykerk-Kauffman and her students found the first thrust fault in the Coyote Mountains ("T" in Figure 51). This is also a zone of accommodation for the western detachment fault (Figure 51). **BU** shows where Alverson with White and Red Fanglomerates form a buttress unconformity with White Cross Gneiss. Just beyond the gate formed by the Viejo marine sediments and Green Fanglomerate (lower part of photo) is the right-lateral strike-slip fault that is the young continuation of the western detachment fault (Figure 51). Buttes A and B on the skyline are Viejo marine sediments over Viejo freshwater Green Fanglomerate. **Mav**-Alverson Formation basalts, **amph**-amphibolite, **Jt**-Jurassic tonalite. Up this drainage is another of Mace's (1981) paleomagnetic sites and this site shows no rotation while sites in the northern Alverson Canyon area have 111 degrees of clockwise rotation.

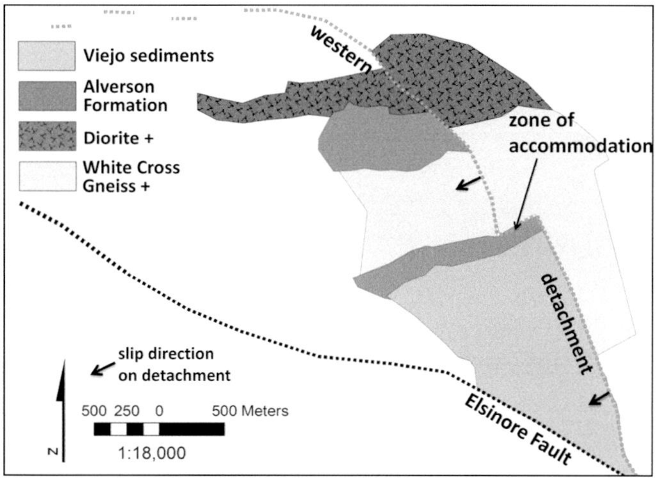

Figure 50. Simplified geological reconstruction of Alverson Canyon before right-lateral faulting reactivated the western detachment fault and before thrusting. The western detachment fault has a 3m wide fault zone between the White Cross Gniess and the Alverson basalts north of the zone of accommodation. Elsinore fault likely did not exist but is used as the boundary for the area. In the map the Alverson Formation includes the Red and White Fanglomerates. The Viejo includes the Green Fanglomerate. There are outcrops of Viejo sediments and Alverson Formation without Red and White Fanglomerates, overlying the diorite and White Cross Gneiss.

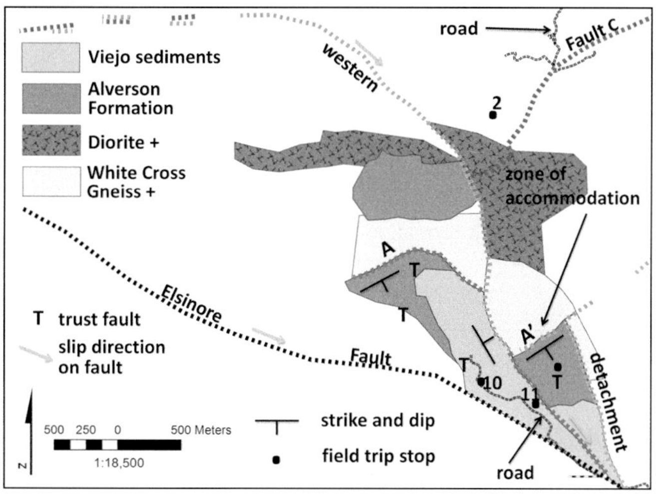

Figure 51. Simplified geologic map of Alverson (Fossil) Canyon after thrusting and right-lateral faulting. The Alverson Formation includes the Red and White Fanglomerates. The Viejo includes the Green Fanglomerate. There are outcrops of Viejo sediments and Alverson Formation without Red and White Fanglomerates on the diorite and White Cross Gneiss. **WDF** represents the western detachment fault. A' is the area of accommodation between the two branches of the **WDF**. The fault at point A (continuing past Stop 11), south of the zone of accommodation, is a strike-slip fault that connects the northern and southern branch of the **WDF** and is the continuation of the reactivation of the northern branch of the **WDF** as a right-lateral strike slip fault. This fault's zone is about 8+ cm wide. Steely et al. (2008) dates the transition from detachment faulting to strike-slip faulting at about 1 Ma. The fault at point **A** appears to be strike-slip fault with a thrust component. The thrusting appears to be restricted to the areas south of **A** and south of **A'**, using the strike-slip faults as zones of accommodation within these areas. No thrusting has yet been found north of these points. The areas of point **A** and point **A'** appear to have been offset right laterally approximately 1000 m. The contact between the diorite and White Cross Gneiss also appears to be offset right laterally.

the underlying Viejo freshwater Green Fanglomerate.

Dibblee (1954) and Bell (1980) considered the Green Fanglomerate to be a part of the Imperial. Winker and Kidwell (1996) combined it with the Garnet and the other unnamed fanglomerates under the Imperial Group and above the Alverson Formation, and put this group of fanglomerates with the Alverson and the Red and White Fanglomerates in the Split Mountain Group. We have mapped the Green Fanglomerate as a fresh-water member of the Viejo Formation. A more detailed description of the Green Fanglomerate can be found in Morgan and Morgan, this volume.

Turn around and backtrack down Alverson Canyon.

STOP 11

Viejo marine sediments interfingering with the Green Fanglomerate.

15.8 MILES *This is a no hammer outcrop.* Walk a short way up a small side canyon and look at Viejo marine interfingering with Viejo freshwater Green Fanglomerate (Figure 52). Note the Viejo marine marker Dark Cuesta high on the north side at the mouth of the canyon. Many of the fault zones in this section of the Green Fanglomerate are characterized by thin brown cataclasites. Here are Viejo fossils and thin sandstone interfingering with Viejo freshwater Green Fanglomerate. Some of the fossils appear to be in place. Marine fossils are absent up section.

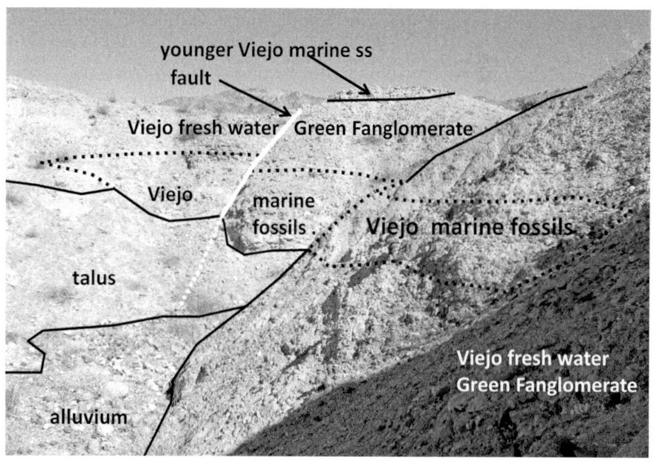

Figure 52. Looking north at fresh-water Viejo Green Fanglomerate with interfingering marine Viejo sandstone and basal conglomerate.

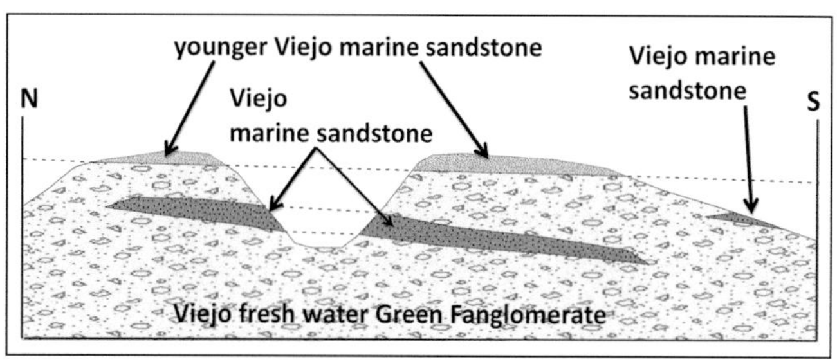

Figure 53. Simplified cross section of the Viejo marine sandstone interfingering with the freshwater Viejo Green Fanglomerate. Not to scale.

Unconformably younger Viejo marine sandstone overlies the Green Fanglomerate (Figures 52 and 53). Just to the east of the outcrop is the right-lateral strike-slip fault that is the continuation of the western detachment fault. On the eastern side of this fault is Red Fanglomerate and basalts of the Alverson Formation.

There are other areas in the Alverson Canyon that show Viejo marine sediments interfingering with the Viejo freshwater Green Fanglomerate. We have not found any other Viejo freshwater fanglomerates with interfingering Viejo marine units.

Return to vehicles.

Continue driving south down the canyon.

15.9 MILES Turn west through a narrows. As we turn note the amphibolite on the south side of the canyon. Blaettler (1991), using thin-section petrography, found that this amphibolite and other amphibolites in the southern Coyote Mountains were basalts at one time. The relic pillows we have found in one of the amphibolites in the southern Coyote Mountains confirm Blaettler's findings.

16.3 MILES Back on the "paved" road. Cross under power lines. Note an outcrop of the Dark Cuesta marker bed on the descending ridge to the east.

16.4 MILES Turn onto the power line dirt road. The first canyon to the north is Chuckawalla Canyon (Bell, 1980). Bell reports sand dollar fossils in "the middle" of what we call the Viejo freshwater Green Fanglomerate. This road for the most part will stay on Deguyños mudstone and younger Quaternary units. The road will parallel, for a distance, the Elsinore fault.

16.9 MILES Turn north at the intersection east of the terrace to the south. The low hills just north of the road have outcrops of Viejo Green Fanglomerate with and without capping Viejo marine sediments. On the skyline is the Dark Cuesta marker bed (Pimb). The green sediments under this marker bed are the Viejo freshwater Green Fanglomerate.

17.2 MILES On the western side of the wide wash turn north. At the next two intersections take the west forks.

STOP 12

Green Fanglomerate overlain by Alverson Formation volcanics.

17.5 MILES Climb up on a Deguyños mudstone hill and look at the Viejo Green Fanglomerate capped by Alverson volcanics. No marine fossils have been found in this outcrop of the Green Fanglomerate (Figures 54 and 55). The Viejo Green Fanglomerate outcrops previously noted to the west contain marine fossils. This implies that some of the Viejo sediments are older than some of the volcanics of the Alverson Formation.

Return to the vehicles.

Drive to next stop. Look north. Note the stripped dip slope of the Dark Cuesta marker bed.

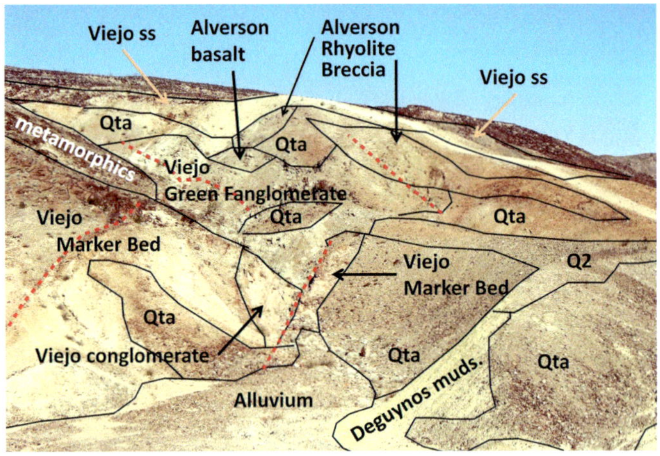

Figure 54. Looking north at the Viejo freshwater Green Fanglomerate capped by a basalt flow and a rhyolite breccia of the Alverson Formation. **Qal**-alluvium, **Qta**-talus, **Q2**-Terrace 2, Viejo marine ss (sandstone) is the unit under the Viejo marine Dark Cuesta marker bed-(**Pimb**).

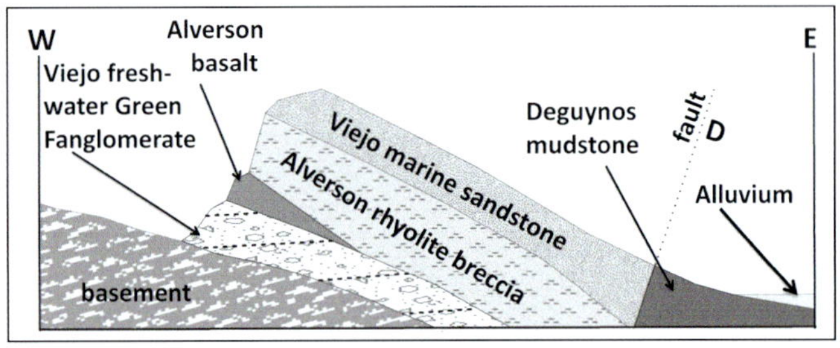

Figure 55. Simplified cross section of the Green Fanglomerate below basalts and rhyolite of the Alverson Formation. Not to scale. The dotted lines in the Green Fanglomerate represent bedding planes, dipping 30° south. The Alverson Formation and Imperial sediment dip to the east.

STOP 13

Woodring's outcrops. Imperial interfingering with the Alverson Formation.

18.3 MILES *This is a no hammer outcrop.* Woodring's outcrops are the only place in the Coyote Mountains where Viejo marine sediments interfinger with basalt flows of the Alverson (Figures 56, 57, and 58). Woodring (1931) described these outcrops while looking at the Tertiary sediments in the Salton Trough. He did not produce a map that could be used to reacquire these outcrops. The last two field trip stops and Bell's Chuckawalla Canyon outcrops support Woodring's observations that Viejo marine sediments do interfinger with the volcanic Alverson Formation. Woodring's outcrops are small: even with the *no hammer* restriction they may not last long. These outcrops contain ~60 cm+ thick Viejo siltstone that interfingers with volcanic basalt flows that may be in the upper section of the Alverson Formation. The fossils in the siltstone are broken and not in place. We have identified fossil turritella, sand dollars, sea urchin spines, oysters, barnacles, and clams. There are no baked zones associated with the volcanic flows. Note the green tint of the Woodring basalts. The gray and red basalts up slope are the next unit below the Woodring basalts (Figure 57).

Figure 56. Looking north at the northern outcrop of Imperial (Viejo marine sediments) between two basaltic flows of the Woodring basalts. There is no baked zone.

Figure 57. Looking at Woodring's southern outcrop. There is no baked zone associated with the basalt.

A short walk (0.1 mile) to the northeast and we can look at Butaca Canyon and the fault contact between the Alverson basalts and mudstones of the Deguyños. Up the canyon are paleomagnetic sites from Mace (1981). The clockwise rotation ranges from 43 to 75 degrees. Mace also has a K-Ar date of 17.1±0.5 Ma from a basalt near the mouth of the canyon.

This ends Field Trip A.

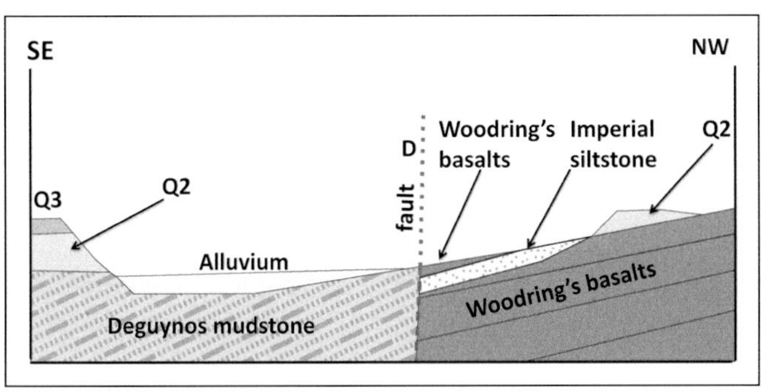

Figure 58. Simplified cross section of Woodring's Outcrops. **D**-down side of fault, **Q2**-Terrace 2, **Q3**- Terrace 3

REFERENCES

Axen, G. J., and Fletcher, J. M., 1998, Late Miocene-Pliocene extensional faulting, northern Gulf of California, Mexico, and Salton Trough, California: International Geological Review, V 40, p. 217–244.

Bell, Patricia J., 1980, Environment of Deposition, Pliocene Imperial Formation Southeast Coyote Mountains, Imperial County, California: M.S. thesis, San Diego State University.

Christensen, A. D., 1957, Part of the geology of the Coyote Mountains area, Imperial County, California: Unpublished M.S. thesis, University of California at Los Angeles, 188 p.

Dibblee, T. W., 1954, Geology of the Imperial Valley region, California, in Jahns, R. H., ed., Geology of southern California: California Division Mines Bulletin, v. 170, p. 21–28.

Dibblee, T. W., 2003, Geology of the Coyote Mountains, Imperial and San Diego Counties, San Diego. In, M.L. Murbach and M.W. Hart (eds.), Geology of the Elsinore Fault Zone, San Diego Region: San Diego Association of Geologists, San Diego, South Coast Geological Society, Santa Ana, California, p. 117–123.

Dorsey, R., 2006, Stratigraphy, Tectonics, and Basin Evolution in the Anza-Borrego Desert Region, in Jefferson, G., and Lindsay, L., eds., Fossil Treasures of the Anza-Borrego Desert (Chapter 5). San Diego: California State Parks and Sunbelt Publications.

Link, M. H., and Abbott, P. L., 1991, Eocene sedimentary history, San Diego, California: overview and Field trip stops, in Abbott, P. L., and May, J. A., eds., 1991, Eocene geological history San Diego Region, Pacific section SEPM, Vol. 68, p. 1–26.

Mace, N., 1981, A paleomagnetic study of the Miocene Alverson Volcanics of the Coyote Mountains, western Salton Trough, California: Master's Thesis, San Diego State University.

Magistrale, R., and Rockwell, T. K., 1996, The central and southern Elsinore fault zone, southern California, in Bulletin of the Seismological Society of America, Volume 89, No, 6, pp. 1793–1803, December.

Marshall, M., Mace, N., Erskine, B., and Jamsen, E., 2006, What regional and local paleomagnetic data tells us about the tectonics of the Coyote Mountains, southwestern Imperial Valley, southern California, in Bloom, D. M. and Farquharson, P. T., eds., Geology and history of southeastern San Diego, California, p. 55–62.

Miller, R, H., and Dockum, M. S., 1983, Ordovician conodonts from Metamorphosed carbonates of the Salton Trough, California: Geology, v. 11, p. 410-412.

Morgan, G. J., and Morgan, J. R., 2005, The Painted Gorge Wash Fault: recent left-lateral faulting, east side Coyote Mountains, Imperial County, California: Bloom, D. M., ed. Geology and history of southeastern San Diego County, California, San Diego Association of Geologists, p. 189–204.

Morgan, G. J., and Morgan, J, R., 2010, Post Pliocene detachment on the eastern side of the Coyote Mountains, Salton Trough, southern California:, Abstract and Programs, v. 42, n. 4, p. 94.

Morgan, G, J., and Morgan, J. R., 2013, Recent mapping in the Coyote Mountains, western Salton Trough, Imperial County, southern California, USA: GSA, Abstract and Programs, v. 45, No. 6, p. 61.

Morgan, G. J., and Morgan, J, R., 2014, A 2014 update of an Overview of the Geology and Geological Problems of the Coyote Mountains, Salton Trough, Imperial County, southern California. In: Coast to Cactus, Geology and Tectonics San Diego to Salton Trough, California, Morton, J. B., Editor, p. 135–148.

Morgan, J. R., Morgan, G. J. and Pecha, M., 2012, U/Pb dating of tuffs from the Alverson Volcanics in the Fossil Canyon and Painted Gorge areas of the Coyote Mountains, western Salton Trough, California, GSA, Abstract and Programs, Vol. 44, No. 3, p. 22.

Morton, P. K., 1977, Geology and mineral resources of Imperial County, California: California Division of Mining and Geology, County Report No.7.

Ruisaard, C. I., 1979, Stratigraphy of the Miocene Alverson Formation, Lmperial County, California: Masters of Science Thesis, San Diego State University, 125p.

Spencer, J. E., Patchett, P.J., Peartree, P. A., House, P. K., Sarna- Wojcicki, A. M., Wan, E., Roskowski, J. A. and Faulds, J. E., 2013, Review and analysis of the age and origin of the Pliocene Bouse Formation, lower Colorado River Valley, southwestern USA: Geological Society of America, Geosphere.gsapups.org, p. 444–459.

Steely, A. N., Janecke, S. U., Dorsey, R. J., and Axen. G. J., 2009, Early Pleistocene initiation of the San Felipe fault zone, SW Salton Trough, during reorganization of the San Andreas fault system: Geological Society of America Bulletin, v. 121, P. 663-687, doi. 10.1130/B26239.1

Watkins, R., 1990, Paleoecology of a Pliocene rocky shoreline, Salton Trough Region, California: Society for Sedimentary Geology, Palaios, Research Letters, V. 5, p. 167–175

Watkins, R., 1990b. Pliocene channel deposits oyster shells in the Salton Trough region, California. Palaeogeography Paleoclimatogy Palaeoecology v. 79, p. 249–262

Winker, C. D., and Kidwell, S. M., 1996, Stratigraphy of a marine rift basin: Neogene of the western Salton Trough, California, in Abbott, P. L., and Cooper, J. D., eds., Field Conference Guidebook and Volume for the Annual Convention: San Diego, California, American Association of Petroleum Geologists, Pacific Section, p. 295–336.

Woodring, W. P., 1931, Distribution and age of the Tertiary Deposits of the Colorado Desert: Carnegie Institution of Washington Publication 148, p. 1–25

COMPLEX PALEOTOPOGRAPHY AND FAULTING NEAR THE ELSINORE FAULT, COYOTE MOUNTAINS, SOUTHERN CALIFORNIA

M. J. Brenneman and A. Bykerk-Kauffman

Geological and Environmental Sciences, California State University, Chico

The Coyote Mountains of southern California are bounded on the southwest by the Elsinore fault, an active dextral fault within the San Andreas fault zone. According to Axen and Fletcher (1998) and Dorsey and others (2011), rocks exposed in these mountains comprise a portion of the hanging wall of the east-vergent Salton Detachment Fault, which was active from the late Miocene-early Pliocene to ca. 1.1–1.3 Ma. Detachment faulting was accompanied by subsidence, resulting in deposition of a thick sequence of marine and nonmarine sedimentary rocks. Regional detachment faulting and subsidence ceased with the inception of the Elsinore fault, which has induced uplift of the Coyote Mountains. Detailed geologic mapping in the central Coyote Mountains supports the above interpretation and adds some intriguing details. New discoveries include a buttress unconformity at the base of the Miocene/Pliocene section that locally cuts across strata at an angle so high that it could be misinterpreted as a fault. We thus conclude that the syn-extension strata were deposited on a surface with very rugged topography. We also discovered that locally-derived nonmarine gravel deposits exposed near the crest of the range, previously interpreted as part of the Miocene Split Mountain Group by Winker and Kidwell (1996), unconformably overlie units of the marine Miocene/Pliocene Imperial Group and must therefore be Pliocene or younger. The presence of such young gravel deposits on the crest of the range provides evidence for its rapid uplift. Additional new discoveries flesh out details of the structural history of the range. We mapped just two normal faults, both of which were relatively minor, thus supporting Axen and Fletcher's assertion that the hanging wall block of the Salton detachment fault had not undergone significant internal deformation during extension. We found abundant complex synthetic and antithetic strike-slip faults throughout the area, some of which offset Quaternary alluvial deposits. We interpret these faults as Riedel shears of the Elsinore fault that distribute dextral strain over an area at least 2 km wide. Finally, our mapping of the Elsinore fault itself reveals two releasing bends that are superimposed on the overall transpressive regime in the area.

References

Axen, G.J. and Fletcher, J.M., 1998, Hall Volume, GSA, p. 365-392.

Dorsey, R.J., Housen, B.A., Janecke, S.U., Fanning, C. M., Spears, A.L.F., 2011, GSA Bulletin, v. 123, p. 771-793.

Winker, C.D. and Kidwell, S.M., 1996, Field Conference Guide, Pacific Section AAPG/SEPM, Book 80, p. 295-336.

Originally published by the American Geophysical Union,
Fall Meeting 2012, abstract #T23D-2714, December 2012.
Retrieved from http://adsabs.harvard.edu/abs/2012AGUFM.T23D2714B

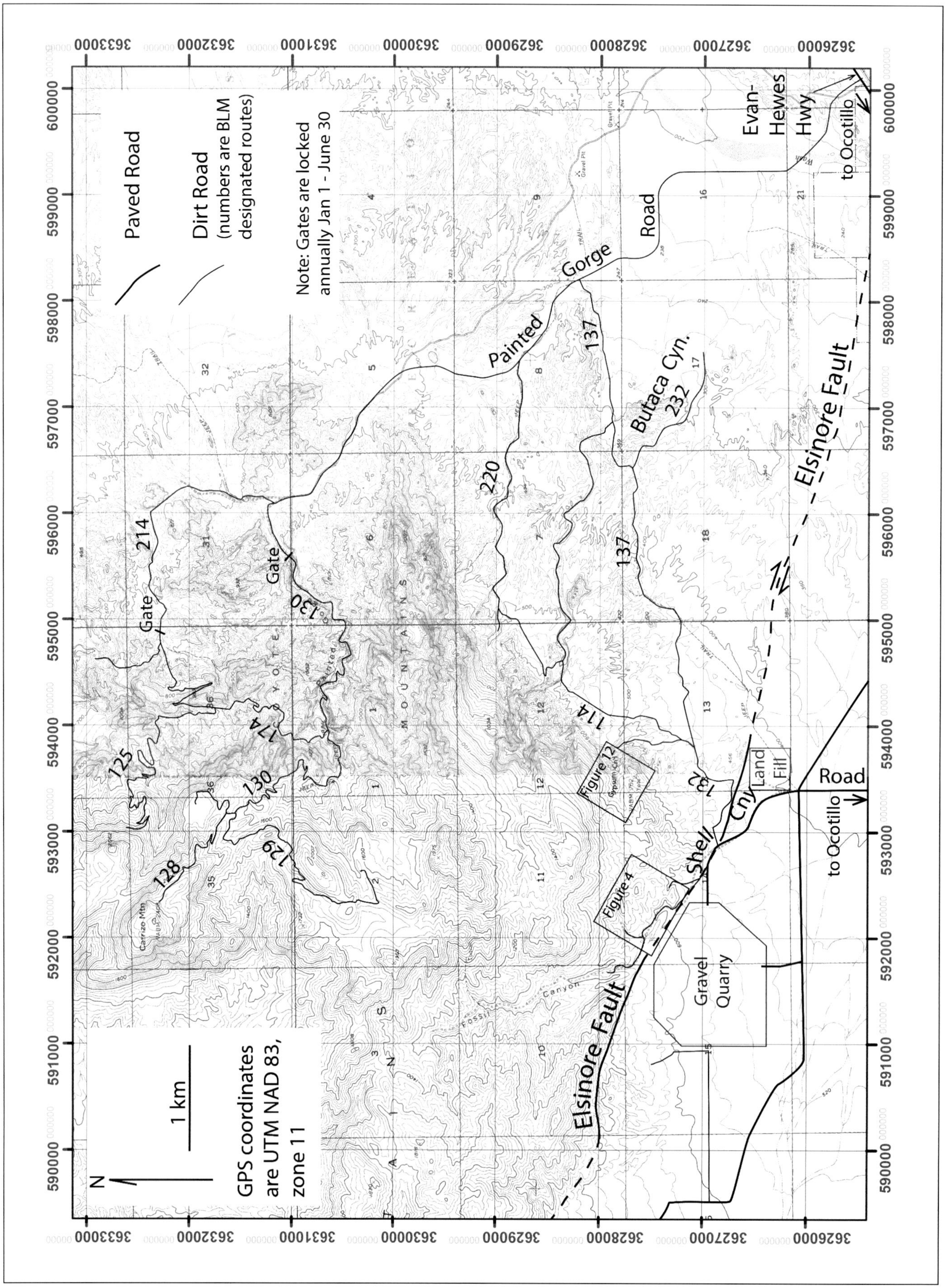
Paved Road
Dirt Road
(numbers are BLM designated routes)
Note: Gates are locked annually Jan 1 - June 30
N
1 km
GPS coordinates are UTM NAD 83, zone 11
Painted Gorge Road
Evan-Hewes Hwy
to Ocotillo
Butaca Cyn.
Elsinore Fault
Gravel Quarry
Land Fill
Road
Shell Cny
Figure 4
Figure 12
Gate
Gate
125
128
129
130
130
174
214
220
232
137
137
114
132
590000
591000
592000
593000
594000
595000
596000
597000
598000
599000
600000
3626000
3627000
3628000
3629000
3630000
3631000
3632000
3633000

FIELD TRIP B: Fossil Canyon to Gypsum Canyon

ANN BYKERK-KAUFFMAN
California State University, Chico

INTRODUCTION

THE COYOTE MOUNTAINS HAVE A REMARKABLY rich, complex, and—as Becky Dorsey says—"world class" geology. I have been mapping in the range since 1991, but until 2008 that mapping was limited to a tiny (3 km^2) postage stamp near the end of Shell Canyon Road, where I taught field camp for California State University, Chico. In 2008, I started venturing out of that little postage stamp and, together with my students, I mapped the Butaca Canyon, Canyon Sin Nombre, and upper Fossil Canyon areas (Figure 1), becoming more and more intrigued. Finally, in 2011, I started a serious research program on the range, focusing on the faulting and the basic geometry of the rock units. I took a sabbatical year in 2011–2012 to map the bulk of the range and have continued this effort since, albeit on a more limited scale. During this time, I have mentored the research efforts of five graduate and two undergraduate students. I can't pretend to fully understand this complex range; what follows is a summary of my current understanding of its geology—a progress report.

PUBLISHED GEOLOGIC MAPS OF SIGNIFICANT PORTIONS OF THE RANGE

The Coyote Mountains were first mapped by Dibblee (1996) in 1943; his published maps cover the entire range but include no topographic base. Christensen (1957) mapped the eastern two thirds of the range on a topographic base (contour interval of 50 and 100 ft) at a scale of 1:12,000. Winker and Kidwell (1996) published a simplified geologic map of the range, without a topographic base. Morton (1966) included the eastern two-thirds of the Coyote Mountains on his 1:125,000 scale geologic map of Imperial County. Todd (2004) included the western two-thirds of the Coyote Mountains on her 1:100,000 scale geologic map of the El Cajon 30′ × 60′ Quadrangle.

Figure 1. Road map of the eastern Coyote Mountains

OVERVIEW OF THE GEOLOGY OF THE COYOTE MOUNTAINS

The Coyote Mountains consist of Paleozoic/Jurassic metamorphic basement overlain by nonmarine lower Miocene sedimentary and volcanic rocks that were primarily deposited in paleovalleys carved into the basement. All of these units are overlain along a locally rugged buttress disconformity by a nonmarine-to-marine-to-nonmarine sequence of upper Miocene to upper Pliocene sedimentary rocks. Pleistocene and Holocene gravels locally overlie all of these units on a profound buttress angular unconformity. All of these lithologies are cut by a complex network of primarily strike-slip faults, dominated by the active range-bounding right-lateral Elsinore fault. Table 1 and Figures 2 and 3 depict additional details of this basic geology.

LITHOLOGIES

There is no consensus on the nomenclature for the various lithologies of the Coyote Mountains. I have adopted the nomenclature of Todd (2004) for the metamorphic basement and that of Winker and Kidwell (1996) for the sedimentary and volcanic cover rocks.

Paleozoic and Jurassic Metamorphic Basement

The oldest rocks in the Coyote Mountains are marbles, fine-grained schists, and metaquartzites that have been interpreted as metamorphosed Paleozoic miogeoclinal sedimentary rocks (Todd, 2004). These rocks dominate the high elevations in the eastern Coyote Mountains. They display a steeply dipping foliation and locally a dip-subparallel lineation. On the southern flank of the range between Fossil and Butaca Canyons, these rocks are isoclinally folded on a map scale (Figure 4). The only published age date from these metasedimentary rocks is based on Ordovician conodont fossils extracted from marble just south of Carrizo Peak (Miller and Dockum, 1983). My student, Kathi McCarthy, and I collected more than a dozen samples from the same area and elsewhere in the range, but failed to find any conodonts.

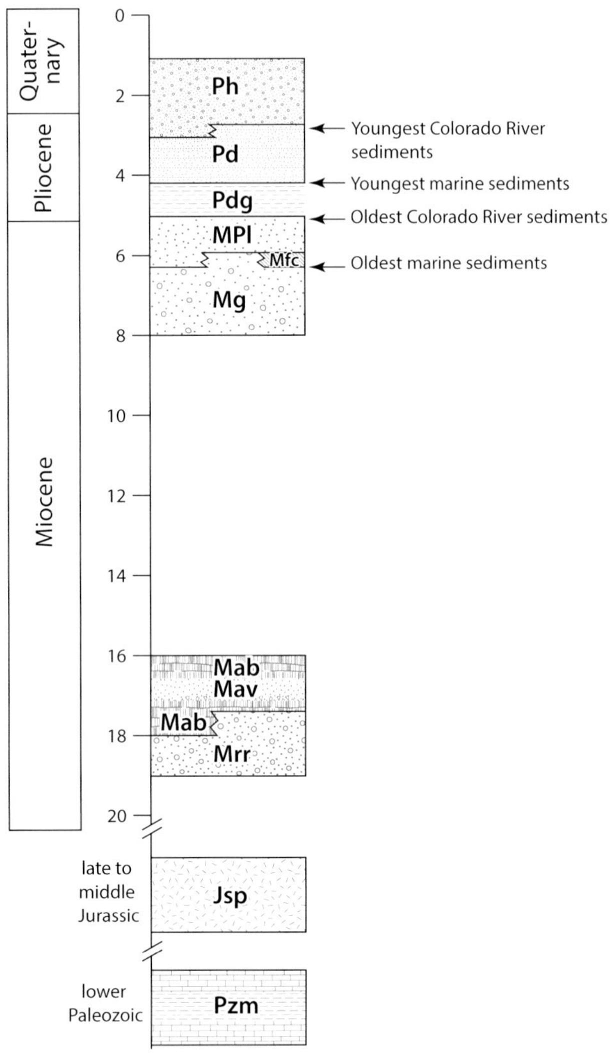

Figure 2: Chronostatigraphic column for the rock units in the southern Coyote Mountains. Rock unit names are from Winker and Kidwell (1996) and Todd (2004). Ages of the basement units are from Todd (2004). Age of the Red Rock Formation from Walsh (George Jefferson, personal communication, 2011); age of the Alverson Formation is from Morgan and others (2012); ages of the younger sedimentary are from Dorsey and others (2011).

Ph: Pliocene-Pleistocene Hueso Formation. Nonmarine fluvial sandstones and conglomerates, locally derived.

Pd: Pliocene Arroyo Diablo Formation. Nonmarine fluvial deltaic fine sandstones and mudstones derived from the Colorado River.

Pdg: Pliocene Deguyños Formation. Coarsening-upward marine deltaic sequence of mudstone and sandstone derived from the Colorado River. Includes oyster shell coquina beds.

MPl: Upper Miocene-Pliocene Latrania Formation. Locally derived fossiliferous shallow marine sandstones and conglomerates.

Mfc: Upper Miocene Fish Creek Gypsum. Flaggy turbidite sandstones with interlayered gypsum beds.

Mg: Upper Miocene Garnet Formation. Locally derived nonmarine alluvial fan conglomerates.

Mab: Basalt flows of the ~17 Ma Alverson Formation

Mav: Volcaniclastic layers of the ~17 Ma Alverson Formation. Includes tuff, tuff breccias, and debris flow deposits.

Mrr: Miocene Red Rock Formation. Braided stream conglomerates and sandstones.

Jsp: Upper and middle Jurassic migmatitic schist and gneiss of Stephenson Peak. Foliated granitoids and various inclusions, including amphibolites and marbles.

Pzm: Metamorphosed miogeoclinal sediments of probable Paleozoic age, including marbles, fine-grained schists, and metaquartzites.

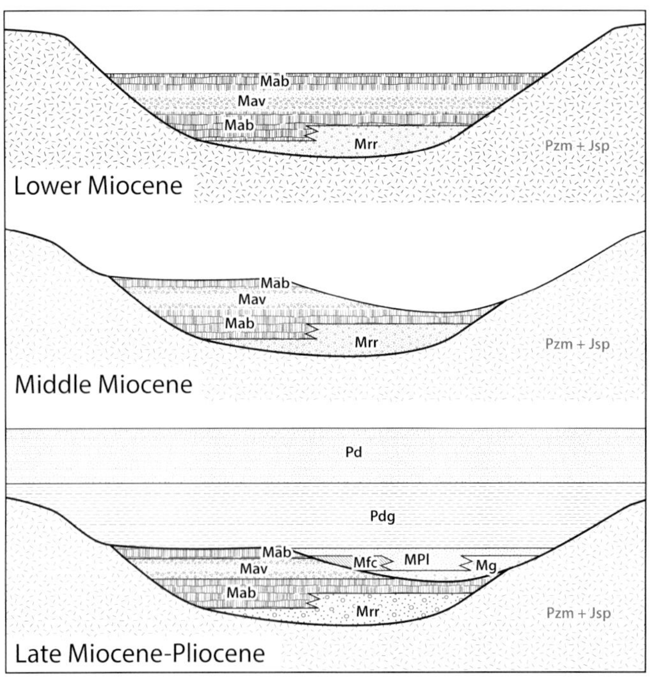

Figure 3: Schematic diagram showing the Miocene-Pliocene depositional and erosional events recorded in the rocks of the eastern Coyote Mountains. Rock unit names are the same as those in Figure 2.

The miogeoclinal metasedimentary rocks are intruded by variably foliated granitoids with large (building-sized plus) inclusions of foliated marble and amphibolite, all of which are intruded by pegmatite dikes. Todd (2004) designated these rocks as the upper and middle Jurassic migmatitic schist and gneiss of Stephenson Peak (Jsp), a unit that also crops out extensively in the Jacumba Mountains to the southwest. The Jsp dominates the central and western Coyote Mountains. Morgan and Morgan (2015, this volume) report U-Pb zircon ages ranging between 160 and 164 Ma for this unit.

Lower Miocene Sedimentary and Volcanic Rocks

Red Rock Formation: The Red Rock Formation is the oldest supracrustal unit in the Salton Trough area; it is named after Red Rock Canyon in the southern Fish Creek Mountains, 11 km north of the Coyote Mountains (Winker and Kidwell, 1996). In the Coyote Mountains, the Red Rock Formation is up to 40 m thick and consists primarily of a basal white coarse pebbly arkose that grades upward into a red cobble conglomerate (Figure 5). These sediments were deposited by braided streams in discrete paleovalleys carved into the underlying basement rocks (Winker and Kidwell, 1996). One 2 hectare exposure of the Red Rock Formation in the eastern Coyote

Mountains[1] includes eolian sandstones similar to more extensive eolian sandstones in the Red Rock Canyon area. Samples of the fluvial Red Rock Formation, collected in the Coyote Mountains by Stephen Walsh in 2005, yielded lower to middle Miocene rodent teeth (George Jefferson, personal communication, 2011). The upper part of the Red Rock Formation is interbedded with the volcanic Alverson Formation, which has yielded several U-Pb zircon ages, all statistically identical at ~17 Ma (Morgan and others, 2012; Kimbrough, personal communication, 2013).

Alverson Formation: The Red Rock and Alverson Formations are both included in the Split Mountain Group (Winker and Kidwell, 1996). The Alverson Formation consists of andesite flows and tuffs (Figure 8), basalt flows (Figure 5), volcanic breccias (Figures 9, 10 and 11), and rare fluvial conglomerates. It is as much as 120 m thick (Todd, 2004) and crops out primarily in the eastern and central Coyote Mountains, but there are a few isolated small outcrops near the western tip of the range. It also crops out in the southern Fish Creek and northern Jacumba Mountains (Todd, 2004). Many Alverson flows occupied and at least partially filled the paleovalleys which had been floored by the Red Rock Formation; thus much of the Alverson Formation is deposited directly on basement.

Strangely, the Alverson Formation is not associated with rifting or any other deformational event; all faults that cut the Alverson Formation also cut younger units. The buttress unconformity between the Alverson Formation and the younger sedimentary units is a disconformity, not an angular unconformity (Figure 11), indicating that erosion but no tilting occurred before the deposition of the younger units.

Dickinson (1997) proposed that the numerous short-lived mid-Miocene magmatic events in southern California, including the event that produced the Alverson Formation, were caused by asthenospheric upwelling into the slab window created when the East Pacific rise collided with the west coast subduction zone, forming the San Andreas Fault system. In this model, upwelling would have occurred once for a short period of time in any particular place, as the edge of the slab window passed by the area, resulting in a single short-lived episode of magmatism. This model nicely explains why all of the U-Pb age dates on the Alverson Formation are statistically identical at ~17 Ma and why this episode of magmatism was not accompanied by deformation.

Upper Miocene to Pliocene Sedimentary Rocks

Garnet Formation: A commonly subtle but locally rugged buttress disconformity separates the Alverson Formation from the nonmarine(?) fanglomerates of the upper Miocene Garnet Formation, the uppermost formation of the Split Mountain Group (Winker and Kidwell, 1996) (Figure 11). The Garnet Formation (Mg) resembles the Elephant Trees Formation of the Split Mountain area, the base of which has been interpreted as ~8 Ma in age by Dorsey and others (2011); Dorsey and others further interpret these fanglomerates as representing the initiation of regional subsidence due to normal-slip offset on the West Salton detachment fault, which structurally underlies the Coyote Mountains (Axen and Fletcher, 1998).

Geology students from CSU Chico have collected sand dollars from the upper portions of the Garnet Formation, demonstrating that at least part of this unit was deposited in a marine setting.

Table 1: Summary of the Geologic History of the Coyote Mountains. All specific age dates are from Dorsey and others (2011, 2012) except for the 17 Ma age for the volcanism which is from Morgan and others (2012).

TIME	EVENT
~1.2 Ma to present	Motion on the West Salton detachment fault and accompanying subsidence ceased. Inception and continued motion on the right-lateral Elsinore fault caused complex faulting, uplift, tilting, rotation, erosion, and fanglomerate sedimentation in the Coyote Mountains.
~2.8 Ma to ~1.2 Ma	Continued subsidence caused by motion on the West Salton detachment fault is recorded by locally derived fanglomerates in the western Coyote Mountains; no record of this sedimentation remains (if it ever existed) in the eastern Coyote Mountains.
~4.2 Ma to ~2.8 Ma	Continued subsidence caused by motion on the West Salton detachment fault and continued sediment influx from the Colorado River but delta progradation caused the delta environment to become nonmarine.
~5.3 Ma to ~4.2 Ma	Continued subsidence caused by motion on the West Salton Detachment fault. This was accompanied by sediment influx from the Colorado River into the marine environment. The Coyote Mountains started out in a pro-delta setting but the delta gradually prograded into the area, resulting in a coarsening-upward sedimentary sequence.
~8 Ma to ~5.3 Ma	Continued subsidence caused by motion on the West Salton Detachment fault. This caused flooding by marine waters and deposition of locally-derived fossiliferous conglomerates and sandstones.
Upper Miocene (~8 Ma)	Inception of the West Salton Detachment fault—located structurally below the Coyote Mountains—initiated subsidence and alluvial fan deposition.
~17 Ma to ~8 Ma	Erosion carved a rugged topography.
~17 Ma	Bimodal volcanic activity, which was in part contemporaneous with the tail end of the fluvial deposition. This volcanic activity at least partially filled some of the paleovalleys.
Lower Miocene	Carving of basement paleovalleys and deposition of fluvial conglomerates and sandstones (and, locally, eolian deposits) on the valley floors.
Cretaceous(?)	Metamorphism and deformation, possibly contemporaneous with the intrusion of the nearby Peninsular Ranges batholith
Middle to upper Jurassic	Intrusion of granitoids
Paleozoic	Deposition of miogeoclinal sediments

[1] UTM NAD 83 coordinates: 11S, 593382E, 3630782 N

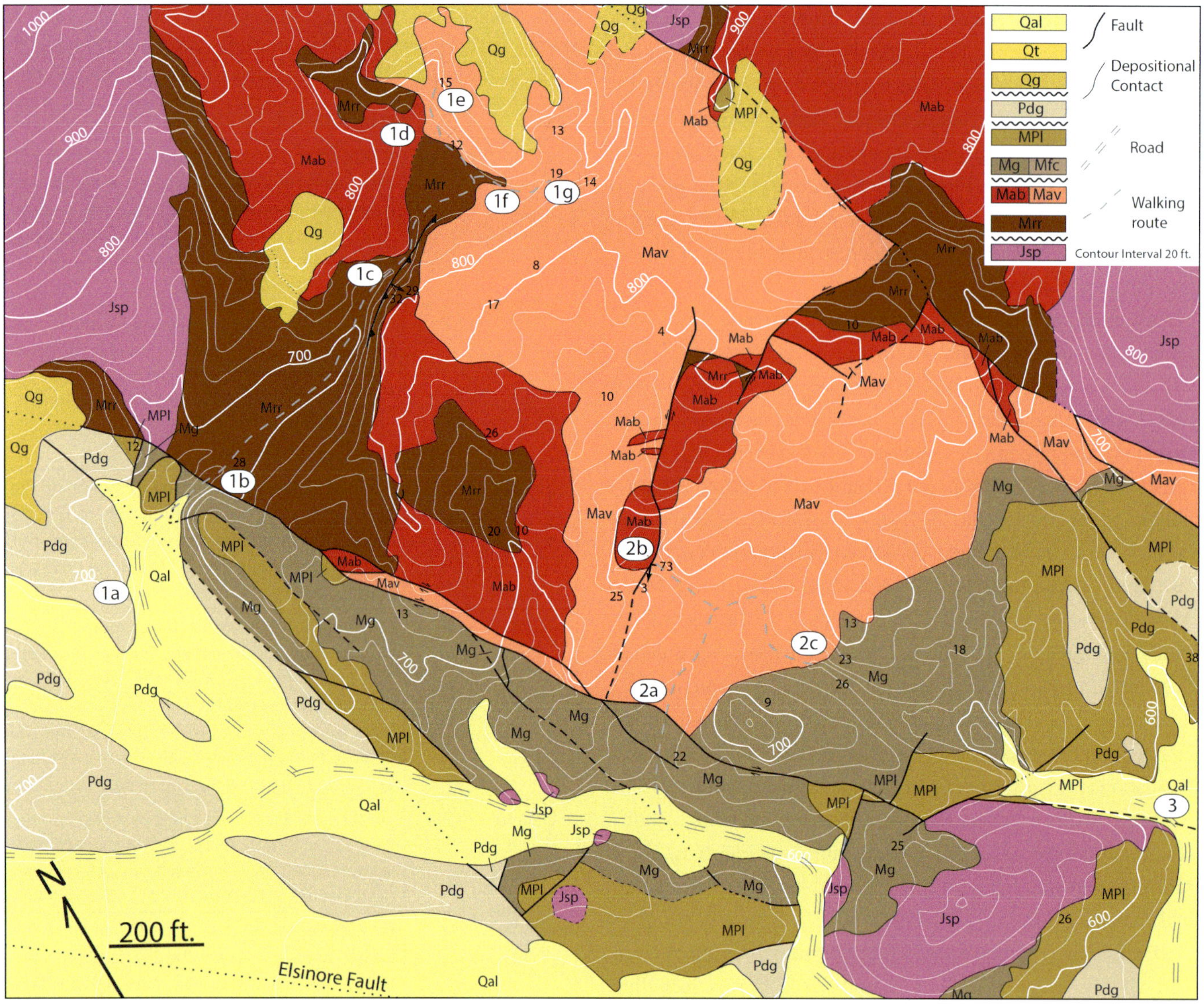

Figure 4: Geologic Map of Stops 1-3 and surrounding areas

The fanglomerates of the Garnet Formation grade upward and laterally into the Latrania Formation which is part of the marine Imperial Group (Winker and Kidwell, 1996).

Imperial Group: Winker and Kidwell (1996) defined the Imperial Group as the marine portion of the upper Miocene to Pliocene sedimentary section of the Salton Trough. It consists of the Fish Creek Gypsum, Latrania Formation, and Deguyños Formations.

The basal formation of the Imperial group, the Fish Creek Gypsum, crops out in just one location in the Coyote Mountains (Figure 12) where it appears as flaggy turbidite sandstones with sparse interlayered gypsum beds (Figure 13); its total outcrop area is less than 1 hectare in size. At this locality, the Fish Creek Gypsum overlies the Alverson Formation and underlies the Latrania Formation.

In the Coyote Mountains, the Latrania Formation consists of fossiliferous marine conglomerates, sandstones, and sparse limestones. All of its detrital sediment is locally derived and it reaches a maximum thickness of 60–70 m (Winker and Kidwell, 1996). It is much more widespread than is the underlying Garnet Formation. Where the Garnet Formation is absent, the Latrania Formation rests unconformably on the Alverson Formation or on metamorphic basement. Wherever the top of the Latrania Formation is preserved, it is disconformably overlain by the Pliocene Deguyños Formation.

The Latrania-Deguyños contact, dated at 5.3 Ma by Dorsey and others (2011), marks the cessation of locally derived sedimentation and the beginning of Colorado River derived sedimentation. Where the Latrania Formation is absent in the Coyote Mountains, the Deguyños Formation overlies the Alverson Formation or the metamorphic basement along a buttress unconformity that is locally so profound that it resembles a fault (Caldwell and Bykerk-Kauffman, 2015).

The Deguyños Formation is by far the thickest formation in the Imperial Group, reaching a thickness of over 1000 m in the Split Mountain area (Dorsey and others, 2011). The Deguyños Formation is a coarsening upward deltaic sequence of greenish yellow claystone, siltstone, and fine sandstone with oyster-shell coquina interbeds (Winker and Kidwell, 1996). The marine Deguyños Formation and the overlying nonmarine Arroyo Diablo Formation, "represent a single, very large depositional system, the Colorado River delta, which spanned the width of the Salton Trough, and spread across the mosaic of environments existing in pre-deltaic times." (Winker and Kidwell, 1996)

The Pliocene-Pleistocene Arroyo Diablo Formation is the basal member of the nonmarine Palm Spring Fm Group and the only member of this group present in the central and eastern Coyote Mountains. Its thickness in the Coyote Mountains has not been determined, but it is over 1000 m thick in the Fish Creek-Vallecito Basin (Dorsey and others, 2011). The Arroyo Diablo Formation crops out extensively in the foothills of the Coyote Mountains, but not near the crest of the range. The contact with the underlying Deguyños is conformable and gradational (Winker and Kidwell, 1996) and was dated at 4.2 Ma by Dorsey and others (2011). The Arroyo Diablo Formation consists of light pink to buff fine-grained well-sorted very quartz-rich sandstone and reddish mudstone (Winker and Kidwell, 1996). The sandstones of the Arroyo Diablo Formation are easily distinguished from those in the Latrania Formation by their high quartz content, high degree of sorting, and the presence of hematite-coated pinkish and orange quartz grains that are indicative of sand derived from the Colorado River (Winker, 1987). Deposition of the Arroyo Diablo Formation ceased at 2.8 Ma (Dorsey and others, 2011).

Pleistocene and Holocene gravels

Pleistocene and Holocene locally-derived boulder gravels up to 10 m thick locally overlie all older rock units along an angular unconformity of 20–30°. These gravels all post-date the ~1.2 Ma inception of oblique motion along the Elsinore fault and the contemporaneous inception of uplift of the Coyote Mountains (timing of inception from Dorsey and others, 2012). The tops of all but the oldest of these gravel layers form planar gently-dipping terrace or pediment surfaces. Although most of these gravel layers generally cross-cut the many faults in the range, a significant number are demonstrably offset along those faults, especially the oldest layers.

STRUCTURAL DEFORMATION

Bedding in the Pliocene and older units generally dips gently southeast on the southern flank of the range and gently northeast on the northern flank, suggesting that the Coyote Mountains are folded into a large east-plunging anticline (Dibblee, 1996). South of the Elsinore fault, however, where the Arroyo Diablo Formation dominates, bedding is commonly steeply dipping, strike direction is variable, and smaller folds are common.

Faulting in the Coyote Mountains is dominated by the right-lateral Elsinore fault, which is largely responsible for the uplift of the range. According to Fletcher and others (2012), the Coyote Mountains segment of the Elsinore fault has had a mean slip rate of 1.6 ± 0.4 mm per year over the past ~40,000 years. Total offset for the Coyote Mountains segment of the Elsinore fault has not been estimated but, about 40 miles NW of the Coyote Mountains, the Elsinore fault has a total bedrock offset of 1–2 km (Dorsey and others, 2012).

West of Shell Canyon Road, the Coyote Mountain range front trends northwest, following the trace of the active Elsinore fault; east of Shell Canyon Road, the range front makes an abrupt turn to the northeast where it follows a series of left-lateral faults that can be interpreted as rotated R' shears associated with the Elsinore fault.

Faulting within the Coyote Mountains affects all lithologies and most of this faulting can be interpreted as being kinematically coordinated with right-lateral distributed shear associated with the Elsinore fault (Figure 7). Right-lateral synthetic R and left-lateral antithetic R' faults abound. Many of these faults are oblique, such as the major right-lateral normal fault on the northeast wall of Fossil Canyon (the extension of this fault crops out at Stops 1b and 2a on Figure 4) and several minor left-lateral normal faults in the upper Fossil Canyon area. One minor thrust fault and associated hanging-wall anticline has been identified in the Coyote Mountains (Stop 1c and Figures 4 and 6); its orientation is compatible with the expected direction of compression in a zone of right-lateral distributed shear parallel to the Elsinore fault.

I have found no faults in the Coyote Mountains that can be unambiguously interpreted as predating the Elsinore fault. However, since the Coyote Mountains were on the hanging wall block of the ~8–1 Ma West Salton detachment fault, one would expect to see evidence of normal faulting contemporaneous with the deposition of the upper Miocene-Pliocene section. There is indeed a network of conjugate 020°–050°-striking normal faults on the north flank of the central Coyote Mountains (in and near the Domelands area) that can be interpreted that way (Ewing, 2014). They cut units as young as the Deguyños Formation and thus must post-date ~5.3 Ma but they are not crosscut by any younger units; thus there is no upper limit to their age and they can also be interpreted as having formed during local transtension within a zone of right-lateral distributed shear associated with the Elsinore fault. This interpretation would, however, require 27°–57° degrees of clockwise rotation of these normal faults since the inception of this faulting. Additional detachment-related or older faults may very well exist in the Coyote Mountains but the Elsinore-related faults are so pervasive that they have obscured and probably reactivated the vast majority of these older structures.

In addition to the largely brittle younger deformation in the Coyote Mountains, the metamorphic basement underwent a significant episode of ductile deformation and metamorphism between the intrusion of the upper and middle Jurassic migmatitic schist and gneiss of Stephenson Peak (Jsp) and the deposition of the lower Miocene Red Rocks Formation, as evidenced by pervasive foliation, common lineation, and local map-scale isoclinal folding. No analysis of this deformation has been published and I have not yet ventured to attempt one—stay tuned!

DRIVING/WALKING DIRECTIONS

STARTING POINT Corner of the Imperial Highway (S-2) and Shell Canyon Road in Ocotillo. Set your odometer to zero here. See Figure 1 for a map of the roads. See Figure 2 for a chronostratigraphic diagram of the rock units of the eastern Coyote Mountains; see Figure 3 for an idealized schematic diagram showing the spatial relationships among these rock units. See Figure 4 for a geologic map of Stops 1–3 and surrounding areas.

Take Shell Canyon Road north toward the Coyote Mts.

1.5 MILES Enter Figure 1 from the south, just before this point. Cross an unnamed paved road (be sure to stop at the stop sign and watch out for fast-moving gravel trucks). Continue north; the road soon curves to the northwest.

2.0 MILES The low hill on your right is located just southwest of the Elsinore Fault. It consists of Quaternary alluvial gravels overlying fluvial deposits of the Pliocene Arroyo Diablo Formation (Pd) which is part of the Palm Spring Group. The mature fine sands and clays of the Arroyo Diablo Formation were deposited in the nonmarine portion of the Colorado River delta ~4.2–3.6 Ma (Dorsey and others, 2011). We will examine these outcrops at Stop 4.

2.2 MILES Immediately after passing under a double row of power lines, continue straight, leaving the pavement (The paved road veers left toward a gravel quarry). This stretch of road follows the trace of the Elsinore Fault. The yellow mounds of clay on the right (just northeast of the Elsinore Fault) are weathered outcrops of the basal Coyote Clay of the Pliocene Deguyños Formation (Pdg), part of the marine Imperial Group. The Coyote Clay was transported into the Salton Trough area by the Colorado River and was deposited ~5.1–4.8 Ma (Dorsey and others, 2011) by suspension settling in a distal, sand-starved pro-delta marine setting (Winker, 1987, p. 287). Progradation of the Colorado River delta into the Salton Trough followed, as recorded by the coarsening-upward sequence of marine clays and sands of the Deguyños Formation (above the Coyote Clay) as well as the nonmarine sands and muds of the Arroyo Diablo Formation.

2.4 MILES Outcrops of the Miocene-Pliocene Latrania Formation (MPl) on the right. In the Coyote Mountains, this marine sandstone and conglomerate disconformably underlies the Deguyños Formation and is composed exclusively of locally derived sediment with no sediment input from the Colorado River (Winker and Kidwell, 1996).

2.5 MILES You are entering the area depicted in Figure 4, a color geologic map. The road takes a sharp turn to the right; the cliff on the right consists of metamorphic basement—specifically, the upper and middle Jurassic migmatitic schist and gneiss of Stephenson Peak (Jsp), which "apparently originated as a partial melt of predominantly metasedimentary rocks that are lithologically similar to the Julian Schist and the rocks of Jacumba Mountains" (Todd, 2004).

2.6 MILES The road curves to the left, resuming its NW course. Outcrops on both sides of the road consist of conglomerate beds of the Miocene Garnet Formation (Mg), which represents the initiation of regional subsidence due to normal-slip offset on the West Salton detachment fault ~8 Ma (Dorsey and others, 2011). The nonmarine alluvial fan deposits of the Garnet Formation grade upward into the marine Latrania Formation (Winker and Kidwell, 1996). Geology students from CSU Chico have collected sand dollars from the upper portions of the Garnet Formation, demonstrating that at least part of this unit was deposited in a marine setting. Despite the gradational conformable contact between the Garnet and Latrania Formations, Winker and Kidwell (1996) placed the Latrania Formation in the marine Imperial Group and the Garnet Formation in the underlying nonmarine Split Mountain Group.

2.7 MILES Small outcrops of pegmatite, part of the migmatitic schist and gneiss of Stephenson Peak, poke out from underneath the Garnet Formation, demonstrating the rugged topography on which the Garnet Formation was deposited. Several right-lateral faults (R Riedel shears to the Elsinore Fault) juxtapose the Garnet, Latrania and Deguyños Formations.

2.8 MILES Turn right (north) into a major wash flanked by mounds of weathered Coyote Clay of the Deguyños Formation.

3.0 MILES Southern Boundary of the Coyote Mountains Wilderness Area. Vehicular traffic prohibited beyond this point. Park here for Stop 1.

STOP 1

Observe the steep cut in the Deguyños Formation on the west side of the parking spot.

STOP 1A (UTM 0592019, 3627726)

This is a rare unweathered outcrop of the Coyote Clay layer at the base of the Deguyños Formation. The cementing agent for this claystone is calcite; dropping dilute hydrochloric acid on this rock dissolves the cement and leaves a fine clay residue.

To get to the next stop, walk up the wash as it becomes a short narrow rocky canyon through the Garnet Formation and overlying Latrania Formation. Cross a major right-lateral fault as the canyon opens up again. The trace of this fault trends 303°, which is 15° clockwise from the 288° trend of the Elsinore Fault—precisely the predicted trend for an R Riedel shear to the Elsinore Fault.

STOP 1B (UTM 0592103, 3627768).

The unit exposed northeast of the fault is the fluvial Miocene Red Rock Formation of the Split Mountain Group. The Red Rock Formation is the oldest supracrustal unit in the Salton Trough area; its upper part is interbedded with the volcanic Alverson Formation, which has yielded several U-Pb zircon ages, all ~17 Ma (Morgan and others, 2012; Kimbrough, personal communication, 2013). Samples of the Red Rock Formation, collected 1.3 km NW of here by Stephen Walsh in 2005, yielded lower to middle Miocene

Figure 5: View east from Stop 1b. Braided stream deposits of the Miocene Red Rock Formation overlain by basalt flows of the 17 Ma Alverson Formation. Note that the contact is much higher on the south side of the canyon than it is on the north side. This offset is due to a north-vergent thrust fault exposed on the south wall of the canyon (Stop 1c).

Figure 6 (Stop 1c): Ellie Spangler (January, 2008) is pleased to see a sheared boulder that serves as a shear sense indicator revealing this gently dipping fault as a thrust.

rodent teeth (George Jefferson, personal communication, 2011). The Red Rock Formation was deposited by braided streams in discrete channels carved into the underlying basement rocks. In this locality, the Red Rock Formation consists of a basal white coarse arkose that grades upward into a red cobble conglomerate. In this area, all of the clasts in the Red Rock Formation consist of metamorphic and plutonic basement rocks.

The view up-stream (east) shows the following features (Figure 5):

1. On the distant skyline: Latrania Formation (MPl) overlying the Garnet Formation (Mg) and basalt flows of the Alverson Formation (Mab).
2. In the middle distance: isoclinally folded and metamorphosed Paleozoic miogeoclinal sedimentary rocks (Pzm) intruded by the migmatitic schist and gneiss of Stephenson Peak (Jsp).
3. Immediately beyond the canyon gap: a thick sequence of tuffs and tuff breccias of the Alverson Formation (Mav).
4. In the foreground: basalt flows of the Alverson Formation (Mab) overlying the Red Rock Formation (Mrr).

Note that the Alverson/Red Rock contact is much higher on the south side of the canyon than it is on the north side. The fault responsible for this offset is a north-vergent south-dipping thrust fault which is exposed on the south wall of the canyon at Stop 1c. To get to this stop, walk upstream to where the canyon becomes a narrow slot canyon.

STOP 1C (UTM 0592267, 3627808).

Observe the gently dipping fault surface exposed on the south canyon wall. The fault surface has a strike and dip of 087°, 32°; the slickenlines have a rake of 65°E and a plunge and trend of 29°, 148°. The exposed trace of the fault is entirely contained within the Red Rock Formation. A sheared boulder in the footwall block indicates thrust offset (Figure 6); the rake of the slickenlines suggests an additional left-lateral component of motion.

This thrust is, in general, kinematically compatible with distributed shear within the right-lateral Elsinore Fault zone (Figure 7). The average trend of the Elsinore Fault trace in the field trip area is 288°. Thus one would expect the compression axis of the stress regime within the Elsinore Fault zone to be 153°/333° (45° clockwise from the trend of the fault trace). The slickenlines on the thrust fault deviate only 5° from this trend. But the 087° strike of the thrust fault is only 66° from the expected compression axis, not 90° as one might expect, which probably explains the oblique component of motion on this fault.

To get to Stop 1d, continue walking upstream until the canyon widens and branches; follow the north (left) branch to an exposure of the top of a basalt flow.

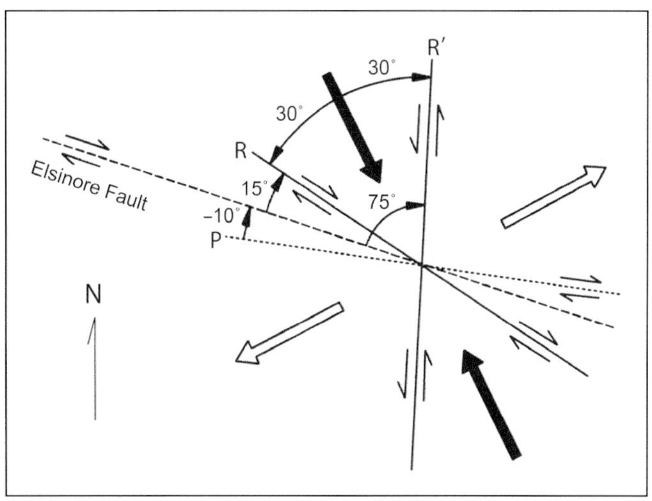

Figure 7: Expected attitudes of the compressional and extensional stress axes and initial (pre-rotation) attitudes of Riedel Shears near the Elsinore Fault in the vicinity of the north end of Shell Canyon Road, Coyote Mountains, CA. Mean trend of the Elsinore Fault in this area is 108°/288°. Modified from Figure 6.145 on p. 339 of Davis and others, 2012.

Figure 8 (Stop 1e): Michael Glas (2008) maps the 17 Ma andesitic tuff within the volcaniclastics of the Alverson Formation.

Figure 9: Unusually thick section of volcaniclastics of the Alverson Formation (Mav) directly overlying the Red Rock Formation (Mrr). Note that the two units are locally interbedded.

STOP 1D (UTM 0592343, 3627883): Pahoehoe basalt flow in the Alverson Formation

Observe the pahoehoe texture beautifully preserved in this vesicular basalt outcrop. This flow is overlain by a thick series of andesitic tuffs, tuff breccias, and debris flows.

To get to Stop 1e, scramble toward the northeast, up the steep drainage, to a white tuff bed.

STOP 1E (UTM 0592351, 3627905): Tuff bed within the Alverson Formation (Figure 8).

This is one of the tuffs within the Alverson Formation that yielded a 17 Ma U-Pb age (Kimbrough, personal communication, 2013).

To get to Stop 1f, return to the main wash and follow the south branch to an exposure of the Alverson-Red Rock contact.

STOP 1F (UTM 0592365, 3627836): Alverson-Red Rock contact

At this locality, the Alverson volcaniclastics (Mav) directly overlie the Red Rock formation (Mrr) with no intervening basalt flow (Figure 9). This is the only locality within the Coyote Mountains with this stratigraphic relationship.

To get to Stop 1g, continue upstream to where car-sized blocks have fallen off of the north-facing cliff.

STOP 1G (UTM 0592404, 3627808): Block of the "Hoodoo" bed.

Observe a displaced block of a high energy debris flow bed from the Alverson volcaniclastics; this particular block displays dramatic slickensided surfaces. This debris flow bed, which is up to 10 m thick, contains andesitic boulders up to 1 m in diameter and rip-up clasts up to 2 m in diameter. My students and I have informally dubbed this bed the "Hoodoo" bed because it has the tendency to form hoodoos. The Hoodoo bed has been observed as far west as Fossil Canyon and as far east as Gypsum Canyon. We will see it in place at Stop 2a.

To get to the next stop, return to the vehicles.

DRIVING TO STOP 2

Backtrack 0.25 miles to UTM 0592258, 3627436 (past the pegmatite outcrops but before the cliff of metamorphic rocks). The odometer should read 3.3 miles. Park here and walk up the small narrow side canyon to the northeast.

STOP 2

STOP 2A (UTM 0592293, 3627492): Subtle outcrop of the R Riedel shear fault

For the first 200 ft. of your traverse, on your way to this stop, you pass through outcrops of the Garnet Formation (Mg) which rest directly on the metamorphic basement with no intervening Red Rock or Alverson Formations. After crossing the right-lateral R Riedel shear fault (the same one that you crossed at Stop 1b), the canyon widens as you pass through thick beds of the Alverson volcaniclastics. This dramatic contrast in stratigraphy across the fault signals its strike-slip nature; it also highlights the spotty distribution of the Red Rock and Alverson Formations, which were deposited primarily in discrete paleochannels carved into the metamorphic basement. The best displaced match for this particular paleochannel crops out in Fossil Canyon approximately 1.4 km to the northwest.

The fault itself is easily overlooked at this location because it juxtaposes two conglomerate units. These units are best distinguished by their clast composition: The Garnet Formation contains both basement clasts and volcanic clasts; the Alverson volcaniclastics—with rare exception—contain only volcanic clasts.

Continue walking up the drainage. At 450 ft from the vehicles (UTM 0592352, 3627531), the drainage branches. Take the left (west) fork, walking through outcrops of the Hoodoo debris flow bed which caps the Alverson volcaniclastics. On your right, observe the large hemispherical depression in the rock with the remains of a rip-up clast at the bottom of it. In about 150 ft., the drainage branches again. You are now standing on the contact between the Hoodoo bed and an Alverson basalt flow. Stop here.

Figure 10: View west-southwest from Stop 2b, looking along the strike of the left-lateral R' Riedel shear fault.

STOP 2B (UTM 0592334, 3627576)

Look to the WSW and observe (Figure 10)....

1. A large rip-up clast within the Hoodoo bed of the Alverson Formation, which attests to the high-energy nature of the debris flow that formed this bed.

2. A fault displacing the Hoodoo bed. This fault displays a beautiful overhanging fault surface with a strike and dip of 040°, 73° and slickenlines with a plunge and trend of 3°, 220°. The offset of the Hoodoo bed indicates a left-lateral sense of slip on the fault, antithetic to the slip on the Elsinore Fault. The strike of this left-lateral fault is rotated 112° clockwise relative to the 288° trend of the Elsinore fault. The initial attitude of an R' Riedel shear should, in theory, deviate only 75° clockwise relative to the main fault (Figure 7). Therefore, I have interpreted this fault as an R' Riedel shear fault that has rotated ~37° clockwise as a result of distributed right-lateral shear within the Elsinore Fault zone.

To get to Stop 2c, walk back to the fork in the drainage (at UTM 0592352, 3627531) and then continue southeast, taking the other fork this time, and crossing a very subtle divide to where the wash drains southeast instead of southwest. Continue southeast until the wash dramatically narrows to form a slot canyon.

Figure 11: Buttress disconformity between the Alverson volcaniclastics (Mav) and the Garnet Formation (Mg).

STOP 2C (UTM 0592399, 3627464)

The buttress disconformity between the 17 Ma Alverson volcaniclastics (Mav) and the ~8 Ma Garnet Formation (Mg) crops out here. A careful examination of the contact reveals that bedding in the two units is parallel, but the contact itself is at an oblique angle to bedding (Figure 11). This geometry suggests that erosion, but no significant deformation, occurred in the area between 17 and 8 Ma.

The ~8 Ma inception of deposition of the Garnet Formation signals the beginning of regional subsidence caused by motion on the West Salton detachment fault, which crops out northwest of here; the Coyote Mountains are in the hanging wall block of this fault (Dorsey and others, 2011). The 17 Ma Alverson Formation significantly predated this tectonic episode and does not seem to be associated with any rifting event. Dickinson (1997) proposed that the numerous short-lived mid-Miocene magmatic events in southern California, including the event that produced the Alverson Formation, were caused by asthenospheric upwelling into the slab window created when the East Pacific rise collided with the west coast subduction zone, forming the San Andreas Fault system. In this model, upwelling would have occurred once for a short period of time in any particular place, as the edge of the slab window passed by the area, resulting in a single short-lived episode of magmatism. This model nicely explains why all of the U-Pb age dates on the Alverson Formation are statistically identical at 17 Ma (Morgan and others, 2012; Kimbrough, personal communication, 2013).

To get to the next stop, return to the vehicles.

Continue to backtrack another 0.25 miles to where a major wash comes in from the northeast (UTM 0592440, 3627141). Turn left into this wash and drive another 0.1 mile to where the wash branches; turn left and park here (UTM 0592531, 3627292).

STOP 3

Observe the large outcrop of pegmatite to your south; it is part of the upper and middle Jurassic migmatitic schist and gneiss of Stephenson Peak (Jsp). The sharp wall-like northeastern boundary of this pegmatite body is the same R Riedel shear fault that crops out at Stops 1b and 2a. The displaced half of this pegmatite body crops out 400 ft. southeast of here.

To view the displaced portion of the pegmatite body, walk southeast along the strike of the fault, directly crossing the main wash, and then climbing the low ridge of eroded Deguyños mudstone on the other side (UTM 0592585, 3627233). At the crest of this low ridge, you have a good view of a southwest-facing wall of pegmatite that closely matches the northeast-facing one near the vehicles.

Return to the vehicles and observe the buttress nonconformity between the sandstones of the Latrania Formation and the pegmatite on the southeast-facing point of the pegmatite hill

(UTM 0592546, 3627263). Although this contact is disturbed by the adjacent R Riedel shear fault, it is still possible to see that the Latrania Formation was deposited on a rugged surface carved into the pegmatite.

Turn to the north side of this wash and observe a small mound of the Deguyños Formation disconformably overlying the Latrania Formation (UTM 0592548, 3627317). Both units dip gently southeast. Note the sharp transition from very coarse locally derived sand to fine Colorado River derived clay. The erosion-resistant fossiliferous coarse sandstone bed capping the Latrania Formation is a unique laterally extensive bed that thickens toward the northwest. In some places, this bed constitutes the entire thickness of the Latrania Formation. One km northwest of here, this erosion-resistant bed forms a distinctive 5 km^2 flat iron which is stained so dark by desert varnish that it resembles basalt from a distance. We will visit the edge of this flatiron at Stop 5. Note how angular and poorly sorted the sand grains are; note also the high proportion of metamorphic lithic grains. We will contrast this locally-derived sandstone with Colorado River-derived sandstones at Stop 4.

DRIVING TO STOP 4

Continue to backtrack until the pavement begins and you reach the power lines.

Mile 4.0: Turn left onto the power line service road (BLM route 132) and park. Walk toward the low hills to the south.

STOP 4 (UTM 0593008, 3626788)

This stop is on the Elsinore Fault; it is not shown on the Figure 4 geologic map. Pale yellow massive mudstones of the marine Deguyños Formation on the north side of the fault are juxtaposed against fluvial light pink fine well-bedded sandstones and mudstones of the Arroyo Diablo Formation on the south side of the fault (overlain by Quaternary alluvial gravels). Both rock units are composed of sediment brought into the area by the Colorado River during Pliocene time. Bedding in the Deguyños Formation dips gently southeast; bedding in the Arroyo Diablo Formation has a generally east-west strike and a near vertical dip. Observe the poorly consolidated fine sandstones and mudstones of the Arroyo Diablo Formation. Note how fine, well sorted, and well rounded the sand grains are; note also the high quartz content and the presence of hematite-coated pinkish and orange quartz grains that are indicative of sand derived from the Colorado River (Winker, 1987).

DRIVING TO STOP 5

Follow the power line service road (BLM route 132) another 0.7 mile: Park at the sharp left turn in the road (UTM 0593518, 3626714).

STOP 5

This stop is also not on the Figure 4 geologic map. Walk in a southeast direction, toward the tan erosion-resistant southeast-dipping bed (UTM 0593518, 3626714). This is a coquina bed at the base of the Yuha (middle) member of the Pliocene Deguyños Formation. These coquina beds consist primarily of accumulations of fragmentary *Lopha vespertina* oyster shells, deposited by lateral accretion in meandering channels of the Colorado River delta (Watkins, 1990; Winker and Kidwell, 1996).

The Elsinore fault trace is immediately south of this outcrop, north of the county landfill.

DRIVING TO STOP 6

Continue to follow the power line service road (BLM Route 132) another 0.4 mile to where a major wash comes in from the north. Pause to observe the flat iron of Latrania to the north; we are headed that way.

Turn left onto BLM Route 114, traveling upstream (north) within the wash. After just 0.1 mile, take the left fork to follow the less major western tributary of the wash straight north between ridges of Deguyños Formation (the other route heads slightly northeast, following the western edge of a large relatively flat plain). Continue another 0.4 mile to the end of the road. Park here (UTM 0593823, 3627811); you are now just within the geologic map shown in Figure 12.

STOP 6

The outcrops west of the wash consist of the locally derived fossiliferous sandstones of the Latrania Formation; the outcrops east of the wash consist of the Colorado River-derived mudstones of the Deguyños Formation. Walk upstream to where the sand wash becomes a bedrock canyon.

STOP 6A (UTM 0593772, 3627887): Excellent water-scoured outcrops of the fossiliferous Latrania Formation

Continue walking/scrambling upstream as the canyon takes a turn to the left and widens out.

STOP 6B (UTM 0593729, 3627866): Fish Creek Gypsum

The Fish Creek Gypsum underlies the Latrania formation and is exposed here on both sides of the canyon. The approximately one hectare area of Fish Creek Gypsum shown on Figure 12 is the only outcrop of this unit in the Coyote Mountains. Here there are just a few 1-2 ft thick beds of gypsum interbedded with flaggy mica-rich turbidite sandstones. But in the Fish Creek Mountains, this gypsum is so thick and so laterally extensive that it is mined and shipped to Plaster City, located between Ocotillo and El Centro, where it is used to make Sheetrock brand gypsum panels (Wikipedia). In the 1980s, the quarry was estimated to contain a deposit of 25 million tons of gypsum (Todd and others, 1987).

To get to the next stop, continue walking upstream, climbing a makeshift ladder at a small dry waterfall (or, if the ladder is absent, climbing around the dry waterfall).

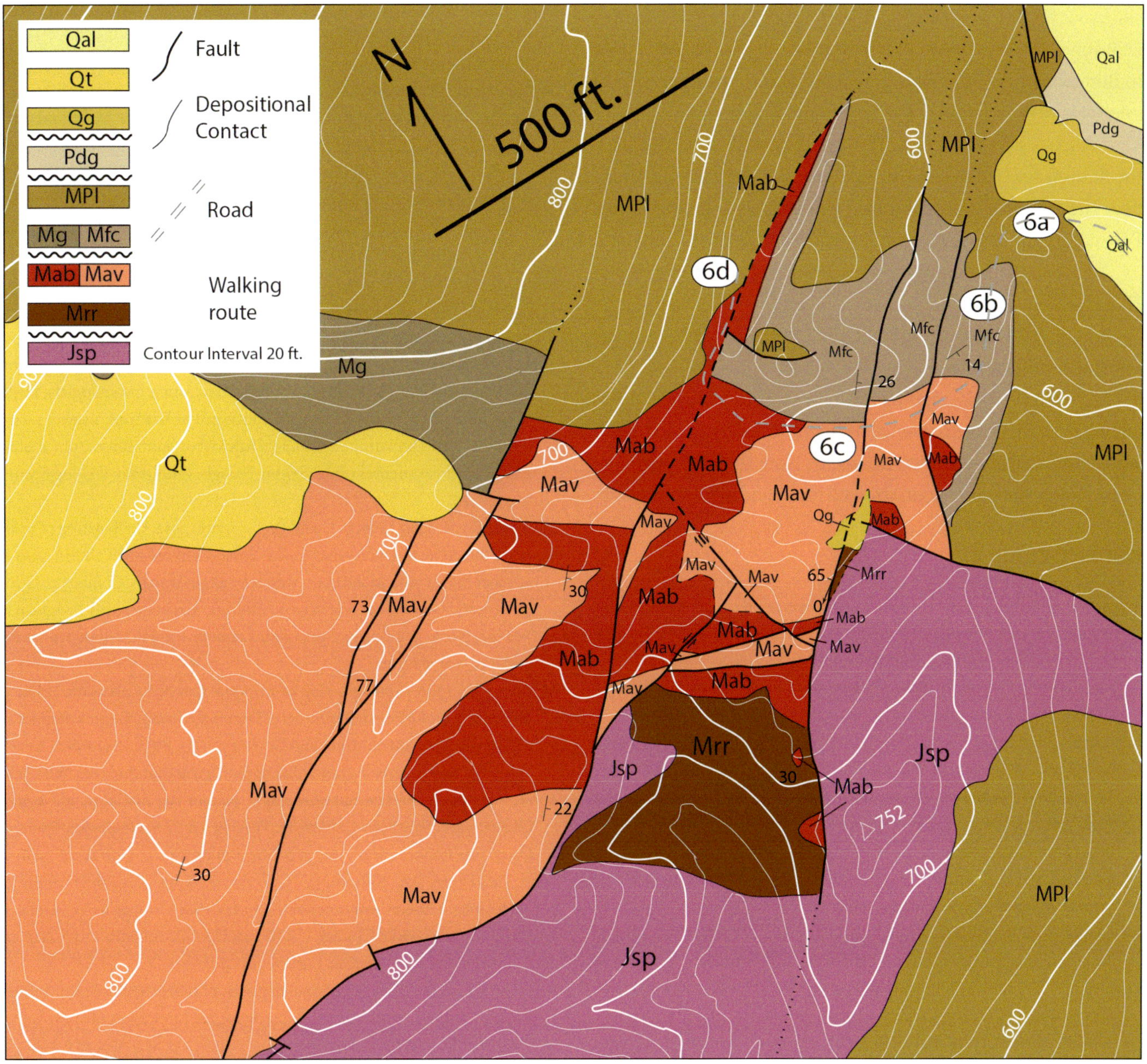

Figure 12: Geologic map of Stop 6 and surrounding area.

STOP 6C (UTM 0593639, 3627840): Left-lateral R' fault and drag fold

A left-lateral fault cuts the Fish Creek Gypsum, forming a dramatic drag fold (Figure 13). Just southwest of here, this fault has a strike and dip of 228°, 63° and the slickenlines are horizontal. The apparent uplift of the southeast side of the fault is due to the fact that the strike of bedding is not perfectly parallel to the strike of the fault; the beds have a slight component of dip toward the southwest (toward the viewer when looking at the drag fold).

The geologic map (Figure 12) reveals that this area is dominated by left-lateral faults whose traces trend approximately ~060°/240°. These northeast-trending left-lateral faults continue toward the east, controlling the trend of the mountain front. The simplest interpretation of these faults is that they are R' Riedel shears that formed at the expected trend of 005° but then rotated approximately 55°

Figure 13: Left-lateral fault and drag fold in the Fish Creek Gypsum at Stop 6c.

clockwise due to distributed right-lateral shear within the Elsinore fault zone. This interpretation is supported by paleomagnetic work in Butaca Canyon (Marshall and others, 2006), which revealed a ~60° clockwise rotation of the Alverson Formation.

Look toward the north-northwest to view the eroded edge of the flatiron formed by the Latrania-capping resistant sandstone bed. The green sedimentary rocks below this layer are the Garnet Formation; the green color comes from the presence of chlorite.

To get to the next stop, continue walking around the hill of Fish Creek Gypsum.

STOP 6D (UTM 0593608, 3627913): Latrania-capping resistant sandstone bed

Here the Latrania-capping resistant sandstone bed occurs as large blocks that have remained as the more easily eroded sandstones underneath were washed away. Note the high degree of cementation, the coarse grain size, and the abundant marine fossils.

REFERENCES

Axen, G.J. and Fletcher, J.M., 1998, Late Miocene-Pleistocene extensional faulting, northern Gulf of California, Mexico and Salton Trough, California, in Ernst, W.G., and Nelson, C.A., eds., *Integrated Earth and Environmental Evolution of the Southwestern United States (Hall Volume)*: Geological Society of America, p. 365–392).

Caldwell, Grant, and Bykerk-Kauffman, Ann, 2015, Evaluation of Late Neogene sediments on the northern flank of the central Coyote Mountains: Paleotopographic setting and sediment sources of the Late Miocene Latrania Formation during initial stages of a marine incursion: Pacific Section of the American Association of Petroleum Geologists 2015 Conference Technical Program. https://www.psaapg.org/2015-conference-technical-program/.

Davis, G.H., Reynolds, S.J., and Kluth, C. F., 2012, Structural geology of rocks and regions (3rd edition): John Wiley and Sons, New York, 839 p.

Dibblee, T.W., Jr., 1996, Stratigraphy and tectonics of the Vallecito-Fish Creek Mountains, Vallecito Badlands, Coyote Mountains, and Yuha Desert, southwestern Imperial Basin, *in* Abbot, P.L., and Seymour, D.C., eds., *Sturzstroms and Detachment Faults, Anza-Borrego Desert State Park, California*: South Coast Geological Society, Inc., Annual Field Trip Guidebook #24, p. 59–79.

Dickinson, W.R., 1997, Tectonic implications of Cenozoic volcanism in coastal California: *Geological Society of America Bulletin*, v. 109, no. 8, p. 936–954.

Dorsey, R.J., Housen, B.A., Janecke, S.U., Fanning, C.M., and Spears, A.L.F., 2011, Stratigraphic Record of Basin Development within the San Andreas Fault system: Late Cenozoic Fish Creek–Vallecito Basin, Southern California: *Geological Society of America Bulletin*, v. 123, p. 771–793.

Dorsey, R.J., Axen, G.J., Peryam, T.C., Kairouz, M.E., 2012, Initiation of the Southern Elsinore Fault at ~1.2 Ma: Evidence from the Fish Creek-Vallecito Basin, southern California: Tectonics, v. 31, TC2006, doi: 10.1029/2011TC003009.

Ewing, Cavan S., 2014, Structural geology of the central Coyote Mountains, Imperial County, California [MS thesis]: California State University, Chico.

Fletcher, K. E. K., T. K. Rockwell, and W. D. Sharp, 2011, Late Quaternary slip rate of the southern Elsinore fault, Southern California: Dating offset alluvial fans via 230Th/U on pedogenic carbonate, J. Geophys. Res., 116, F02006, doi:10.1029/2010JF001701.

Marshall, M., Mace, N., Erskine, B., and Jansen, E., 2006, What regional and local paleomagnetic data tell us about the tectonics of the Coyote Mountains, southwest Imperial Valley, Southern California, *in* Bloom, D.M. and Farquharson, P.T., eds., *Geology and History of Southeastern San Diego County, California:* San Diego Association of Geologists, Field Trip Guidebook for 2005 and 2006, Sunbelt Publications, pages 55–62.

Miller, R.H., and Dockum, M.S., 1983, Ordovician conodonts from metamorphosed carbonates of the Salton Trough, California: Geology, v. 11, p. 410–412.

Morgan, J.R., Morgan, George J., and Pecha, Mark, 2012, U/Pb dating of tuffs from the Alverson Volcanics in the Fossil Canyon and Painted Gorge areas of the Coyote Mountains, western Salton Trough, California, USA [abstract]: Geological Society of America Abstracts with Programs, Vol. 44, No. 3, p. 21. https://gsa.confex.com/gsa/2012CD/finalprogram/abstract_201798.htm

Morton, P.K., 1966, Geologic Map of Imperial County, California (Plate 1 of *Geology and Mineral Resources of Imperial County*, County Report 7, California Division of Mines and Geology, 1977). https://ia700808.us.archive.org/24/items/geologyandminer07mort/geologyandminer07mort.pdf

Todd, V.R., 2004, Preliminary Geologic Map of the El Cajon 30′ x 60′ Quadrangle, Southern California, Version 1.0: U.S. Geological Survey Open File Report 2004-1361, 33 p. http://pubs.usgs.gov/of/2004/1361/

Todd, V.R., Detra, D.E., Kilburn, J.E., Griscom, A., Kruse, F.A., and Campbell, H.W., 1987, Mineral resources of the Fish Creek Mountains Wilderness Study Area, Imperial County, California: U.S.G.S. Bulletin 1711, Chapter C. http://pubs.usgs.gov/bul/1711c/report.pdf.

Watkins, R., 1990, Pliocene channel deposits of oyster shells in the Salton Trough region, California. Palaeogeography, Palaeoclimatology, Palaeoecology, v. 79, p. 249-262.

Wikipedia entry on USG Corporation, https://en.wikipedia.org/wiki/USG_Corporation

Winker, C.D., 1987, Neogene stratigraphy of the Fish Creek-Vallecito section, southern California: Implications for early history of the northern Gulf of California and Colorado Delta [PhD thesis]: University of Arizona, Tucson, 494 pp.

Winker, C.D. and Kidwell, S.M., 1996, Stratigraphy of a marine rift basin: Neogene of the western Salton Trough, California, *in* Patrick L. Abbott and John D. Cooper, eds., 1996, Field Conference Guide 1996, Pacific Section A.A.P.G., GB 73, Pacific Section S.E.P.M., Book 80, p. 295–336.

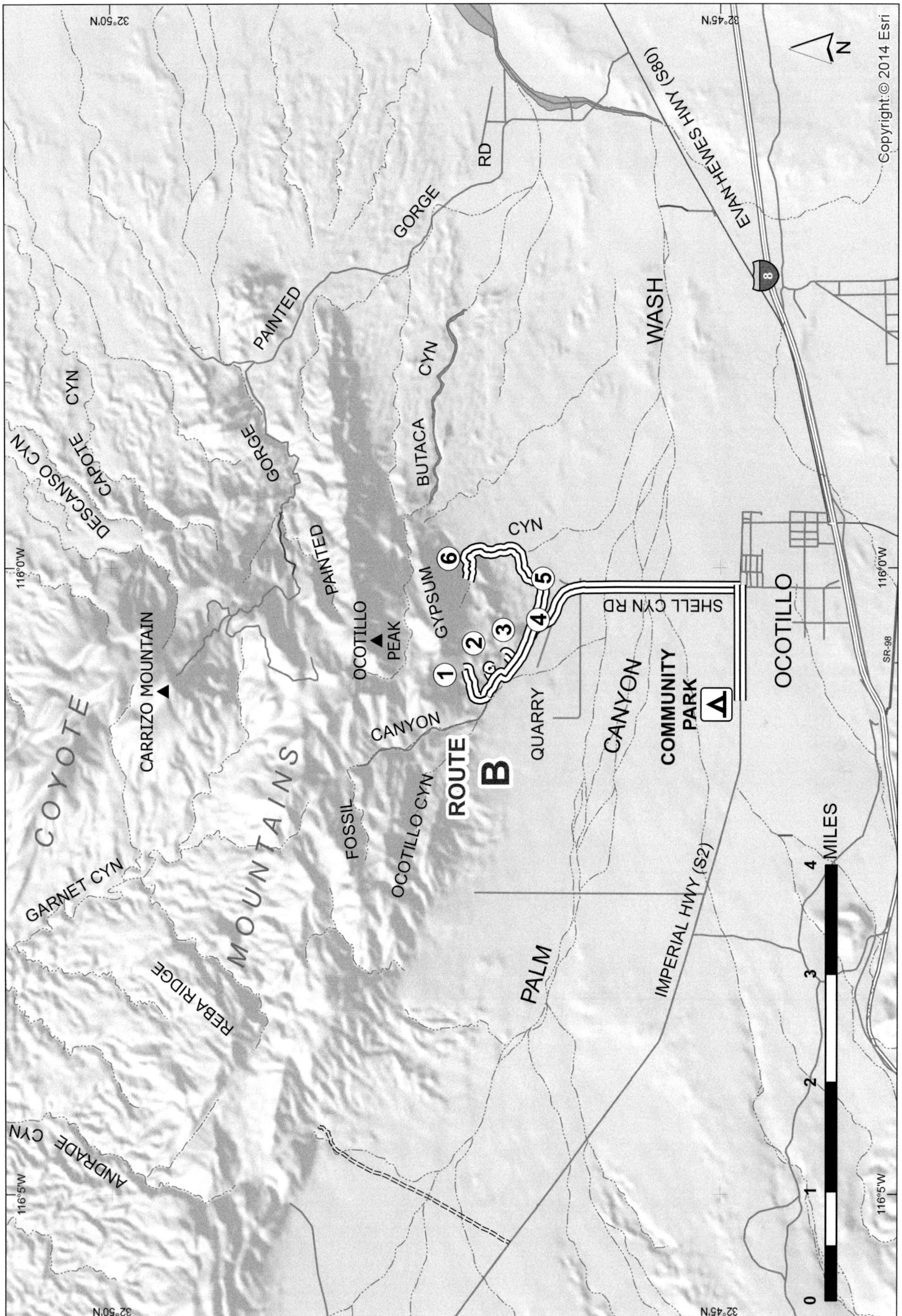

Index Map: Route B

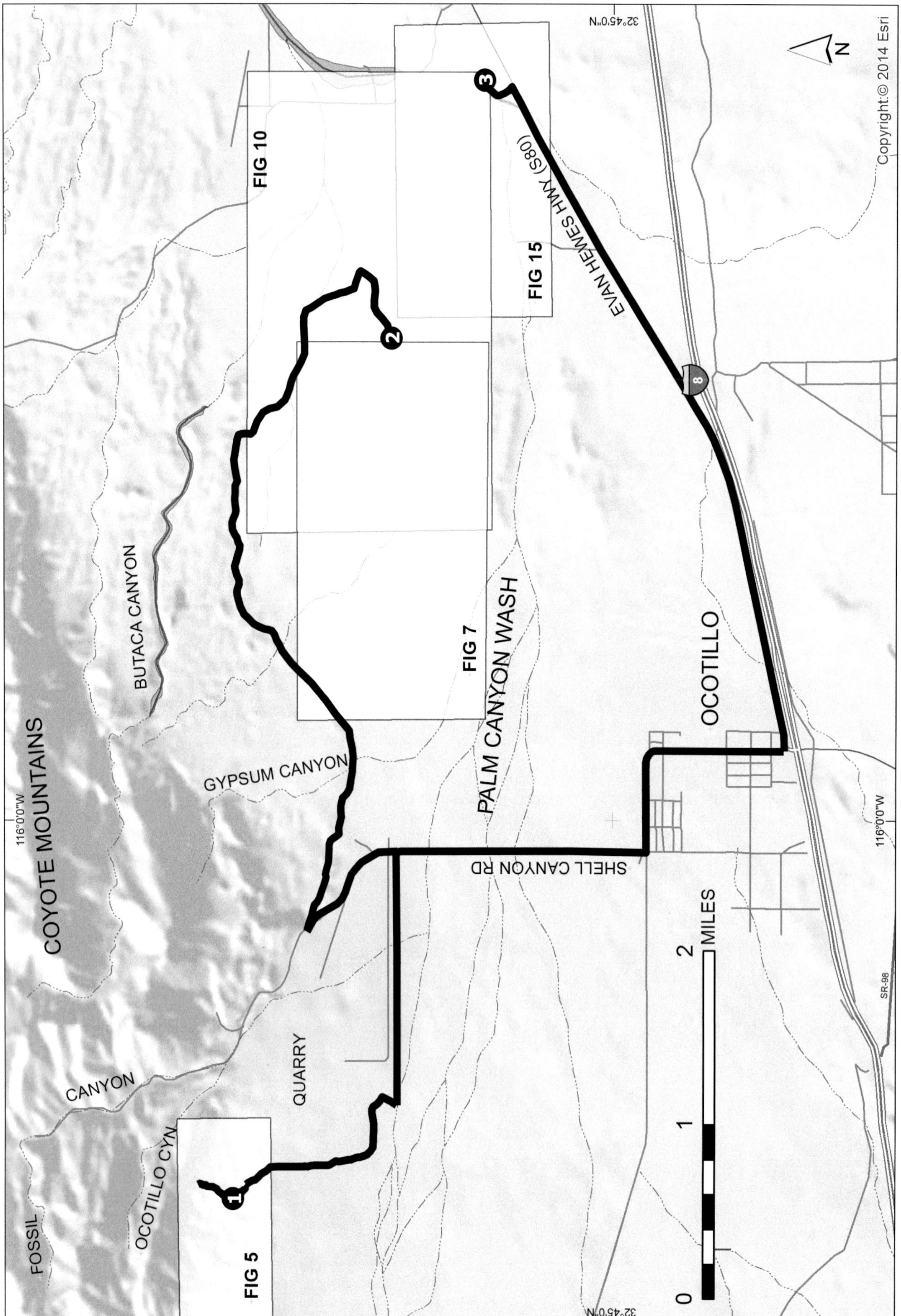

Index Map showing route and stops. Indexed figures are locations of geologic images.

FIELD TRIP C:
Elsinore fault: Evidence from structural analysis of reworked oyster conglomerates:
Southern Coyote Mountains, Imperial County, California Driving Directions to Points of Interest

DAVID ARASH TEIMOORIAN
davidteimoorian@gmail.com

INTRODUCTION

THIS ROAD LOG PROVIDES DRIVING DIRECTIONS and points of interest in connection with a project to analyze the geologic structure of reworked oyster conglomerates in the southern Coyote Mountains near the Elsinore fault.

The project's objective was to geologically map the overly simplified lithology of the southeast Coyote Mountains and map the Elsinore fault further out to the southeast of its currently mapped location, just north of Ocotillo, California. Additionally, the structural responses to the dextrally offsetting Elsinore fault were documented at two outcrops of shell-bearing conglomerate which are interpreted to be offset alluvial fan deposits. Point-count analysis confirms that the two outcrops are similar in composition, although, their provenances stem from different parts of the range; higher concentration of carbonate and metasedimentary clasts documented at the western outcrop are prevalent to the western part of the range while the eastern outcrop displays a higher concentration of volcanic lithics and marble clasts which are prevalent to the east part of the range. Facies variations between outcrops does not allow for them to be used as piercing points but rather two locations along an alluvial fan, distal and proximal to the source, that were offset post deposition via the Elsinore fault.

Cross section analysis revealed that the western outcrop of the shell bearing conglomerate is part of a large scale, overturned anticline with an axial plane trending and plunging to the southeast; the eastern outcrop of the conglomerate responded to localized strain with oblique normal faulting (Riedel shears) which displays dextral shear in the direction of the relative motion induced by the Elsinore fault. It is concluded that the shell-bearing conglomerate outcrops represent two marker points along an alluvial fan which has been offset by the Elsinore fault which extends as far east as Palm Canyon Wash.

STOP 1: Western Exposure of the shell-bearing conglomerate on the south side of the Elsinore Fault

Driving: From the town of Ocotillo, California, head north on Imperial Hwy (S2) and turn right on Shell Canyon Road. Take Shell Canyon Road until it T-intersects with the Ocotillo dump to the right hand corner. At that intersection, zero your odometer and make a left hand turn, continuing westbound for approximately 1.23 miles. To your right you will see an off road path leading to a power line. Take that path for approximately 0.10 mile until you see a wash on your left that will take you around the west side of the mine and directly to a parking spot near the outcrop (Figure 1).

Walking: From the flat-lying parking area, walk across the stream channel to the west and continue up the drainage feature between the flat-lying terrace gravels covered by Quaternary colluvium (Figure 2).

Once you have arrived to this location (32°46′57.79″N, 116°1′53.27″W), the near vertical beds consisting of re-worked oyster shells will be quite apparent. Here the conglomerate displays an easterly dip and overlies the Colorado River (C-class) derived Arroyo Diablo Formation. Hike north approximately 0.23 mile up the most eastern ridge of the conglomerate outcrop to the Elsinore fault, the contact between the Jurassic age basement and the marine pro-delta sediments of the Deguyños Formation. See Figure 3 for a map to the fault location.

As you head up the ridge you will notice the juxtaposing of the Arroyo Diablo Formation and the Deguyños Formation approximately 0.3 mile to the south of the Elsinore fault contact. This juxtaposition of an older unit (Deguyños Formation) adjacent to a younger unit (Arroyo Diablo Formation) results from a fault which runs sub-parallel to the Elsinore fault (Figure 4 and 5).

Figure 1. Offroad path to the western exposure of the shell bearing conglomerate on the north side of the Elsinore fault.

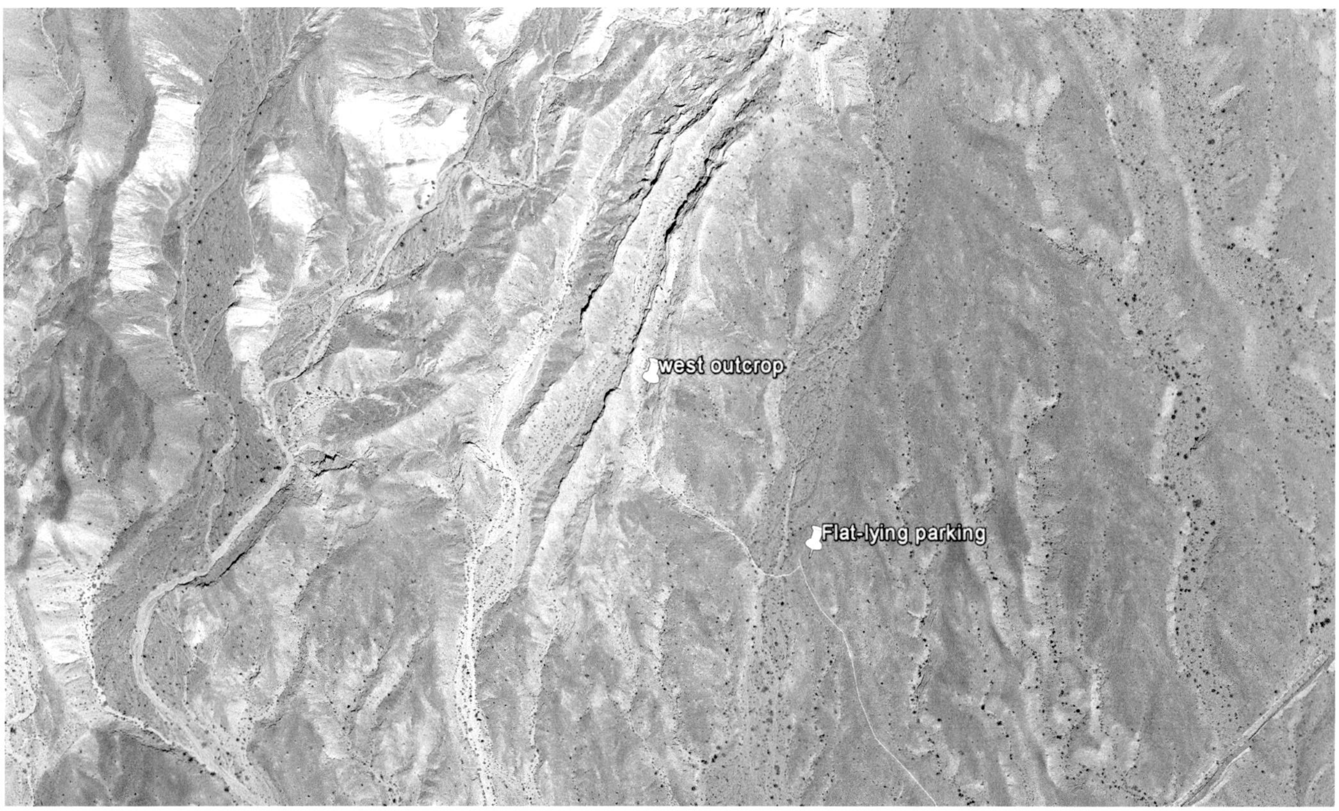

Figure 2. Walking path to the western exposure of the shell bearing conglomerate.

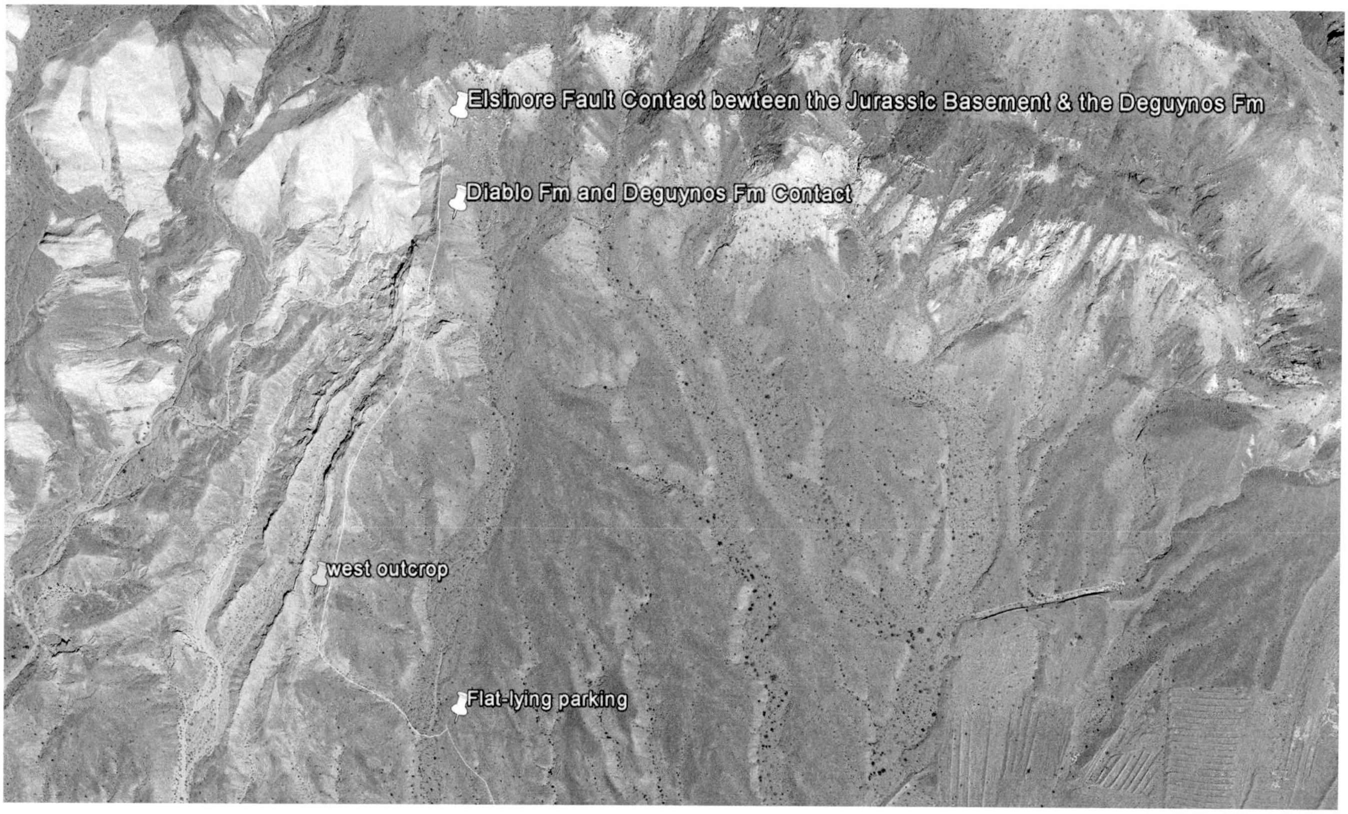

Figure 3. Hiking path towards the Elsinore fault.

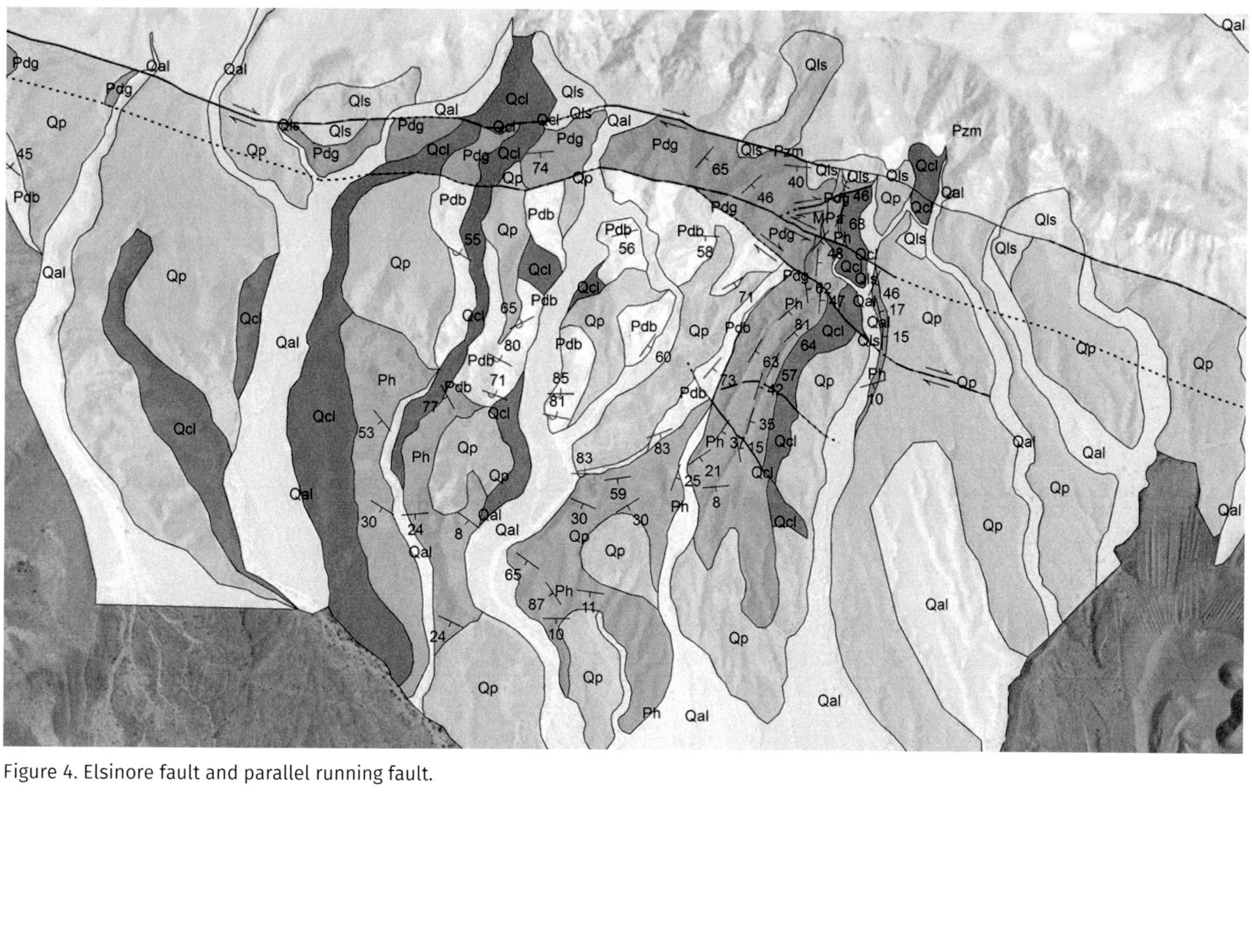

Figure 4. Elsinore fault and parallel running fault.

Figure 5. Geologic map of the western outcropping of the shell bearing conglomerate, Teimoorian 2015.

STOP 2: Eastern Exposure of the Shell-Bearing Conglomerate, north side of Elsinore fault

Driving: Starting from the Ocotillo dump T-intersection, continue north on Shell Canyon Road for approximately 0.6 mile. Turn right and follow the dirt road past the 4×4 park and continue east for 4.0 miles. See Figure 6 for the location of the off road path trailhead.

The 4×4 park is almost entirely underlain by the yellow-colored Deguyños Formation with excellent exposures of the coquina beds. As you continue on the path, just east of the dump, you can see counterclockwise drag folding expressed in the coquina beds of the Deguyños Formation and directly to the south there is a sharp transition into an unconsolidated white sand unit which has counter-clockwise drag folding which is apparent in the topography of digital elevation model (DEM) hillshade overlays (Figure 7); this is interpreted as being the location of the Elsinore fault as well as the location at which this project starts to document the Elsinore fault farther to the southeast than in previous studies in the area (32°46′13.31″N, 115°58′56.02″W). Also in Figure 6, you can see the transition from the older Deguyños Formation into the younger, primarily pink colored Arroyo Diablo Formation but unfortunately the off-road path does not take you to this location.

The 4.0 mile long off-road path to the eastern outcrop is relatively maintained although the path merges into a stream channel which is the continuation of the path for the remaining ~2.0 miles (see arrow on Figure 8). These last two miles have patches of loose sand so it would be wise to engage 4×4 to maintain momentum throughout travel.

The eastern outcrop (32°46′7.58″N, 115°57′35.53″W) is marked by thin, poorly consolidated, flat laying bedding, overlying the Arroyo Diablo Formation. The Arroyo Diablo Formation is commonly undercut, resulting in the conglomerate being perched upon small hills of the Arroyo Diablo Formation (See Figure 9). There are several normal faults with lateral offset al the eastern outcrop which trend toward the Elsinore fault; these normal faults are interpreted to be Riedel shears induced by the Elsinore fault (see Figure 10 and 11). The R shears either terminate into, or, are covered by alluvium.

Figure 6. Trailhead to the off-road path leading to the eastern exposure of the shell bearing conglomerate on the north side of the Elsinore fault.

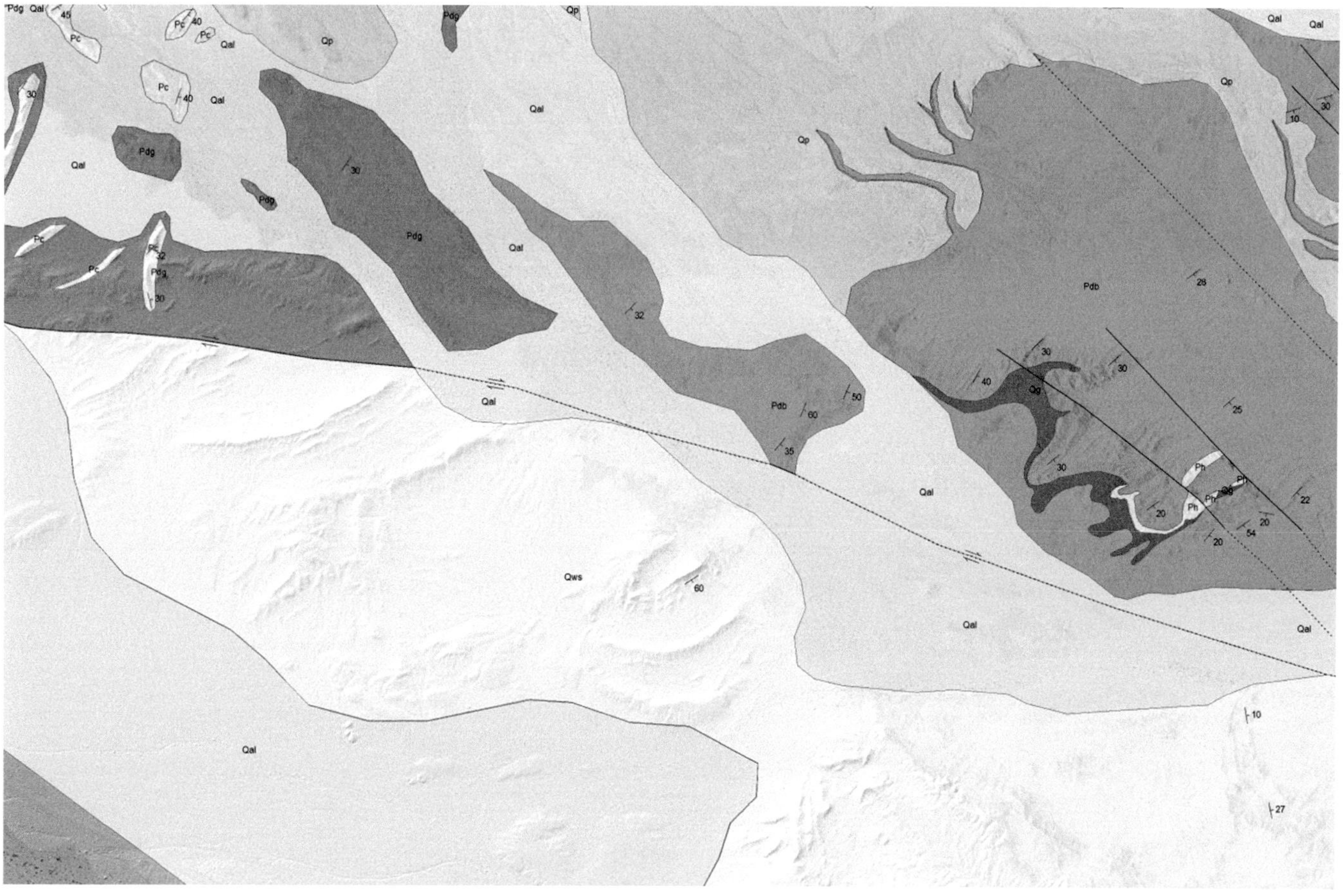

Figure 7. Geologic map showing the Elsinore fault's approximate location between the counterclockwise drag folding of the coquina beds (Deguynos Formation) and the clockwise drag folding of the white sandstone unit, Teimoorian 2015.

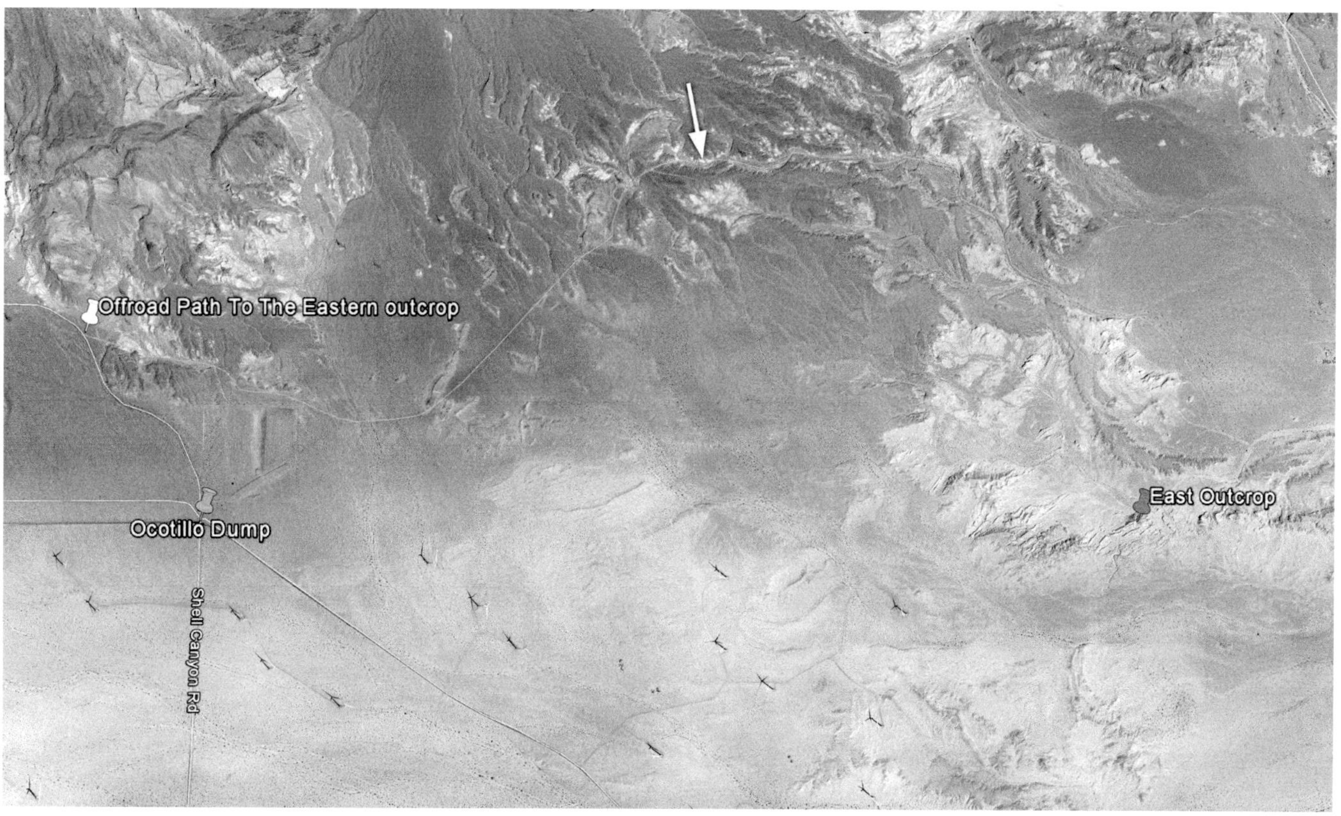

Figure 8. Off-road path to the eastern exposure of the shell bearing conglomerate on the north side of the Elsinore fault.

Figure 9. Shell bearing conglomerate resting atop undercut Arroyo Diablo Formation at eastern outcrop.

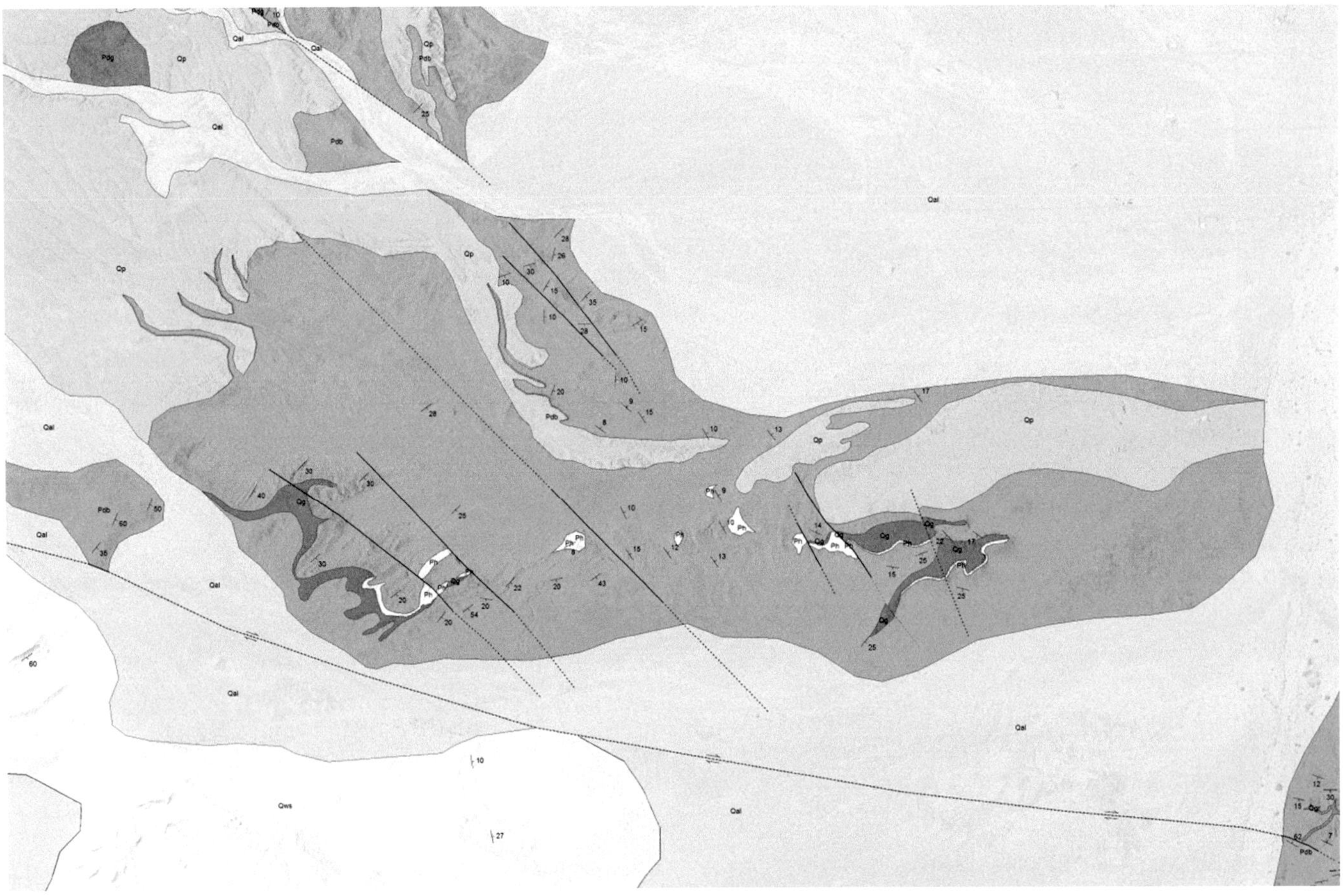

Figure 10. Geologic map of the eastern outcropping of the shell bearing conglomerate, Teimoorian 2015.

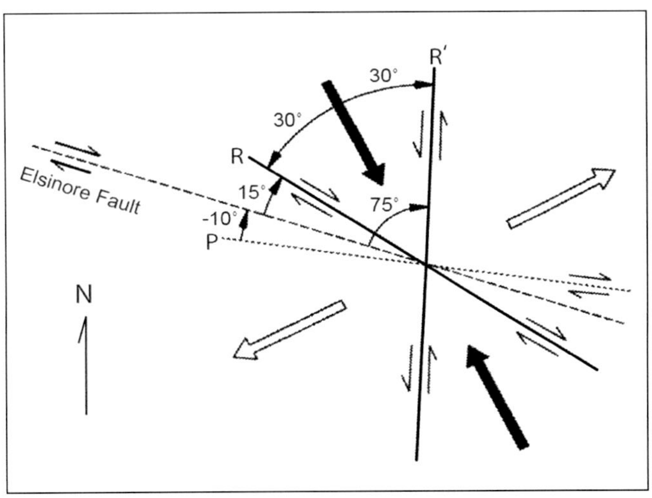

Figure 11. Riedel shears stemming from the dextral Elsinore fault.

STOP 3: Eastern Exposure of the Elsinore Fault

Driving: From the town of Ocotillo, head eastbound on Evan Hewes Highway (road S80) for approximately 3.8 miles (Figure 12). There will be a dirt ramp on the north side of the road which will take you directly to the Elsinore fault via 4×4 trail. You can also park your vehicle on the side of the road and take a short hike to the point of interest (32°45′39.34″N, 115°56′18.96″W), seen in Figure 13.

The location of interest is a canyon carved out of the Arroyo Diablo Formation with overlying Quaternary gravels. This is the farthest the Elsinore fault is mapped in this project terminating in the Quaternary gravels to the south-southeast. The approximate location of the Elsinore fault through the canyon is marked by the opposing bedding attitudes on either side of the canyon (Figure 14 and Figure 15).

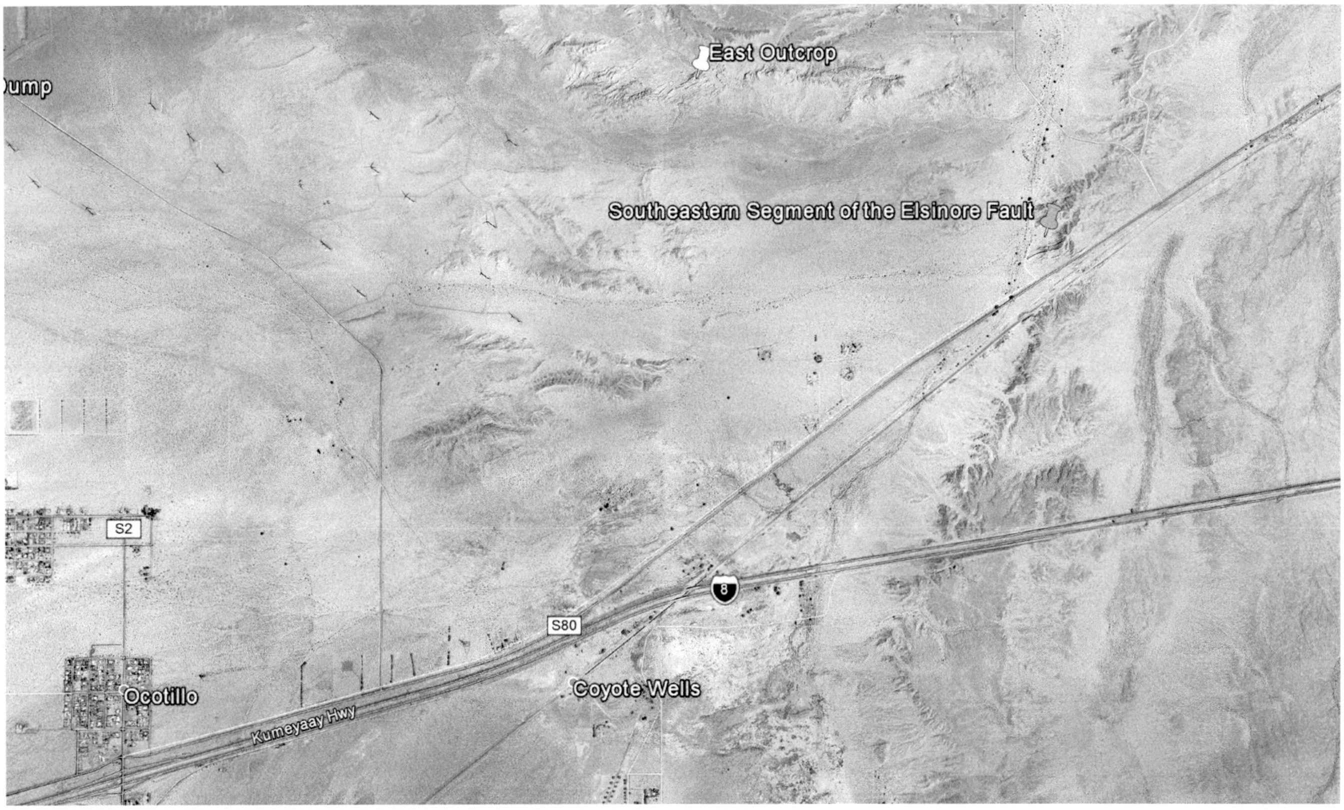

Figure 12. Driving directions to the southeastern segment of the Elsinore fault via S80 road.

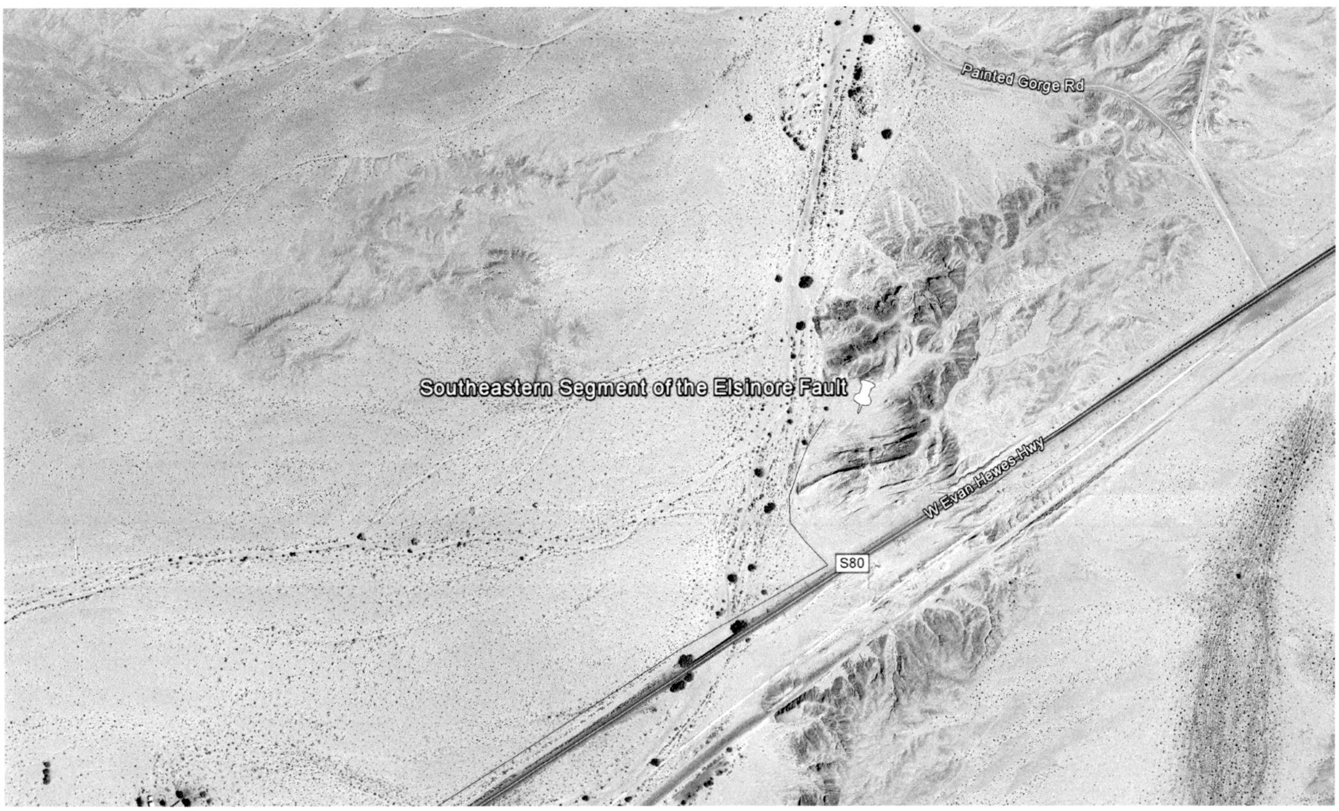

Figure 13. Directions to the southeastern segment of the Elsinore fault from Highway S80.

Figure 14. Picture taken atop a canyon near the southeastern segment of the Elsinore fault, showing the opposing bedding attitudes of the Diablo Formation.

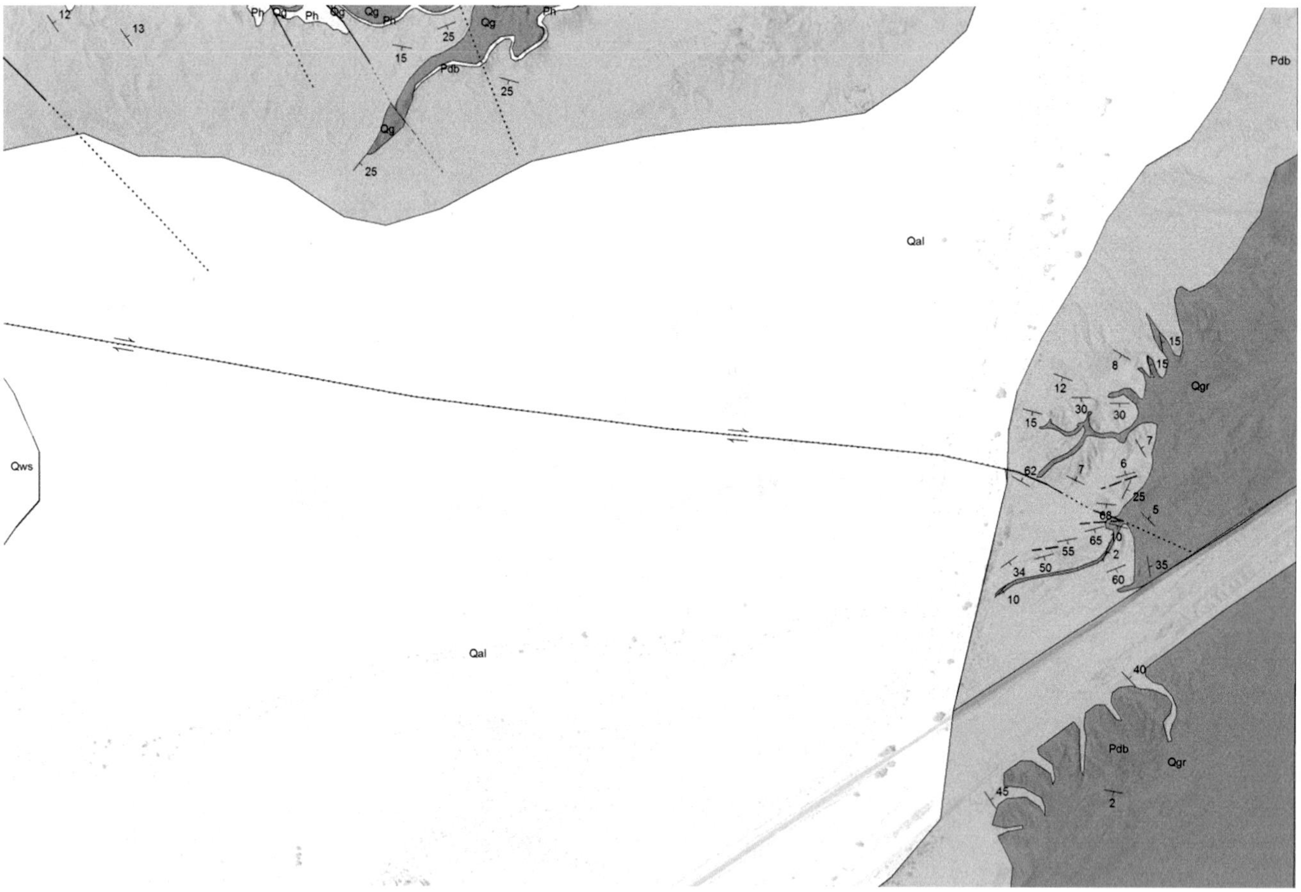

Figure 15. Geologic map of the southeastern segment of the Elsinore fault Teimoorian 2015.

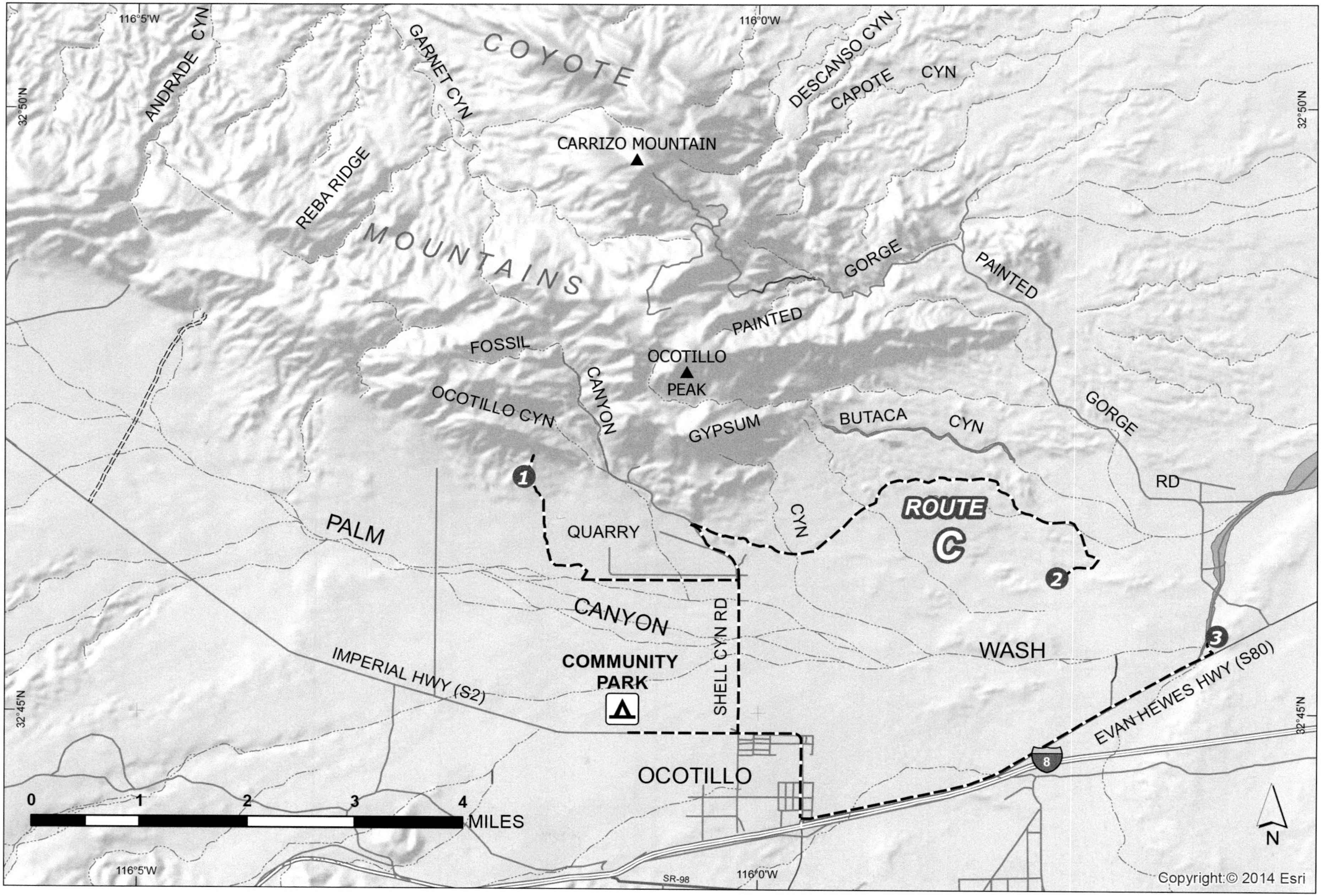

Index Map: Route C

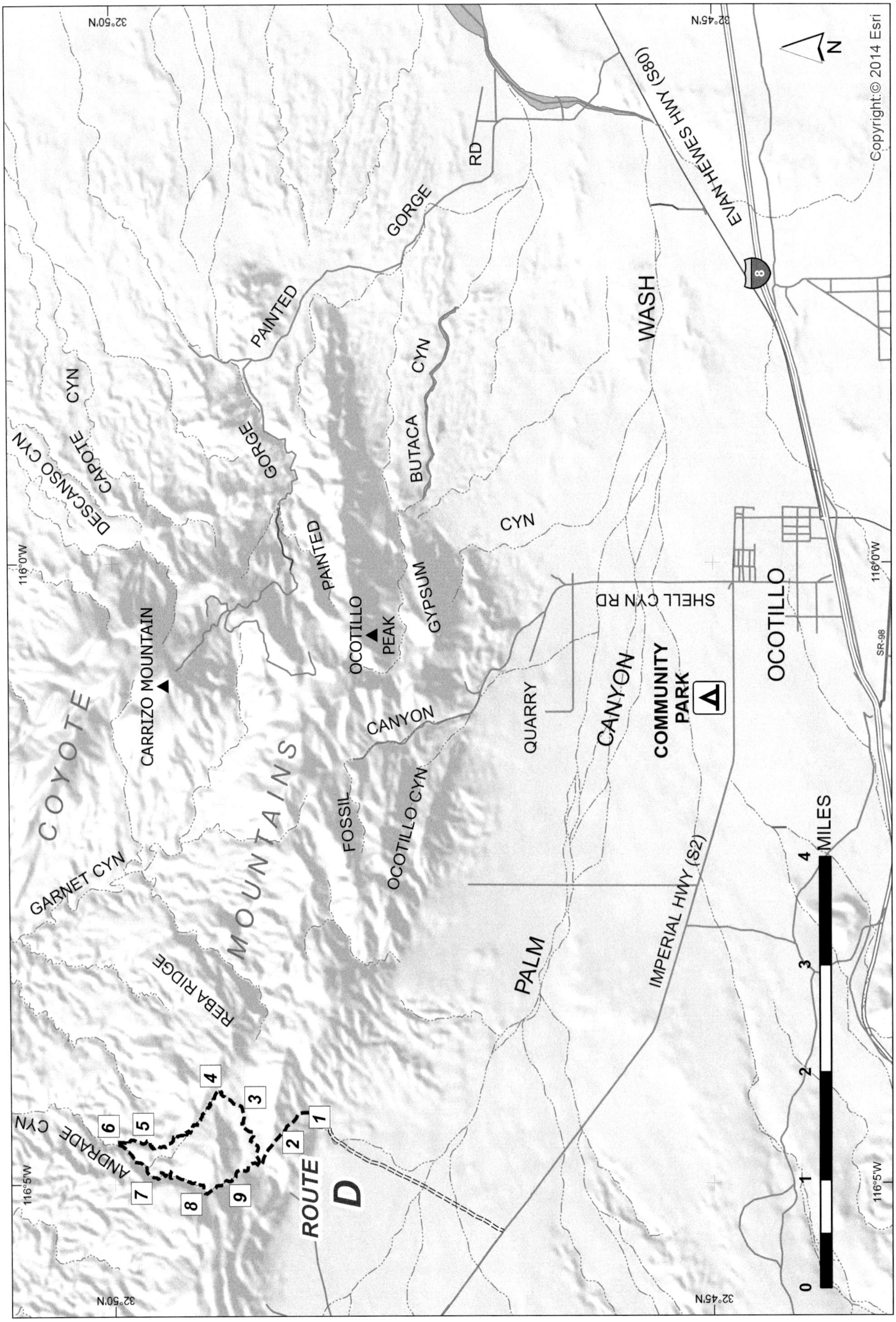

Index Map: Route D

FIELD TRIP D: Andrade Canyon 6 Mile Geological Hike: Slot Canyon and Dry Falls

RANDY WAGNER, *Field Trip Leader*

START TRAIL AT ELEVATION 960 FEET. ASCEND trail toward the Domelands. Instead of turning east up the steep slope to the wind caves, at the saddle which is our high point of 1360 feet, drop north into a dry ravine toward the Carrizo Badlands to our low point of 640 feet. Turn upstream at Andrade Canyon to a dry falls and back to trailhead (Figures 1 and 2).

CAUTION: *Follow this loop counter-clockwise as described; a clockwise route puts the hiker in a dangerous descent down a dry falls.*

This route takes the hiker from the trailhead at the base of the south flank of the Coyote Mountains in Plio-Pleistocene sedimentary strata, up to the spine of range back in time to sandstones from a Miocene sea, then into crystalline basement of Mesozoic granitic intrusions of probable Jurassic age into gneiss, schist, and marbles of Paleozoic protolith. From the saddle, the route goes off the main trail into a dry ravine that becomes a slot canyon with vertical sandstone walls, parts of which have collapsed partially

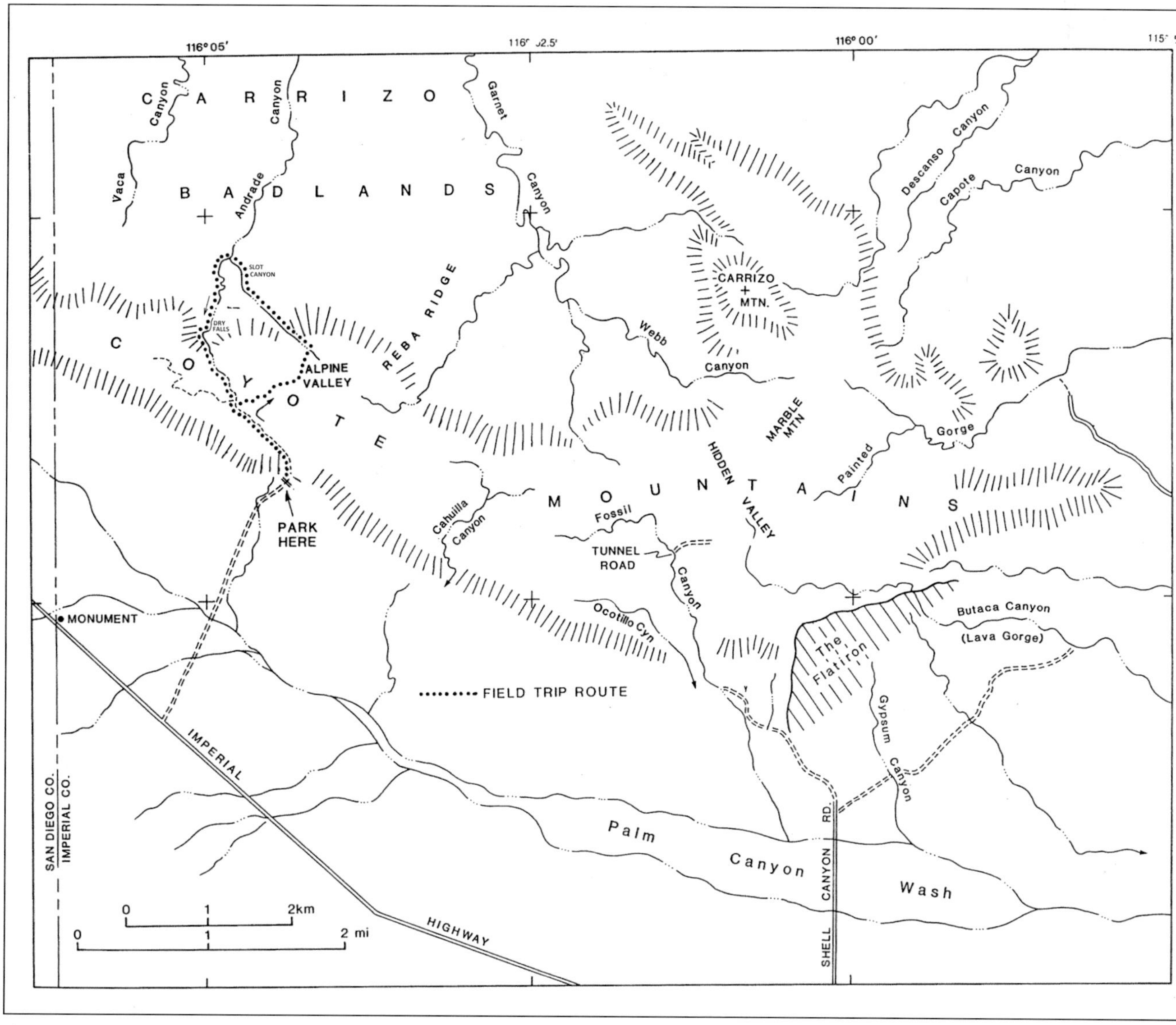

Figure 1. Coyote Mountains index map showing hike route in Andrade Canyon. Modified from Figure 23 of Winker and Kidwell (1996, p. 328).

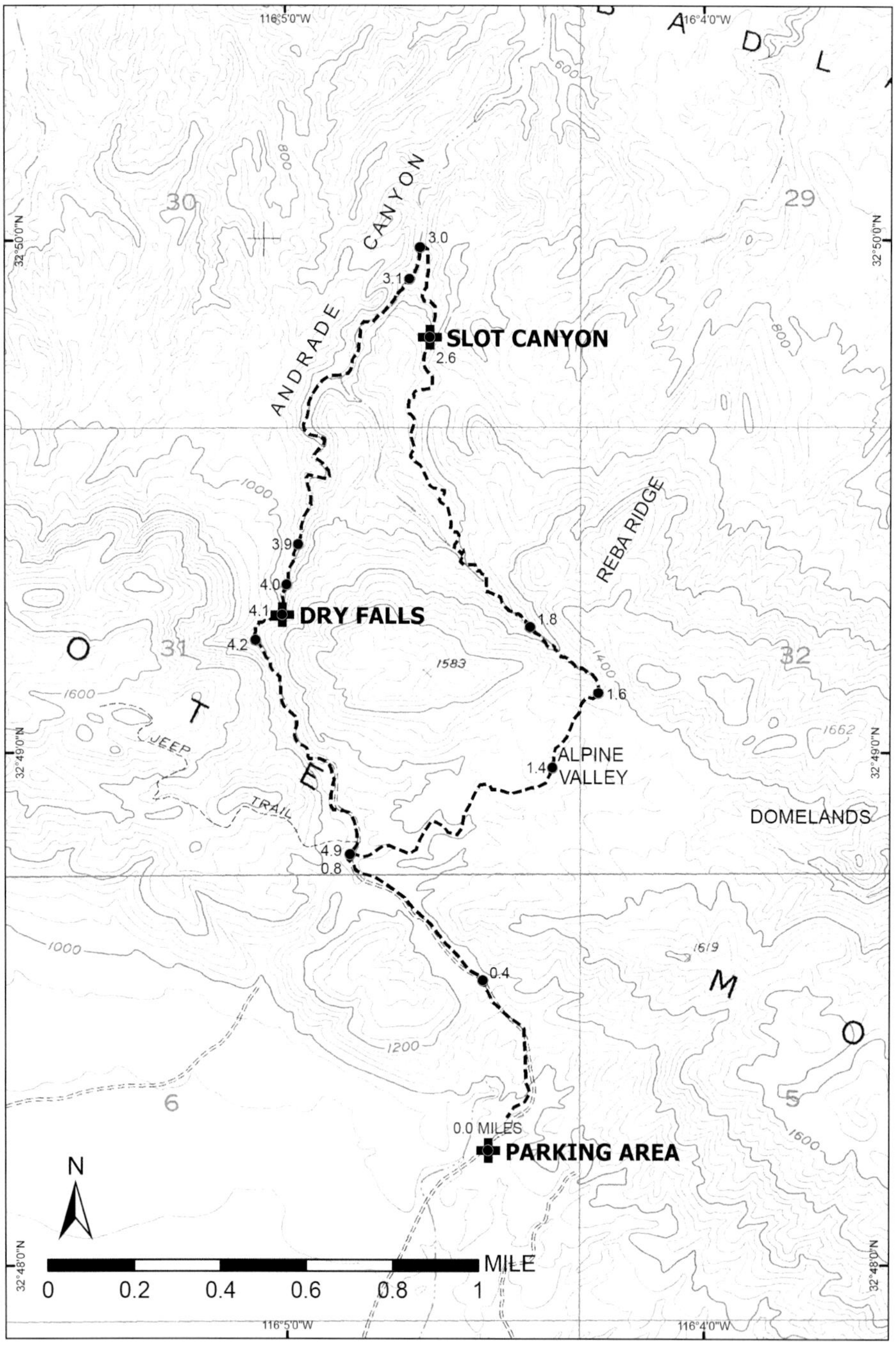

Figure 2. Topographic Map of hike route with dry falls and slot canyon locations.

intact, wedged into the slot, forming a ceiling above the route. This sandstone is the Andrade Member, a shallow facies of a late Miocene sea. At the farthest point north, at a confluence with Andrade Canyon the trail is back into yellow-buff mudstone of the Deguyños Formation. Turn upstream in this canyon to return southbound. From mudstone we return back into sandstone of the Andrade Member of the Latrania Formation. The walls narrow at a dry falls which can be safely climbed with care. At this point crystalline rock of mostly gneiss, schist, and marble dominates. Deguyños mudstone contacts the crystalline basement in a buttress unconformity (see Figure 3). The route rejoins the trail back to the trailhead.

0.0 MILE Parking area and trailhead for the Domelands. (32.804811 N, −116.074562 W, 960'). The main strand of the Elsinore fault is mapped near the south end of the parking area putting pinkish nonmarine fluvial and deltaic beds against old sediment (Figure 3). The muddy hills are free of vegetation, weathering to a distinctive yellow. This late Miocene marine mudstone is distinctively green when fresh. Fossils are scarce but reported within this unit, the Deguyños Formation, the upper part of the Imperial Group. Begin hiking up a fairly steep trail that was once a mining road. As the trail ascends, it takes us generally downsection into older geologic units. At the top of the first incline (0.2 mile), the mudstone gives way to sandstone of the older Andrade Member of the Latrania Formation, the lower portion of the Imperial Group. Shallow marine conditions are represented by high energy and numerous fossils, including snails, clams, oysters, and sand dollars. Continue gradually ascending. The trail crosses a contact with nonmarine fanglomerate, which is shortly truncated by a fault. The trail is back into Deguyños mudstone. The hill (1350′) to the left southwest of the trail is comprised of crystalline bedrock capped by nonmarine fanglomerate of the Garnet Formation.

0.4 MILE The trail gained 220 feet of elevation to reach to the top of the ascent (1180′). The trail descends gradually putting the next hill on the left into view. This hill is comprised of nonmarine Garnet Formation fanglomerate capped by marine Deguyños mudstone.

0.8 MILE Major trail junction at the confluence of a west-southwesterly streambed from the east, a descending streambed to the north-northwest, and a small east-directed drainage from the west. Take the well-traveled trail to the east along the streambed, through faulted Deguyños marine mudstone. These faults are minor parallel and antithetic faults within the larger Elsinore fault zone.

1.4 MILES Alpine Valley. Climb the slope leaving the bottom of the drainage. The valley surface is mantled here by a Quaternary fanglomeratic terrace deposit. The underlying indurated sandstone shortly reveals itself. "In Alpine Valley, Andrade sandstone is thicker than we have seen before and generally rests directly on basement. At Reba Ridge, about 2 miles east of here but hidden from view, the Andrade attains a thickness of 60 m" (Winker and Kidwell, 1996). For the rest of the ascent the trail traverses coarse marine sandstone, in some places concretionary. Nearshore fossils are abundant, but the larger ones, such as sand dollars, are rarely intact.

1.6 MILES Crest of the spine of the Coyote Mountains. This saddle at 1360 feet is the highest point of the hike. The large mountain to the north-northwest on the horizon is Whale Peak. The more distant mountain to its left is Granite Mountain. The Carrizo Badlands, in the middle ground, are an extensive network of dissected yellow mudhills of marine Deguyños Formation and younger pink nonmarine Arroyo Diablo Formation of the Palm Spring Group to the left (Figure 4). Leave the main trail that goes east to the Domelands. Cross north over the saddle ridge taking a faint use trail into the bottom of the dry ravine keeping generally to the right, east, until at the bottom.

1.8 MILES Andrade sandstone faulted against crystalline basement (Figure 5). This canyon is controlled for the next 0.4 mile by this northwest-trending fault. Good exposures of the fault show that its dip varies; farther down the canyon Deguyños mudstone is in spectacular near-vertical fault contact with granitic basement.

Geologic mapping indicates the fault is right-lateral oblique, northeast side up.

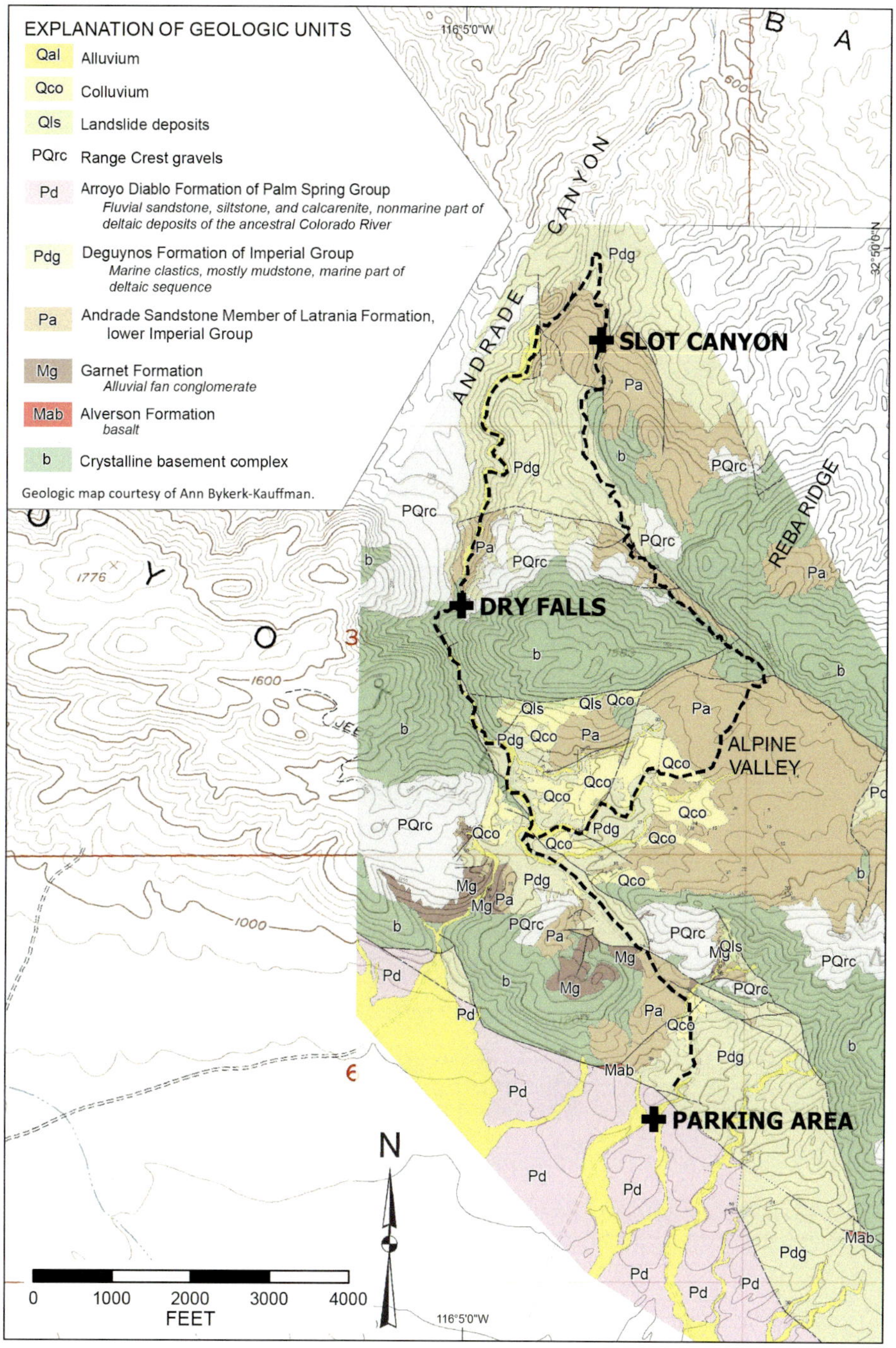

Figure 3. Geologic Map of hike route with dry falls and slot canyon locations.

Figure 4. Carrizo Badlands with Whale Peak in distance from the high point of the hike. Route continues northwest into the ravine, bottom right. (Photo by Dave Bloom.)

Figure 5. Normal fault contact between granitic crystalline basement in the footwall, lower left, and pebbly sandstone of the Andrade member in the hanging wall. (Photo by Dave Bloom.)

Figure 6. Folded gneiss in basement complex. (Photo by Dave Bloom.)

Figure 8. Slot Canyon. (Photo courtesy Ann Bykerk-Kauffman.)

Figure 7: Late Miocene Latrania Formation sandstone above crystalline basement complex. (Photo courtesy Ann Bykerk-Kauffman.)

Figure 9. Path through the slot canyon underneath fallen monolith of conglomeratic sandstone of the Latrania Formation wedged in the slot. (Photo courtesy Ann Bykerk-Kauffman.)

Crystalline basement includes mostly older schist, quartzite, and gneiss of probable early Paleozoic protolith, intruded by granitic rocks including finer grained tonalite and coarser grained pegmatites with feldspar, quartz and large books of muscovite and biotite, with schorl common (Figure 6). Granitic rocks have been zircon-dated as Jurassic (about 160 to 164 Ma; Morgan and Morgan, this volume). Cretaceous rocks have not been reported in the Coyote Mountains.

The metamorphic basement rocks were once oceanic sediments that were deeply buried, heated, and recystallized under directed stress. The protoliths represent sedimentation in an Ordovician sea following rifting of the supercontinent Rodinia that began in the late precambrian (Marshall and Bloom, this volume).

2.6 MILES Slot Canyon. The stream channel turns due north and the walls of ravine narrow to a slot canyon with vertical walls (Figures 8 and 9). Some huge blocks of sandstone have fallen into ravine, the largest ones wedging into the upper part of the canyon creating a roof (Figure 9). This canyon is, in part, fault controlled, cutting through the coarse sandstone of the Andrade Member. The base of the sandstone is depositional directly on crystalline basement (Figure 7), mostly granitic (tonalite). About 150 million years of geologic time is represented by the nonconformity. The nature of the this unconformity should be examined at the entrance and within the slot canyon. It is easy to do because the contact is mainly above the bottom of the drainage. Leaving the slot canyon, the basement rocks disappear beneath the sand. The sandstone walls widen and then diminish as the north dipping strata give way to mudstone.

3.0 MILES Confluence with Andrade Canyon (Figure 10). Turn left, upstream. The mudstone on both sides of the canyon are Deguyños Formation, dipping about 10 to 15 degrees to the northeast.

Figure 10. Confluence of eastern tributary with slot canyon at Andrade Canyon in the badlands formed in mudstones of the Deguyños Formation. Photo courtesy Ann Bykerk-Kauffman.)

3.1 MILES Hike up Andrade Canyon. Coarse sandstone emerges from the bottom of the modern sandy wash. The strata dip northeast toward us. Shortly the walls of the canyon steepen and become nearly entirely Andrade sandstone as Deguyños mudstone disappears over the rim of the canyon (Figures 11 and 12). Gravelly sand beds representing a high energy environment make it easy to observe numerous fault offsets, large and small. A larger fault drops mudstone back into the canyon.

3.9 MILES At this point the canyon crosses from mudstone to sandstone across a fault contact. Erosion steepened the walls and carved fantastic shapes along the rim.

4.0 MILES Crystalline basement rock emerges from the alluvium in the channel bottom (Figures 13 and 14). The overlying gravelly sandstone is in nonconformable contact with the underlying basement, here Jurassic granitic that intruded into strongly foliated schist, ductilely-deformed gneiss, which itself has been intruded by pegmatite dikes and smaller veins.

Figure 11. One of many faults that cut through the Coyote Mountains. Pebbly sandstone of Andrade member of Latrania Formation. (Photo courtesy Ann Bykerk-Kauffman.)

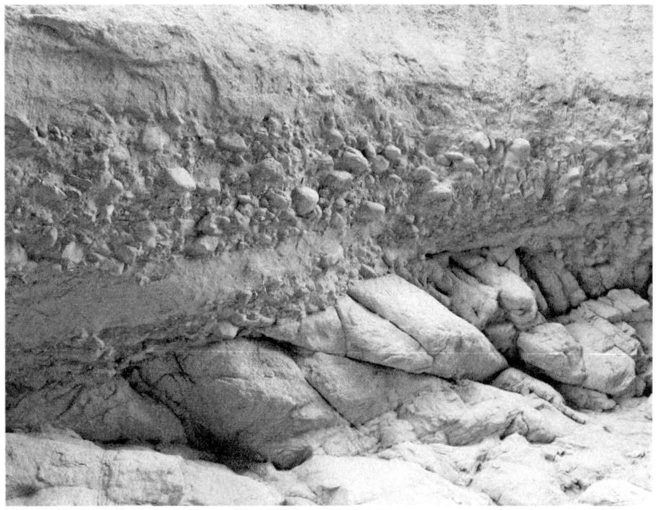

Figure 12. Basal conglomerate in Andrade sandstone member of Latrania Formation in nonconformable contact along ancient erosion surface of underlying Jurassic granitic rock. (Photo by Dave Bloom.)

Figure 13. Relief of underlying granitic rock at nonconformity with Andrade sandstone. (Photo by Dave Bloom.)

Figure 14. Crystalline basement in Andrade Canyon. Strongly foliated schist intruded by a pegmatite dike. (Photo by Dave Bloom.)

Figure 15. Dry falls through crystalline basement complex, Andrade Canyon. (Photo by Dave Bloom.)

4.1 MILES Dry falls (Figure 15). The lower 10 feet is quite steep, but handholds and footholds carefully placed may make for a safe climb to the upper gentler slope to the top, a total of about 30 feet. The dry falls gorge is entirely crystalline basement of various rocks types. Cross-cutting relationships reveal the relative timing of various metamorphic and intrusive events.

4.2 MILES At the top of falls and on the walk southward, note the relationship between the crystalline basement rock and overlying conglomerate and sandstone strata. "Observe the buttress nonconformity of the thin conglomerate and thick sandstone on basement in the floor of the gorge, and the rapid

Figure 16. Buttress unconformity between yellow Deguyños mudstone and crystalline basement. (Photo courtesy Ann Bykerk-Kauffman.)

facies change from conglomerate to sandstone northward away from basement (Winker and Kidwell, 1996, p. 332) (Figure 16).

From here the hike is relatively flat, a gradual rise south-southwest parallel to the strike of a fault that continues back to the main trail. The fault is steeply dipping, juxtaposing basement rock with sandstone or mudstone at various locations.

4.9 MILES Junction of main Domelands trail. Bear slightly left onto Deguyños mudstone and the well-trodden trail we hiked earlier. It is 0.8 mile back to the parking area.

5.7 MILES Parking area.

REFERENCES

Marshall, M., and Bloom, D., this volume, Breaking up is not so hard to do: A history of the geologically brief attachment of southern California to southwestern Arizona, in Wagner, R., ed., Geology of the Coyote Mountains, southern California: San Diego Geological Society, Inc.

Morgan, G., and Morgan, J.R., this volume, Overview of the geology of the southern Coyote Mountains, Salton Trough, Imperial County, Southern California, in Wagner, R., ed., Geology of the Coyote Mountains, southern California: San Diego Geological Society, Inc.

Winker, C. D., and Kidwell, S. M., 1996, Stratigraphy of a marine rift basin: Neogene of the western Salton Trough, California, in Abbott, P. L., and Cooper, J. D., eds., Field Conference Guidebook and Volume for the Annual Convention: San Diego, California, American Association of Petroleum Geologists, Pacific Section, p. 295ñ336.

II. NEW PAPERS

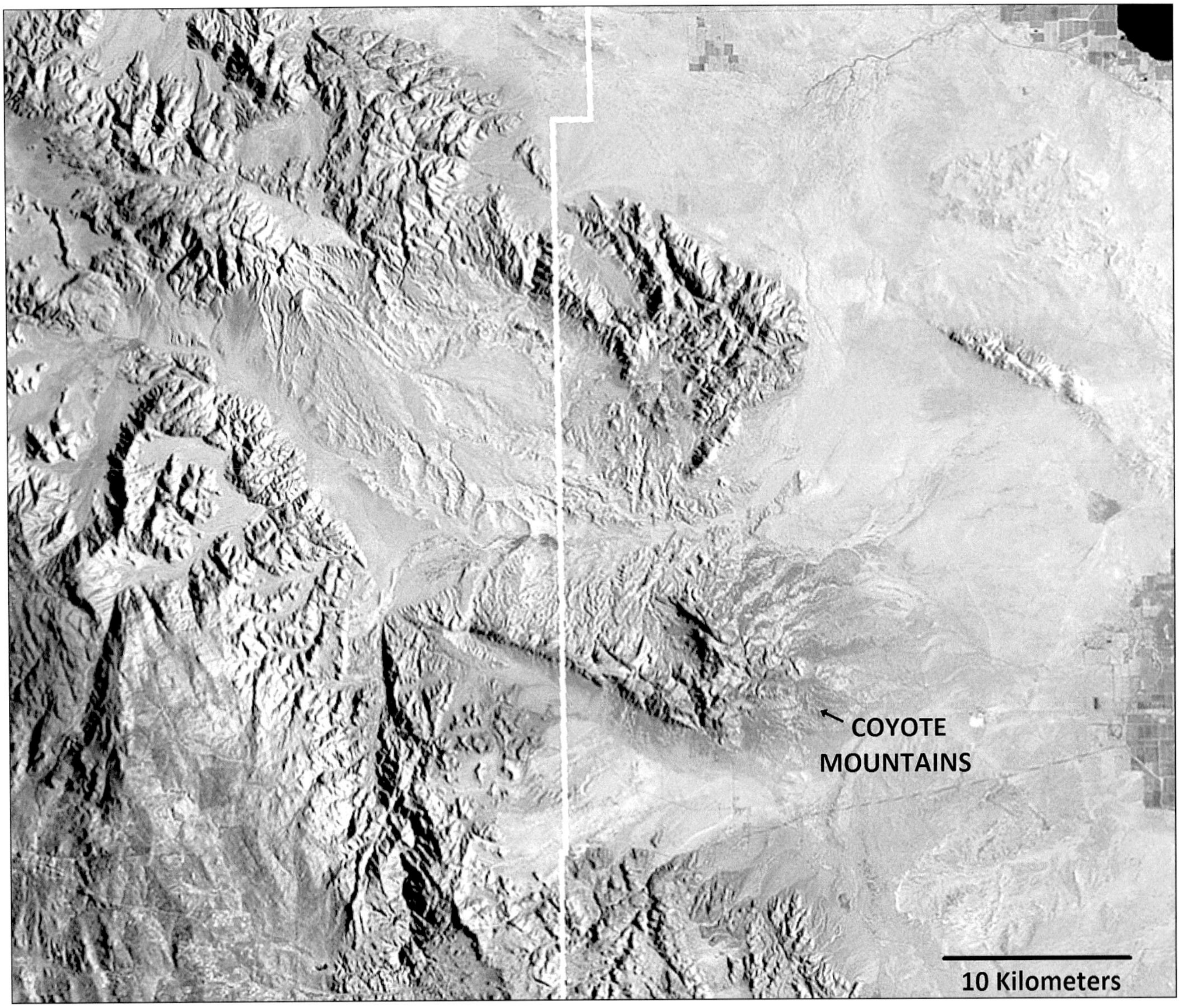

Figure 1. Coyote Mountains and vicinity. The Coyote Mountains are an eastern extension of the Peninsular Ranges at the western edge of the Salton Trough, San Diego and Imperial Counties. County boundary indicated by white line. Landsat 7 data acquired July 1999 to September 2002. Modified from USGS California Landsat Mosaic map created using Tri-Decadal Global Landsat Orthorectified ETM+ Pan-Sharpened data, and draped with National Elevation Dataset data.

Previous page: Enhanced-color view of variegated, predominantly Miocene volcanic strata in the Painted Gorge area. (Photo: Dave Bloom, taken 10-7-2015)

Breaking Up Is Not So Hard To Do: *A History of the Geologically Brief Attachment of Southern California to Southwestern Arizona*

MONTE MARSHALL
Professor Emeritus of Geology and Geophysics
Department of Geological Sciences
San Diego State University

DAVID BLOOM
President, San Diego Geological Society
dave@sandiegogeologicalsociety.org

INTRODUCTION

THE AUTHORS' PURPOSE FOR WRITING THIS PAPER is to set the stage for the people going on this field trip and anyone else interested in this part of southern California. The geology of the Coyote Mountains itself is very interesting—consisting of many different kinds of rocks of many different ages. And the location of these mountains at the western edge of the Salton Trough in the southern Imperial Valley is special (Figures 1 and 2). These mountains lie at the very eastern edge of the Peninsular Ranges and the land is essentially flat to the east, and the next mountain range is the Chocolate Mountains, 50 miles to the northeast. Although covered by sand, fields of alfalfa, and the salty water of the Salton Sea, the geology beneath this flat desert is anything but boring. Buried under thousands of feet of sediments and metasediments lies the northern extension of the transtensional rift zone, or oblique transform, which is currently creating the Gulf of California immediately to the south. This valley is where the major strands of the San Andreas fault system merge into the transtensional rift zone—this flat-floored, innocuous looking valley is in fact an important part of the transform boundary between the North American and Pacific plates (Figure 3).

The main zone of the San Andreas system runs between the Salton Sea and the Chocolate Mountains. Besides occasionally generating large earthquakes, it forms a very special geological boundary in this region. Those of us on this field trip who came from the Pacific coastal areas, like San Diego, passed across sedimentary and igneous rocks no older than about 200 million years (Ma). And most of the rocks in the rift zone to the east of the Coyote Mountains are probably no older than Pliocene—just a few million years old. In fact, there are probably basaltic intrusions that are just a few days old! But, many of the igneous rocks and gneisses in the Chocolate Mountains are 1.8 billion years old (Ga), or 1,600 million years older than most of the rocks to the west. These rocks are correlative with the Vishnu Schist and Zoroaster Granite, seen by many of us lucky enough to raft down through the bottom of the Grand Canyon. These rocks accreted to Laurentia about one billion years ago, and form the basement of much of Arizona and the states surrounding it to the north and east. This is the story of the very old and very new rocks and associated tectonics surrounding the Coyote Mountains that we want to tell the reader in this article.

THE GEOLOGIC HISTORY OF THIS REGION REALLY BEGINS WITH THE BREAKUP OF RODINIA

As seen in the previous section, the Coyote Mountains occupy a special place in the geologic evolution of this region. They are at the eastern edge of the Peninsular Range batholith, between two fault zones of the San Andreas fault system—the Elsinore and San Jacinto, and on the western border of the Salton Trough rift zone (Figures 2 and 3). But even more significantly, the main strands of the San Andreas fault lie 70 km, or 40 miles, to the northeast and on their east side is the same 1.8 Ga igneous and metamorphic crust that underlies Arizona. Aside from some pre-Cambrian rocks in the Transverse Ranges, and Paleozoic to Triassic (?) metasedimentary screens, the plutonic and sedimentary rocks to the west aren't much older than 200 Ma. Behind this great lateral unconformity lies an important chapter in the geological evolution of Western North America.

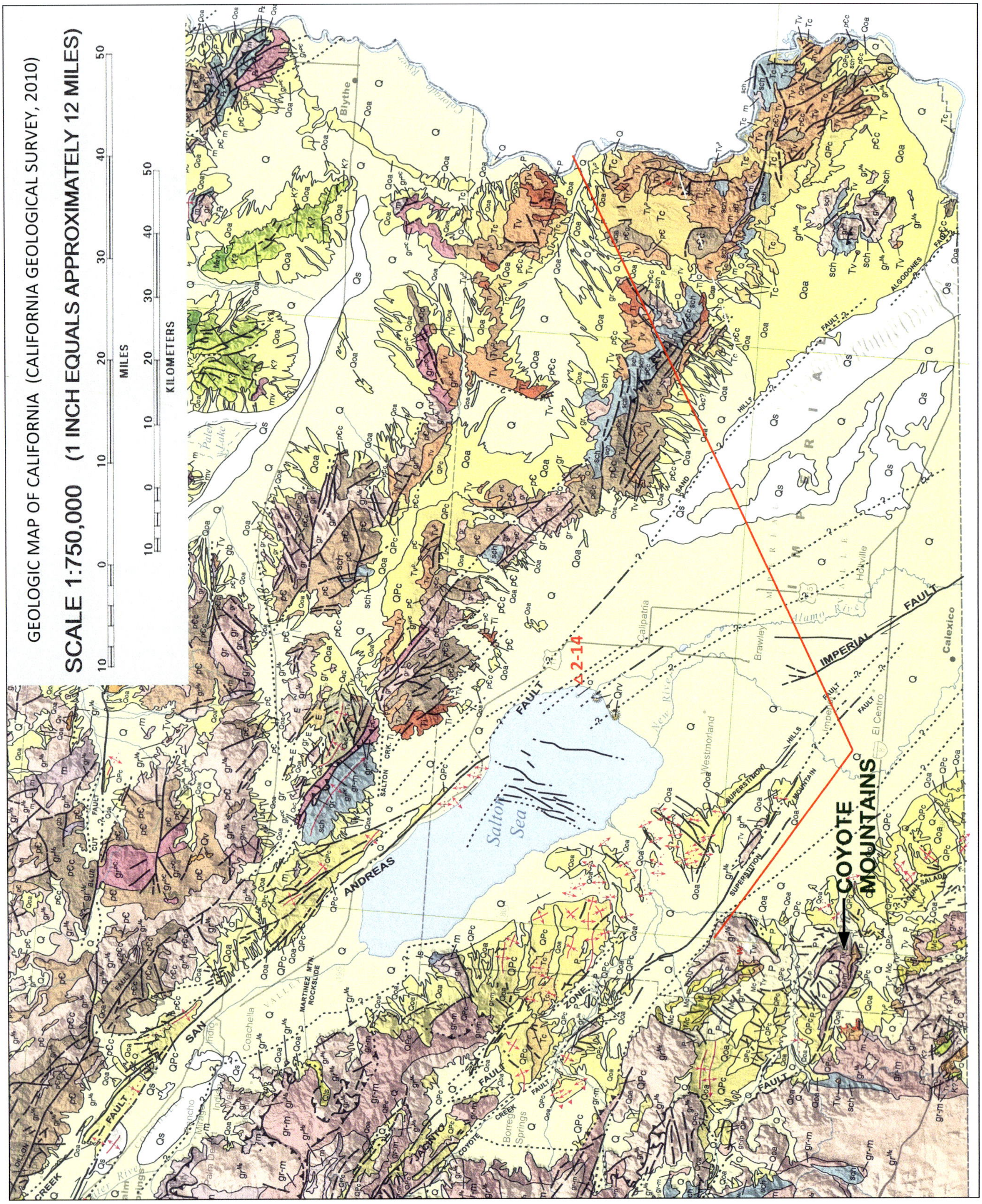

Figure 2 Geologic map of Coyote Mountains and vicinity. Red triangle shows location of Well 2-14. Red line shows location of cross section depicted in Figure 8. In the Coyote Mountains, the basement core is indicated by "gr-m," described as: "Granitic and metamorphic rocks, mostly gneiss and other metamorphic rocks injected by granitic rocks. Mesozoic to Precambrian." Alverson Formation, a mainly volcanic suite on south and eastern flanks, is indicated by "Tv," described as "Tertiary volcanic flow rocks; minor pyroclastic deposits." Marine sedimentary rocks of the Imperial Group on lower slopes indicated by "P," described as "sandstone, siltstone, shale, and conglomerate; mostly moderately consolidated." Younger nonmarine sedimentary rocks of the Palm Spring Group indicated by "QPc," described as "Pliocene and/or Pleistocene sandstone, shale, and gravel deposits; mostly loosely consolidated." Excerpted from "Geologic Map of California" 2010 update of Charles W. Jennings 1977 compilation. California Geologic Data Map Series (No. 2).

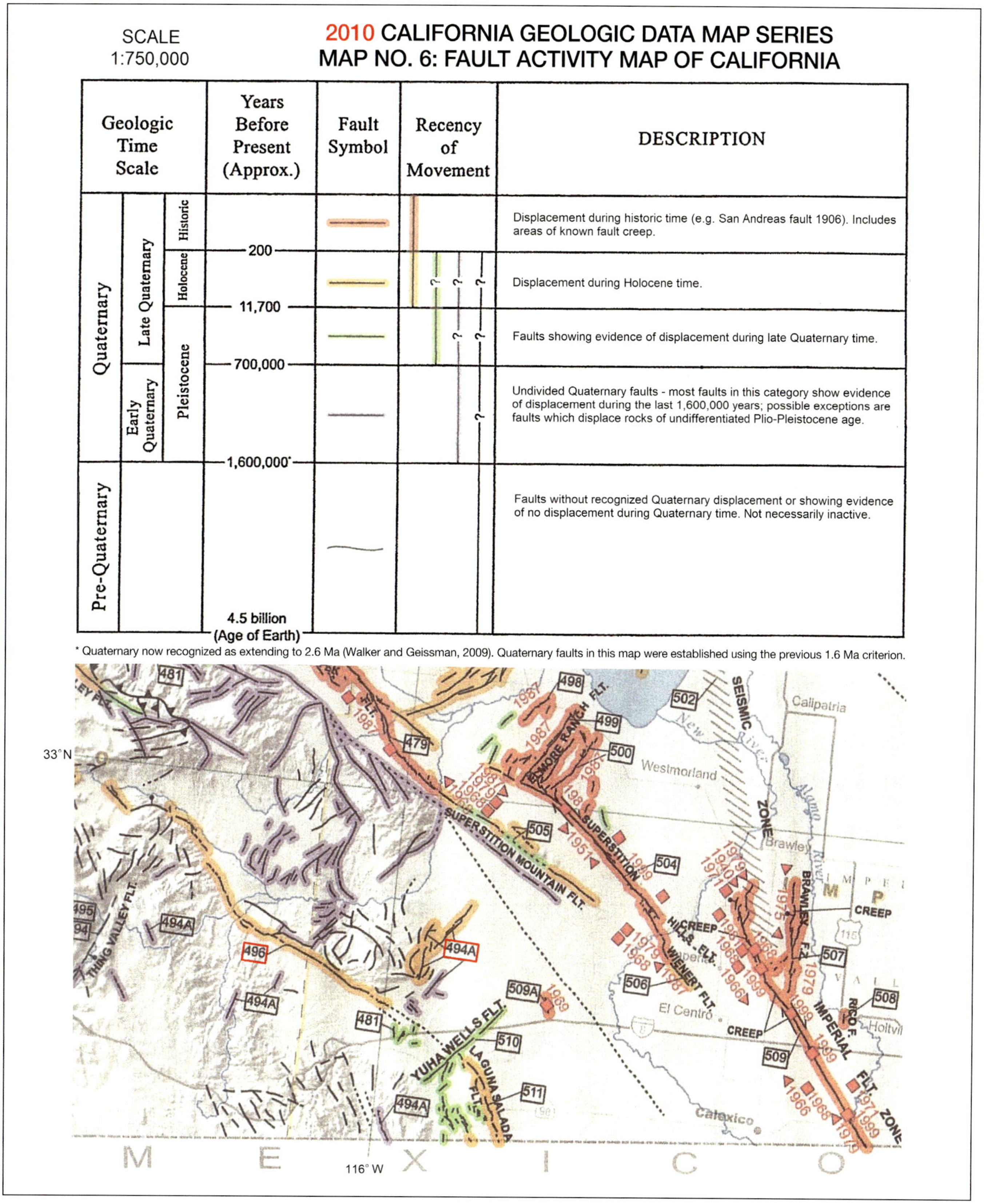

Figure 3. Fault activity map of Coyote Mountains and vicinity. The strand of the Elsinore fault, 496 in red box, is the dominant fault structure in the Coyote Mountain area, bounding the range at base of its southwest slope. This right-lateral fault is west-northwest trending at the Coyote Mountains, and reflects the inherent transpression that is thought to have uplifted the mountains. The Laguna Salada fault may be the southern extension of the Elsinore fault. Numerous northeast trending antithetic faults are expressed in the Coyote Mountains including 494A in red box. See Road Log A, Stop 9, for a visit to the Painted Gorge Wash fault for a detailed discussion. These faults are part of the larger San Andreas fault system and the rift system that formed the Salton Trough. Excerpted from "2010 Fault Activity Map of California," update of Charles W. Jennings and William A. Bryant's compilation. California Geological Survey, Geologic Data Map No. 6.

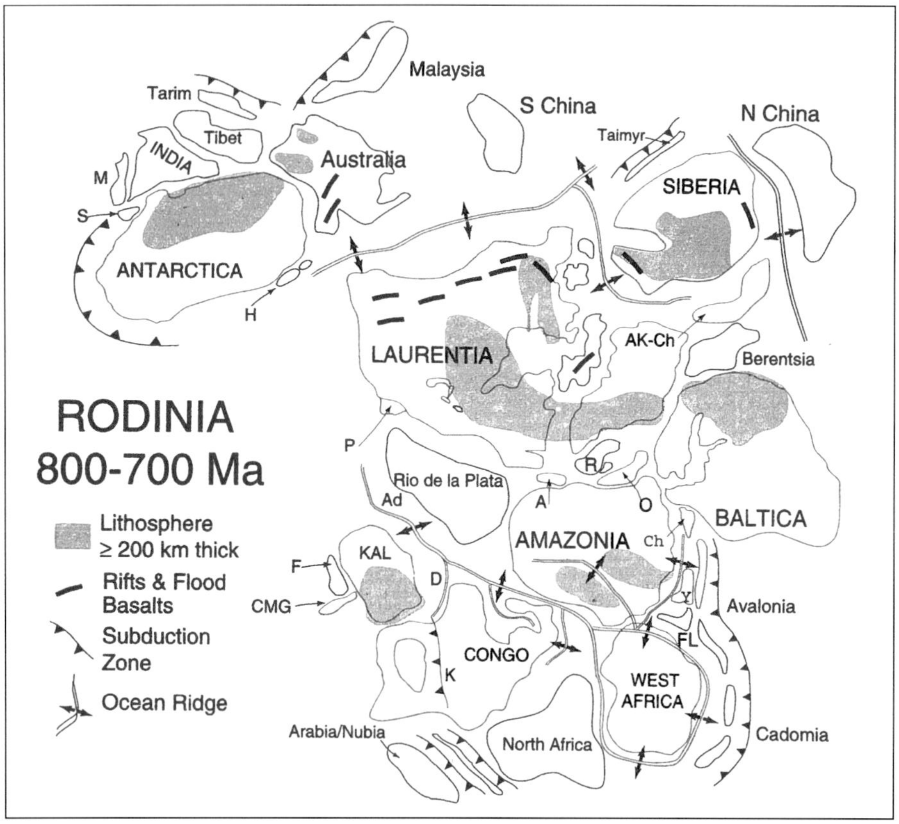

Figure 4. Diagrammatic reconstruction of Rodinia at 800—700 Ma showing possible arc systems/subduction zones, rifts and flood basalts, ocean ridges, and thick Archean lithosphere. Note that north is up in this figure. Laurentia (another name for North America, especially when discussing supercontinents) was in the north-central part of Rodinia and its modern west coast trended east-west and was flanked to the north by the continents of Antarctica, Australia, and S China. The modern east coast of North America was flanked by much of what is now South America. Other symbols are not important to this paper. (From Condie, 2003)

During the past two billion years most of the continental land masses have congregated several times into supercontinents. The last and best known is Pangea that lasted from about 300 to 200 Ma. The one before it, Rodinia, is less well understood and aggregated between about 1300 and 950 Ma and broke up between about 850 and 600 Ma (Condie, 2003). As can be seen in Figure 4, Laurentia/North America occupied the center of Rodinia, with South America and Africa to the south of the present east coast of North America, and Antarctic, Australia, and part of China to the north of the current west coast of North America. The exact location and even identity of some of the circum-Laurentia continents is still debated. Some authors would add Siberia to the row of continents on the northern margin of Laurentia. The reader is recommended to read a more detailed account of the possible surrounding continents in Marshall (2012) and Dickinson (2009). Most important to the geologic history of southern California is the rifting of northern Rodinia along an east-west line beginning about 800 Ma. This is shown beautifully in the landscapes (our Figures 5 and 6) by Blakey and Ranney (2008). Note that the western tier of states didn't exist at that time (800–500 Ma) and that Yuma, Arizona, was probably under a warm, shallow sea that stretched east all the way to present Alaska. You can see in Figure 5 that this shallow sea, underlain by thinned and stretched pre-Cambrian crust, opened first in the east and finally in the west. Dickinson (2009) estimates the eastern section became deep ocean about 750 Ma and the area around Yuma became deep ocean about 600 Ma—this opening was the proto-Pacific. By Cambrian time most of the northern continents were on their way to the southern hemisphere to become part of the Gondwana continents (Figure 6). How far to the north of Yuma was the edge of the continent or the shelf-slope break is unknown, but the present location of San Diego was probably on the deep seafloor, i.e., oceanic crust. Much of the currently western tier of states and the province of British Columbia, and southwest Alaska were probably in the same abyssal situation. As will be seen in the last section of this article, North America rotated to almost its present orientation by the end of the Paleozoic, but Yuma and the southwestern states remained in equatorial waters until the Mesozoic—which helps explain the abundance in the western states of late Proterozoic and Paleozoic carbonates, like the famous Redwall Limestone in the Grand Canyon.

THE CHANGE FROM A PASSIVE/DIVERGENT TO AN ACTIVE/CONVERGENT MARGIN IN THE MESOZOIC ADDS MUCH OF THE CRUST OF SOUTHERN CALIFORNIA

As mentioned above, by the beginning of the Jurassic (200 Ma) North America had rotated almost 60 degrees counterclockwise from its orientation in the Cambrian—the former north coast became the current west coast! But nature had bigger changes in store for us. Far to the east, what was to become the east coast of the future United States began rifting from northwestern Africa as supercontinent Pangea began to break up. As the north-central Atlantic Ocean was born, our continent made way for it by drifting westward. At this time the floor of the east Pacific, just off the coast of Arizona, was about 400 million years old—it was old, cold, and dense—just waiting for an excuse to subduct beneath western North America. Some minor subduction and accretion of island arcs had already begun along the current western margin

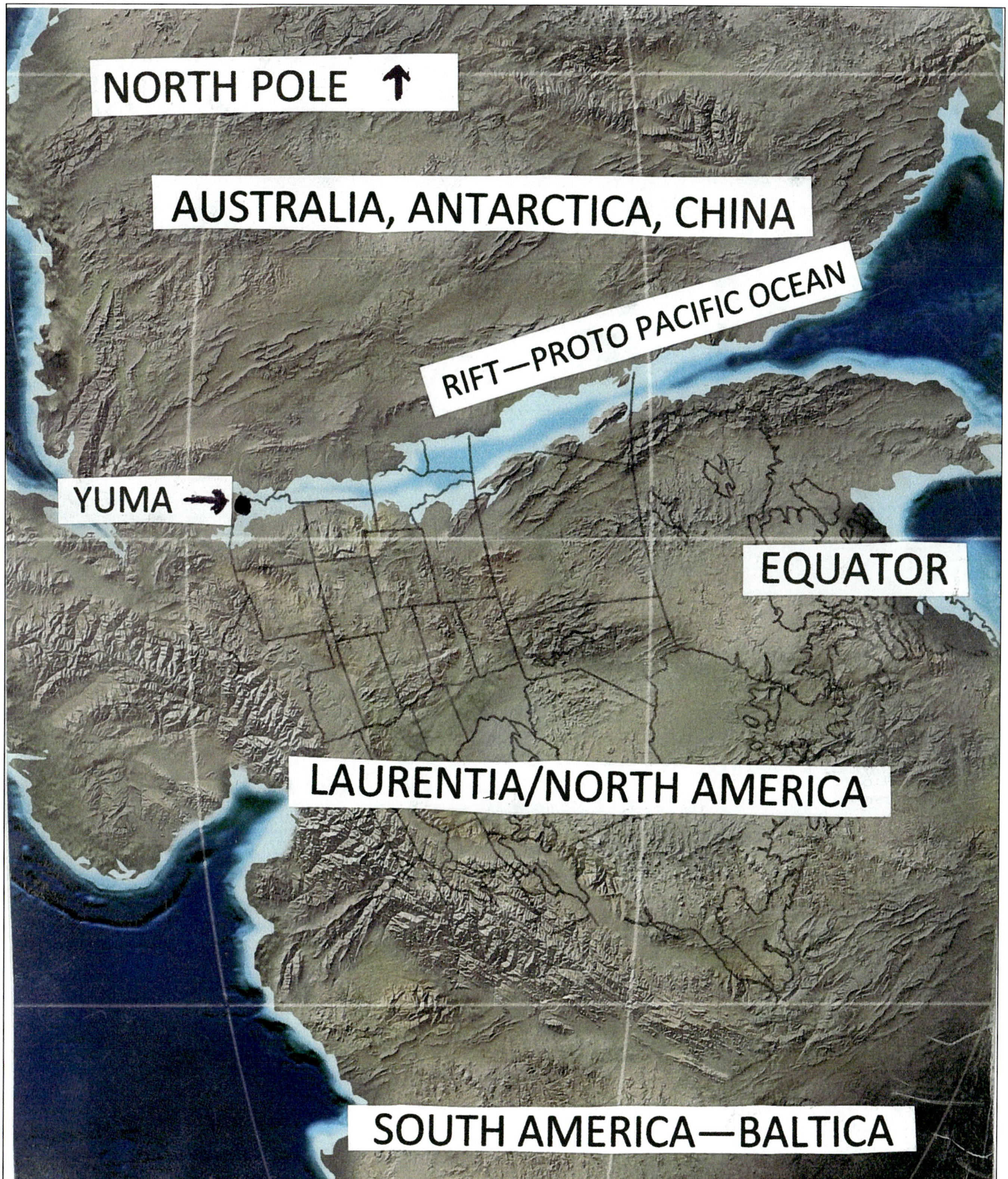

Figure 5. Diagrammatic landscape of the part of the supercontinent of Rodinia, shown in Figure 4, that is of interest in this article, also at about 800 to 700 Ma. Although debated, the giant landmass to the north of Laurentia most likely consisted of at least parts of the current continents of Australia, Antarctica, China, and even Siberia. The dark blue areas are oceanic crust and the light blue is continental shelf and slope, i.e. the continental margins. Note that the area of Yuma is on a continental margin, in more or less shallow sea water. But, this part of Rodinia is starting to rift/break up and soon oceanic crust will form at an east-west spreading center—and the proto-Pacific Ocean will be born. Note especially that most of the western tier of states, and westernmost Canada and Alaska, did not exist at this time. Although not outlined, Greenland was a part of Laurentia (in the lower right of the map). A chain of mountains formed off the present east and northeast coasts of North America by the collision of parts of South America and Baltica. (Modified from Blakey and Ranney, 2008)

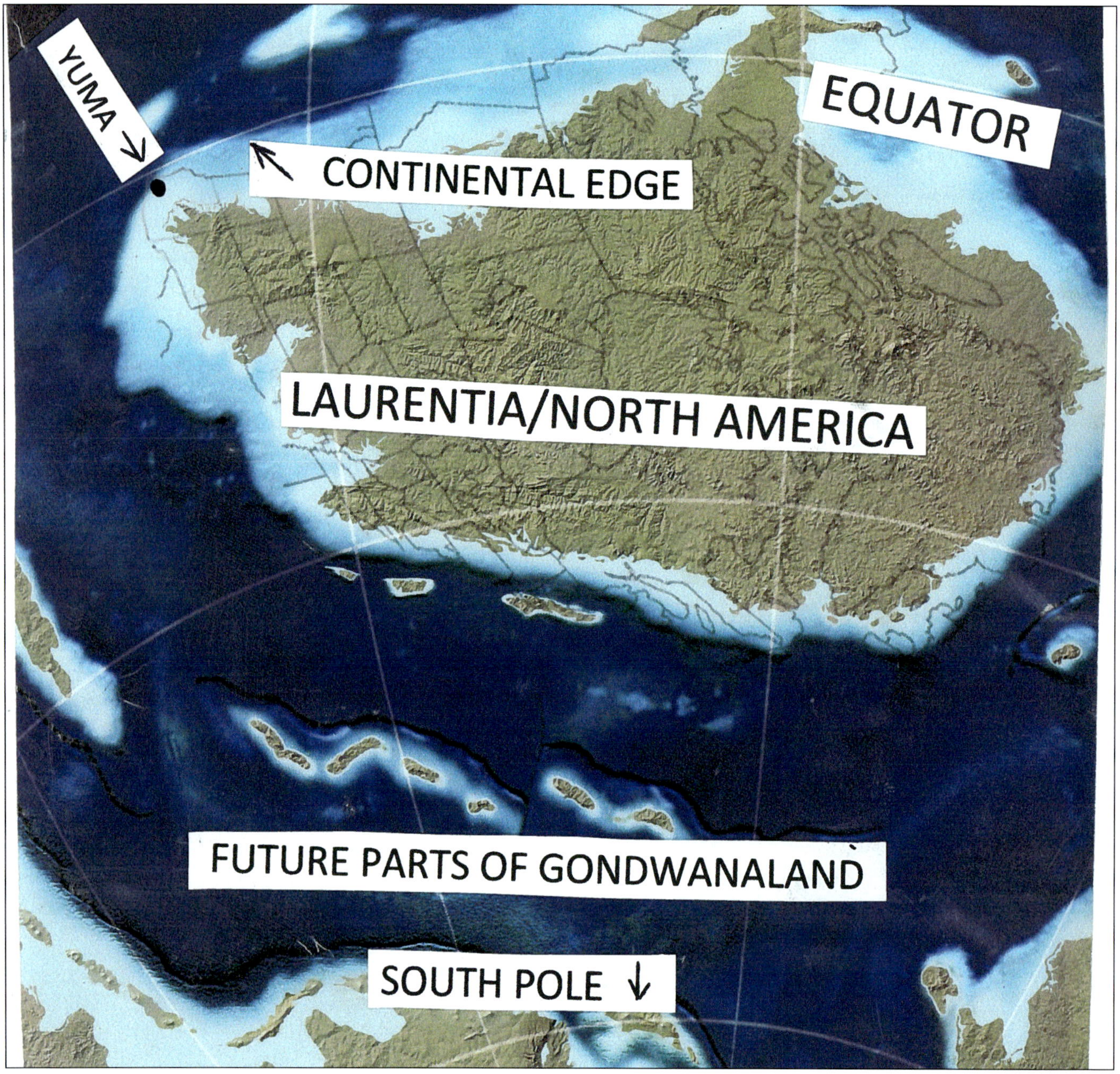

Figure 6. North America at the beginning of the Cambrian, at about 540 Ma. Our continent is still oriented 90 degrees clockwise from its present orientation, and the current west coast still lies along the equator. The Yuma area is still covered by shallow seas, but the continental edge at the shelf-slope break is probably not far to the north. The continent(s) that had been to the north have now made an end-run around North America and are to the south and clustering to form the semi-supercontinent of Gondwanaland. The black, sinuous lines are oceanic trenches above subduction zones that are forming island arcs, one of the main building blocks of new continental crust. The outline of Greenland is now shown on this map. (Modified from Blakey and Ranney, 2008)

in the Paleozoic, but the major pulses of convergent magmatism in the Peninsular Ranges and Sierra Nevada occurred in the Jura-Cretaceous, especially between 120 and 80 Ma (Walawender, 2000; Moore, 2000). For a detailed description of the Pacific plates involved in this subduction and their motions, see Marshall (2012).

Figure 7 shows Mike Walawender's diagrammatic sketches of the subduction geometry involved in, first, the western half of the Peninsular Ranges batholith (Figure 7A), and then that involved in the eastern half (Figure 7B). Although not explicit in Figure 7A, the early magma that formed the volcanic arc and underlying plutons was probably intruded into oceanic crust. A recent synthesis of geophysical data by Langenheim et al. (2014) suggests that the western half of the Peninsular Ranges batholith was emplaced on oceanic lithosphere, whereas the eastern half was emplaced on lithosphere of continental affinity. The age of the plutons decreases towards the east, from about 120 Ma to 90 Ma—the last magma pulse creating the La Posta pluton on the eastern edge of the batholith. This easterly age decrease is seen in other batholiths, like the Sierra Nevada, and is attributed to a shallowing of the subducted plate caused by an increase in the westward velocity of the North

American plate. After 90 Ma the locus of magmatic activity quickly migrated east into the Mojave and Arizona (Dickinson, 2009). The eastern and western halves of the Peninsular Range batholith differ in another respect. Geobarometry studies show that the eastern half has been eroded 3 km, almost 2 miles, deeper than the western half (Walawender, 2000). So, despite being older, the only remnants of the old volcanoes that capped the plutonic rocks, are along the coastal strip—the Santiago Peak volcanics and the newly defined Penasquitos Formation (Kimbrough and others, 2014).

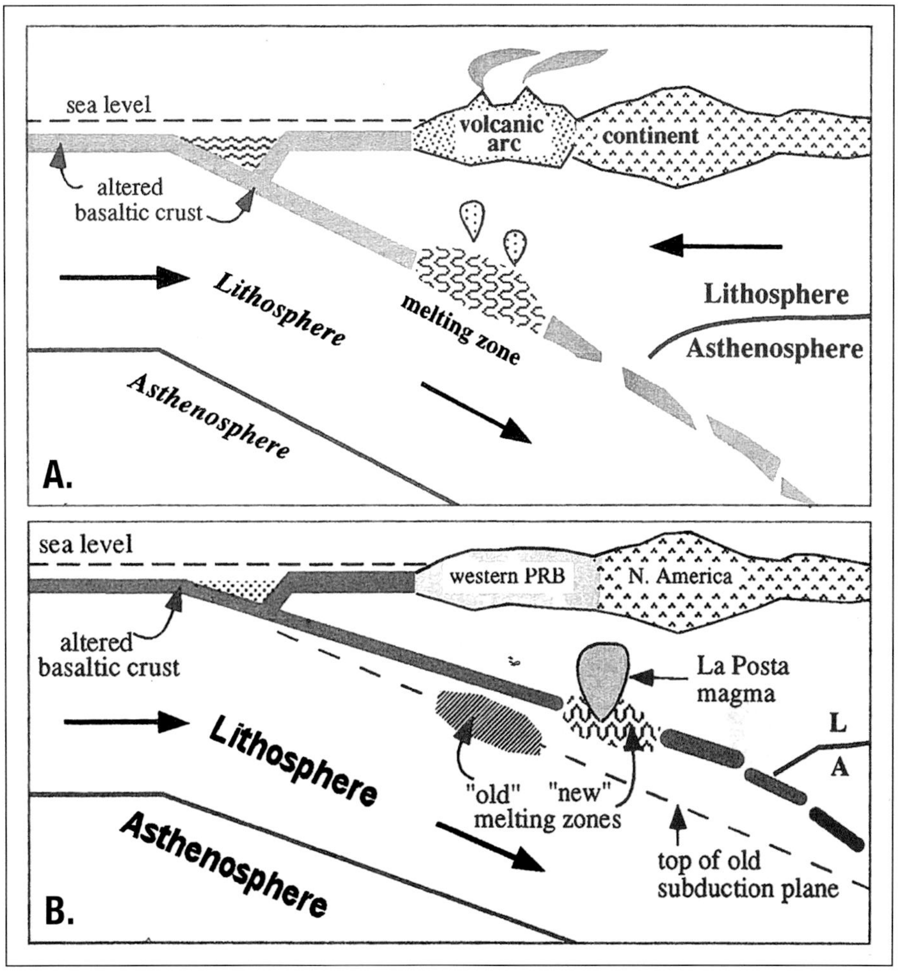

Figure 7. Diagrammatic cross-sections of the subduction zone that existed off the coast of southern California during Jurassic and Cretaceous time that produced the volcanic rocks of the Peñasquitos and Santiago Peak formations and the plutonic rocks of the Peninsular Ranges batholith. **Figure 7A** shows an earlier time when the subduction was producing a volcanic arc (the underlying plutons not explicitly shown), that may have formed at the edge of the North American continent, as shown, or could have formed as an island arc off the coast. One of the current theories for the magma source is the fluxing of part of the down-going crust and overlying mantle wedge (the melting zone) by the release of water from the subducted sediments and clay-altered basalts. This melting typically occurs at a depth of about 100 km. **Figure 7B** shows a later time when the subduction angle has shallowed, displacing the melting zone eastward, forming the La Posta pluton. Intruding between about 100 and 90 Ma, it is the youngest, largest, and easternmost pluton in the Peninsular Ranges at the latitude of the Coyote Mountains. It underlies most of eastern San Diego County and has a surface area of 1400 square kilometers. Its position so far east raises the question of what kind of crust it was intruded into—maybe a very different one than that shown in diagrammatic Figure 7B. (La Posta pluton data and figures are from the late and beloved Mike Walawender, 2000.)

EFFORTS AT CONSTRUCTING A CRUSTAL SECTION ACROSS THIS REGION

One of the most difficult and problematic tasks facing a geologist working in an area is trying to construct what the rocks buried deep beneath his or her study area might be. How can you know what you can't see?! If you have sufficient resources you can drill a well and bring up cuttings or even cores. But what if you want to 'see' all the way through the crust and Moho and into the mantle? The deepest petroleum well in the world is about 10 km, or 6 miles. The deepest scientific well in the world, which is in Russia, is about 13 km, or 8 miles, deep. This depth is about the current limit because the cost goes up exponentially with depth, and the rocks at this depth are so hot that steel alloys begin to lose their strength and the rocks begin to flow plastically and close off the hole.

Sometimes nature itself provides samples of crustal and even mantle rocks, as xenoliths caught up in basalts that rise from the mantle. And there are places where tectonics, like collision zones in convergent margins, can bring slabs of the crust and even upper mantle to the surface. But these places on Earth's surface are special and the normal way to 'see deep' is with geophysics, especially seismology. The basic problem with geophysical surveys, however, is the fundamental principal that all geophysical data are subject to multiple interpretations—there are only geologically reasonable solutions, never unique or certain ones.

With these limitations in mind, let's look at the next two figures in this article. Figure 8 is a somewhat diagrammatic, east-west crustal cross section across the Salton Trough, from the Fish Creek Mountains, just north of the Coyote Mountains, to the Colorado River. It shows the three very different kinds of crust in the region of our study. At the west end, close to the Coyote Mountains, it shows the Mesozoic granitic rocks with septae or screens of metamorphic rocks forming the upper crust. According to the authors, these Mesozoic plutons have intruded and now cap a middle and

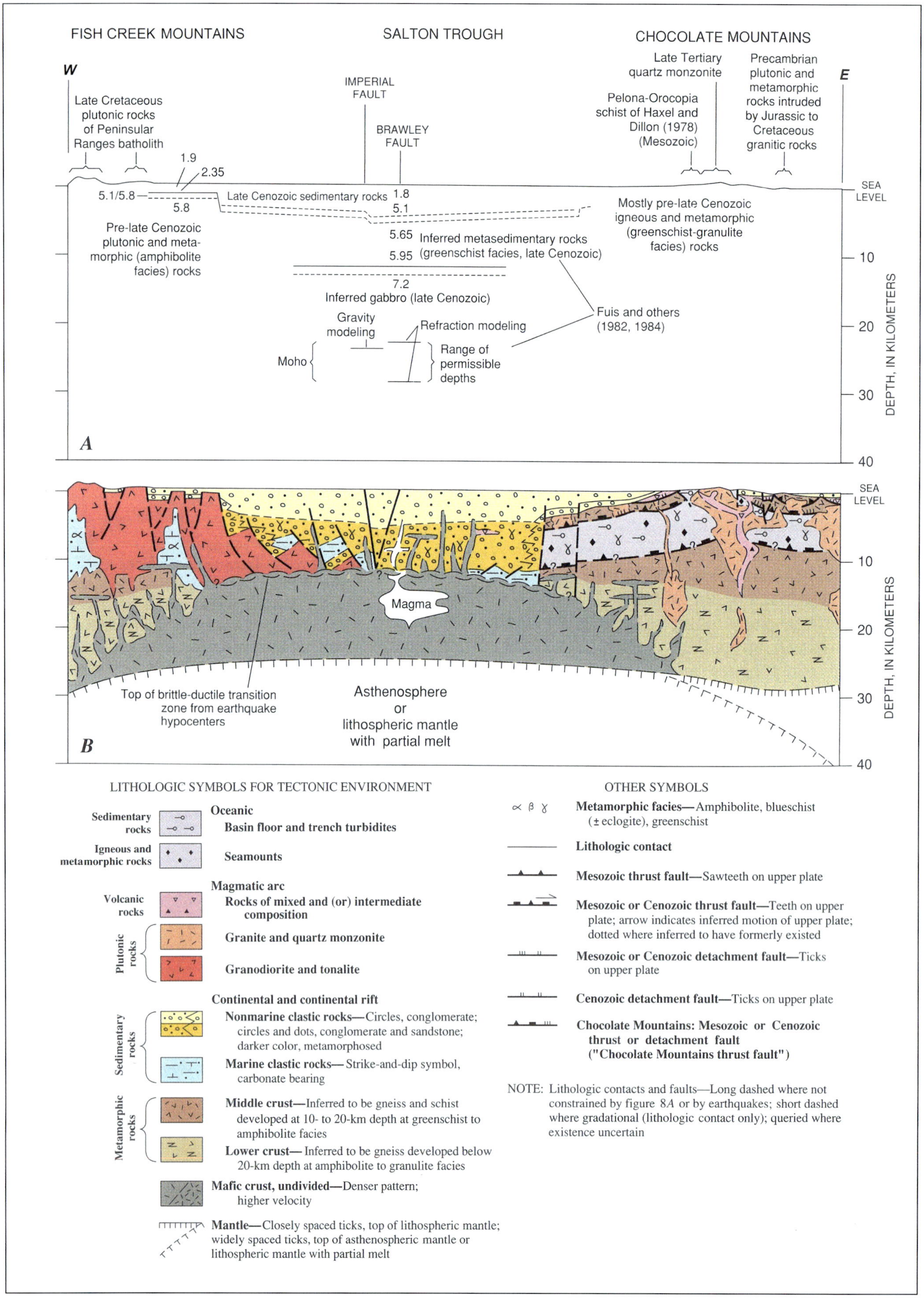
FISH CREEK MOUNTAINS
SALTON TROUGH
CHOCOLATE MOUNTAINS
W
E
Late Cretaceous plutonic rocks of Peninsular Ranges batholith
IMPERIAL FAULT
BRAWLEY FAULT
Late Tertiary quartz monzonite
Pelona-Orocopia schist of Haxel and Dillon (1978) (Mesozoic)
Precambrian plutonic and metamorphic rocks intruded by Jurassic to Cretaceous granitic rocks
1.9
2.35
SEA LEVEL
5.1/5.8
5.8
Late Cenozoic sedimentary rocks 1.8
5.1
Mostly pre-late Cenozoic igneous and metamorphic (greenschist-granulite facies) rocks
Pre-late Cenozoic plutonic and metamorphic (amphibolite facies) rocks
5.65 Inferred metasedimentary rocks
5.95 (greenschist facies, late Cenozoic)
7.2
Inferred gabbro (late Cenozoic)
Fuis and others (1982, 1984)
Gravity modeling
Refraction modeling
Moho
Range of permissible depths
10
20
30
40
DEPTH, IN KILOMETERS
A
Magma
Top of brittle-ductile transition zone from earthquake hypocenters
Asthenosphere or lithospheric mantle with partial melt
B
LITHOLOGIC SYMBOLS FOR TECTONIC ENVIRONMENT
Oceanic
Sedimentary rocks
Basin floor and trench turbidites
Igneous and metamorphic rocks
Seamounts
Magmatic arc
Volcanic rocks
Rocks of mixed and (or) intermediate composition
Plutonic rocks
Granite and quartz monzonite
Granodiorite and tonalite
Continental and continental rift
Sedimentary rocks
Nonmarine clastic rocks—Circles, conglomerate; circles and dots, conglomerate and sandstone; darker color, metamorphosed
Marine clastic rocks—Strike-and-dip symbol, carbonate bearing
Metamorphic rocks
Middle crust—Inferred to be gneiss and schist developed at 10- to 20-km depth at greenschist to amphibolite facies
Lower crust—Inferred to be gneiss developed below 20-km depth at amphibolite to granulite facies
Mafic crust, undivided—Denser pattern; higher velocity
Mantle—Closely spaced ticks, top of lithospheric mantle; widely spaced ticks, top of asthenospheric mantle or lithospheric mantle with partial melt
OTHER SYMBOLS
∝ β γ Metamorphic facies—Amphibolite, blueschist (± eclogite), greenschist
Lithologic contact
Mesozoic thrust fault—Sawteeth on upper plate
Mesozoic or Cenozoic thrust fault—Teeth on upper plate; arrow indicates inferred motion of upper plate; dotted where inferred to have formerly existed
Mesozoic or Cenozoic detachment fault—Ticks on upper plate
Cenozoic detachment fault—Ticks on upper plate
Chocolate Mountains: Mesozoic or Cenozoic thrust or detachment fault ("Chocolate Mountains thrust fault")
NOTE: Lithologic contacts and faults—Long dashed where not constrained by figure 8A or by earthquakes; short dashed where gradational (lithologic contact only); queried where existence uncertain

lower crust consisting of metamorphic rocks, like gneiss, that may have been part of the 1.8 Ga western edge of Proterozoic Laurentia. The crust on the eastern side, east of the active strands of the San Andreas fault, is certainly part of the Proterozoic craton. But its structure is complicated by a thrusted slab or wedge of Cretaceous, Franciscan-like, Pacific seafloor that apparently has been subducted down under the Peninsular Ranges and inserted into the mid and upper crust just to the east of them. Called the Orocopia schist, it is the same unit as the Pelona schist, found in the San Gabriel Mountains, over 200 km (120 miles) to the northwest on the west side of the San Andreas fault. This schist is probably most famous for leading John Crowell, then at UCLA, to make the then very controversial proposal that the San Andreas had over 200 km of dextral slip—a hugh offset which we now take for granted as something that transform faults can do.

The crust in the middle of this cross section, and in the middle of the Salton Trough, is even more special. It's crust that is only a few million years old and is still forming today. The Salton Trough is the northernmost part of the transtensional system of alternating dextral slip faults and rhombochasms that have been forming the Gulf of California and rifting Baja California away from mainland Mexico for the last 6 million years. As the rhombochasms in the crust open they are continually filled by basaltic magma from the mantle, and the depressions at the surface, like the Salton Sea, are continually filled with sediments washed down from the mountains on both sides and periodically brought into this area by the Colorado River. The basaltic magma that slowly cools at depth forms gabbro, whereas the magma that intrudes the sediments forms diabase sills and dikes. The basaltic magma is so hot, probably about 1200°C, that it can melt the sediments and form rhyolite domes, like Obsidian Butte, at the southeast shore of the Salton Sea. Note in the cross section that the sediments in the center of the basin are about 12 km thick. The lower half, the orange colored units, are metasediments because of the high heat flow. A scientific well was drilled at the southeast corner of the Salton Sea in 1986 to a depth of 3200 m, or 10,500 feet (see well 2-14 in Figure 2). The bottom hole temperature was 355° C and the sediments had been metamorphosed to lower amphibolite facies. The fluids were highly saline (mostly alkali chlorides) with 25% total dissolved solids (Elders and Sass, 1988). The high salinity is the main reason why the geothermal resources here are not as highly developed as those south of the border at Cerro Prieto.

It may seem ironic that present-day southern California and Baja California are rifting away from North America apparently along the same 'line' along which a big piece of Rodinia rifted away from Laurentia about 600 Ma, but it may not be coincidental. The late Proterozoic rift may well have provided a zone of weakness in the crust which was then utilized by the San Andreas fault system when it jumped onshore, when the Rivera triple-junction reached the latitude of Cabo San Lucas and Puerto Vallarta. This is the concept that we tried to capture in our title to this article, i.e., zones of crustal weakness/intense faulting can be reoccupied, so that rifting a second time is not so hard to do! The Peninsular Ranges batholith, which extends from Riverside all the way to Cabo San Lucas, and probably continues at least petrologically to the south of Puerto Vallarta, was part of the North American craton for only about 200 Ma and is now rifting away. Maybe there is something strange or divisive about Yuma? Regarding this paper's title, the region involved in these tectonics is so much larger than the area of interest to this field trip, the last part of the title could read "the geologically brief attachment of southern and Baja California to southwestern Arizona and northwestern Mexico."

Before we leave this part on cross sections of the earth's crust, the reader may be interested in seeing what is probably the best cross section of a batholith anywhere in the world—a restored section of the crust in the Southern Sierra Nevada batholith (Figure 9). Jason Saleeby, of Cal Tech, started working in this area before 1990 and he and numerous colleagues have continued this study up to the present day. Miller and Snoke (2009) summarized the main results of this work in a GSA Special Paper and, given the similarities between the Sierra Nevada and Peninsular Ranges batholiths discussed above, this information may well apply to the mountains just to the west of the Coyote Mountains—if not to the Coyote Mountains themselves. Differential uplift, followed by erosion, has exposed about 35 km of this originally vertical crustal section in the southern Sierra Nevada (the upper part of Figure 9). This is an unusually complete section of the crust, since only about 7 km of the lowest crust down to the original Moho are estimated to be missing. The petrology of the lowermost 10 km is a reconstruction based on xenoliths brought up by Neogene volcanic

Figure 8. This is a diagrammatic cross-section of the crustal structure along a roughly east-west traverse across the Salton Trough from the Fish Creek Mountains, 20 km north of the Coyote Mountains, to the Chocolate Mountains and the Colorado River on the east (See Figure 2 for exact location of this transect.). **Figure 8A** lists the surface geology and seismic refraction data from Howell and others (1985). **Figure 8B** is a reinterpretation of this transect by Fuis and Mooney (1990). Starting at the west in the Fish Creek Mountains, the upper crust is late Cretaceous granitic rocks with septa of metamorphosed marine sedimentary rocks (See legend for Figure 8). The middle and lower crust is inferred to be schist and gneiss, ranging from greenschist to amphibolite and granulite facies, whose protolith was possibly precambrian igneous rocks of intermediate composition. The central part is the Salton Trough, composed of entirely new, late Cenozoic crust formed in a rift zone. The lower 13 km of the crust is gabbro. The upper 12 km are sediments and metasediments—the heat flow caused by the basaltic intrusions is so high that sediments buried only 3 km have already been metamorphosed to lower amphibolite facies. The eastern part of this transect is a structurally complex mixture of Proterozoic igneous and metamorphic rocks with a tectonic wedge/slab of Pelona-Orocopia schist—apparently the same Franciscan assemblage found offshore in the continental borderland. No vertical exaggeration.

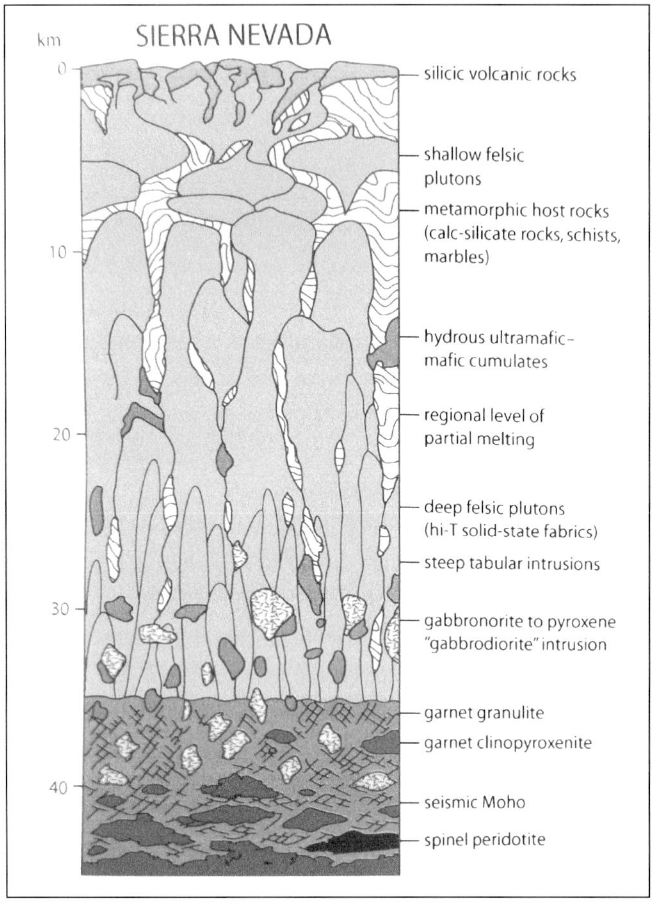

Figure 9. Restored crustal cross section of the southern Sierra Nevada batholith. From the work of Jason Saleeby, as modified by Miller and Snoke (2009).

rocks, and by geophysical data. The seismic Moho is defined as the depth where the p-wave velocity jumps from 7 km/sec to 8 km/sec. The velocity jump here is interpreted to be a transition boundary between plagioclase-bearing and plagioclase-absent rocks. As is the case here, this is not always at the same depth where the petrology changes from mafic to ultramafic.

The petrology changes with depth in Figure 9 are considerably more detailed that those seen in the western end of the transect by Fuis and Mooney (1990). In the upper 15 km narrow screens of metamorphosed volcanic, siliciclastic, and carbonate rocks separate the 100 Ma large felsic plutons. At 15-20 km depth, partial melting of pelitic host rocks occurs. In the shallow to medium depth part of the section, the foliation and lineation are mostly magmatic, but at paleodepths greater than 20 km, the rocks were in the ductile zone and the strong, vertical foliation seen in the plutonic rocks is solid-state deformation, i.e., the plutonic rocks are turned into orthogneiss (similar to Fuis and Mooney's estimate). The metamorphic host rocks at these depths underwent almost total melting, with the exception of the more refractory rocks like quartzite and marble. You can see that, in the lower half of the crust, the plutons become increasing tabular and form thin, intercalated vertical sheets.

In their paper, Miller and Snoke (2009) discussed six exposed Phanerozoic crustal cross sections around the world. Each one was different, and they concluded that "there is no 'typical' Phanerozoic continental crustal section, and the overall crustal composition varies markedly between sections." And they found that only a few sections are in contact (typically faulted) with mantle rocks. In one of the author's (Marshall) experience, the same caveats can be made about ophiolites, or surface exposures of the oceanic crust and upper mantle. Bringing an entire section of either type of crust and upper mantle up to the surface where we can examine it at our leisure takes very special geologic/tectonic conditions.

THE CONTRIBUTIONS OF PALEOMAGNETISM TO OUR UNDERSTANDING OF THE GEOLOGIC HISTORY OF THE COYOTE MOUNTAINS REGION

The science of paleomagnetism has helped us understand the evolution of the region around the Coyote Mountains in many ways and on many levels. On the largest scale, it is one of the most powerful tools in reconstructing former supercontinents. The linear marine magnetic anomalies are how we are able to exactly reconstruct Pangea. We would have a much poorer knowledge of the paleogeography of Rodinia without paleomagnetic data. They were essential in making the landscapes of Blakey and Ranney shown in Figures 4 and 5 not only beautiful but broadly accurate. Going a step further, we show in Figure 10A the paleolatitude and paleo-orientation of southwest Arizona (Yuma!) throughout the Phanerozoic. The details of the paleomagnetic data and kinds of calculations used to make this useful, and unusual, figure are given in the captions, both for this figure and Table 1. You can see that Blakey and Ranney were correct in locating southwestern Arizona at the equator and making the current west coast trend east-west during the Cambrian. And you can also see that North America rotated about 60° counterclockwise during the Paleozoic, while southwest Arizona remained in equatorial waters until at least mid-Triassic (225 Ma). Keeping almost the same orientation, North America then drifted northward almost 40 degrees by the early Cretaceous. Figure 10B shows the actual northward component of the drift rate in cm/y. You can see that the time of fastest northward drift, 5 cm/y, was between the mid-Triassic (225 Ma) and the late Jurassic (160 Ma). Interestingly, this was also the time of the beginning of the breakup of Pangea, whose first rifting was the formation of the north-Central Atlantic as the United States portion of North America moved away from northwest Africa. Although measuring absolute plate motion rates are now possible with very sensitive and expensive GPS instruments, estimates for the geological past have usually relied on the assumption of fixed hot spots in the mantle. Although a northward absolute motion rate of 5 cm/y may seem unusually fast, it is not that unusual. It is midway between the absolute

PERIOD	Age (Ma)	LAP (N)	LOP (E)	PALEO-LATITUDE OF YUMA	PALEOAZIMUTH (°)	RNDS (° /Ma)	NVY (cm/y)
Eocene	45	80	158	33	12	−0.12	−1.4
Early Cretaceous	115	73	191	42	18	−0.30	3.3
Late Jurassic	155	66	155	30	28	0.46	5.1
Late Triassic	215	56	94	2	16	−0.05	−0.6
Late Permian	250	49	117	4	31	−0.07	0.8
Late Carboniferous	315	33	129	−1	49	0.14	1.6
Late Devonian	370	29	121	−8	47	−0.01	−0.1
Late Ordovician	450	22	129	−8	57	−0.30	−3.4
Late Cambrian	500	-2	165	7	97	0.15	1.6
Early Cambrian	535	4	155	2	87		

Table 1. List of ages and the APWP (apparent polar wander path) poles for North America (McElhinny and McFadden, 2000) used in calculating the paleolatitude and paleoazimuth of Yuma, Arizona, in Figure 10. LAP, north latitude, and LOP, east longitude of the paleomagnetic poles; latitude and longitude used for Yuma in these calculations is 33° N and 245° E, respectively; (–) paleolatitude is south; paleoazimuth is degrees clockwise of paleonorth from current north; RNDS is the average rate of rotation for each interval between data points; (–) counter-clockwise rotation; NVY is the northward component of the average speed of the southwest Arizona (Yuma) for each interval; (–) component to the south.

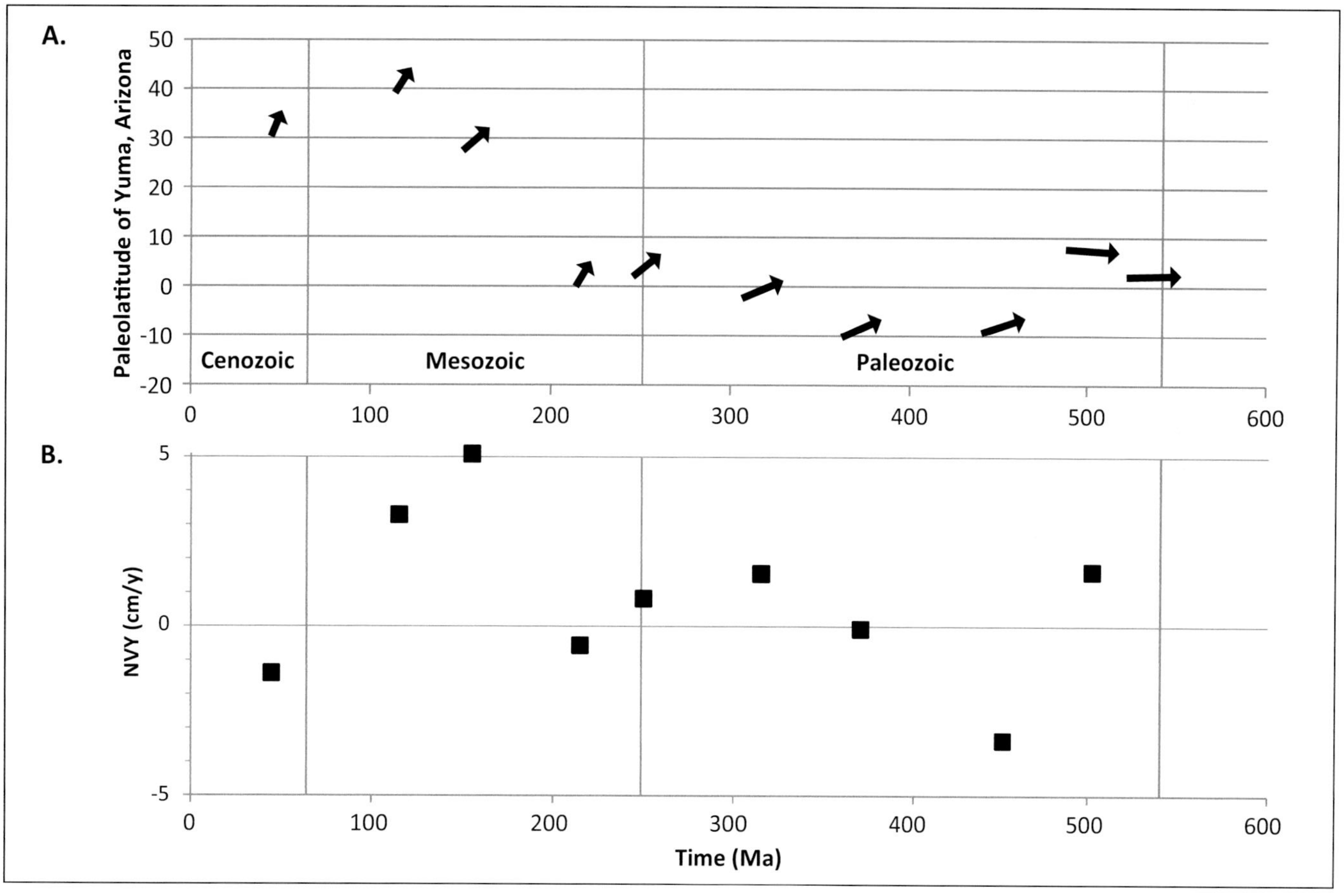

Figure 10A. Paleolatitude and paleoazimuth of this area during the past 600 million years. Calculations were made using the APWP (apparent polar wandering path) for North America (McElhinny and McFadden, 2000) and the standard spherical trigonometry equations used in paleomagnetic studies of this kind. Because it's the best known city of any size whose present location is close to the edge of the stable North American craton and this part of southern California, the current geographical coordinates of Yuma, Arizona, were chosen for these calculations. Note that, as seen in Figures 5 and 6, Yuma/southwest Arizona was close to the equator for all of the Paleozoic. Paleoazimuth is the direction to the North Pole at times in the past. Since north is up in this plot, the arrows show that all of the Phanerozoic compass directions in this part of the world had an easterly component. So, in the early Paleozoic, the coastline of the North American craton ran east-west, as has also already been seen. In fact, the few paleogeographic maps of early Paleozoic North America that show it in the correct orientation and paleolatitude do so because the authors used the same paleomagnetic data and equations used here. The rapid northward drift that occurred from the early Jurassic to the late Cretaceous are discussed in the text. Each data point plots at midpoint of each arrow. **Figure 10B** is a plot of the northward velocity component of the continental drift of the Yuma region (NVY), derived by dividing the latitudinal changes by the times involved. Note that the only significant poleward drift was movement toward the South Pole from late Cambrian to early Devonian of as much as 3 cm/year, and the even more rapid Jura-Cretaceous drift of up to 5 cm/year to the north (data in Table 1).

motion of North America of 2.5 cm/y to the southwest estimated from the Yellowstone hot spot and a rate of 10 cm/y to the northwest for the Pacific plate estimated from the Hawaiian hot spot (Grotzinger and Jordan, 2010).

Another relevant contribution of paleomagnetism was when measuring and identifying the linear magnetic anomalies at the mouth of the Gulf of California dated the opening of the Gulf at about 6 Ma. A still closer example is the results of a paleomagnetic study of basalts in the Coyote Mountains by Neil Mace (1981), who did his master's thesis under the direction of Monte Marshall. His results, along with those of other paleomagnetic studies in the region, are found in Marshall and others (2005, 2006). Mace found an area in north Fossil Canyon that was rotated clockwise by 110°, whereas an area of south Fossil Canyon showed no rotation. The Butaca Canyon area was rotated 60° clockwise, and Painted Gorge had areas that were rotated both clockwise and counterclockwise! As you can see in Figure 3, the Coyote Mountains are highly faulted and crisscrossed by both northwest and northeast striking faults, in addition to the Elsinore fault. Theoretically, the northwest-striking faults should have dextral slip and cause clockwise rotations, whereas the northeast-striking faults should have sinistral slip and cause counterclockwise rotations. But, the amount and complexity of faulting may well render this theory too simplistic, and, along with both senses of rotation, make the Coyote Mountains special.

SUMMARY

The geologic evolution of the Coyote Mountains and the desert to the east is very special and has a narrative that is both unusual and unusually interesting. The story starts over a billion years ago with the assembly of the penultimate supercontinent of Rodinia. The present continent of North America, called Laurentia in this context, occupied the center of this great continental landmass and would've looked strange to our eyes because it was oriented 90 degrees clockwise to its present orientation. Attached to its northern (now western) border was a large land mass that, although still debated, was probably composed of parts of Antarctica, Australia, China, and Siberia. Stranger still, the land/rocks that comprise most of what is now the western tier of states didn't even exist. Supercontinents don't last very long, one or two hundred million years, and between about 750 Ma in the east and 600 Ma in the west, i.e., Arizona, the northern land mass rifted away from Laurentia, forming the proto-Pacific Ocean. The little town of Yuma was close to beachfront property! And the water would've been warm because the whole east-west coastline at that time was near the equator. During the Paleozoic, Yuma kept its warm water beaches, but North America rotated more than 60 degrees counterclockwise and the coastline along current western Arizona and Nevada became a north-south continental margin. During the Mesozoic North America drifted north to almost its present latitudes. More importantly, this pre-Cambrian passive continental margin became an active, convergent margin and subduction of the old, early Pacific seafloor began in earnest. The old oceanic crust that had occupied the western part of the western tier of states began to be replaced by island arcs and, volumetrically more important, gabbroic and tonalitic plutons in the west, at about 120 Ma, to granodiorites in the east, at about 90 Ma. These eastern plutons were immense, both in the Sierra Nevada and our own Peninsular Ranges, but after this magmatic burst the locus of igneous activity migrated east as the subduction plane shallowed even more. By the end of the Mesozoic, the west coast of North America began to look much more like it does today.

Subduction of the Farallon plate continued until about 30 Ma, when the spreading center between it and the Pacific plate started being subducted in the vicinity of Los Angeles. Contrary to what was initially thought, the leading edge of the Pacific plate was also partially subducted and, as it was being pulled back up, caused the large extension and local rotations that formed the continental borderland. But this is a story for another field trip! The contact zone between the Pacific and North American plates lengthened as the northern triple junction migrated up the coast and a southern, the Rivera triple junction, migrated south along the coast of Baja California (Irwin, 1990). For many millions of years the transform motion between the two plates lay off the coast in the continental borderland. But at about 6 Ma, when the Rivera triple junction had reached the latitude of Cabo San Lucas/Puerto Vallarta, most of the transform motion migrated inland to the present location of the San Andreas fault and the oblique transform that is forming the Gulf of California. The localization of the main strand of the San Andreas in the southern Salton Trough and along the axis of the Gulf may be the result of the crustal weakness inherited from the rifting of Rodinia 600 million years ago.

So we come to the Coyote Mountains in the year 2015 and what do we see? The mountains are a block of granitic basement rocks, probably Jura-Cretaceous, with large roof pendants of Paleozoic marble and meta-siliciclastic rocks, partially capped by Neogene and Quaternary sedimentary and volcanic rocks. Most of these rocks were deeply buried until a transpressional bend in the Elsinore fault, which runs along the southwest side of the mountains, forced them up to the surface a million years ago, or so. We can ask ourselves a number of questions about these mountains: Are the granitic rocks just an outlier of the Peninsular Ranges batholith? Most of the young sedimentary rocks were deposited in what is called the proto-Gulf. But what about the protolith of the old metasedimentary rocks in the roof pendants? In what part of the continental margin/miogeocline off of Yuma were the calcareous and siliciclastic sediments deposited — the shallow shelf, the deeper slope, in abyssal depths out on the oceanic crust? What kind of rocks comprise the middle and lower crust here? Have the felsic plutons metamorphosed to orthogneisses, and are they intruded by thin, vertical magmatic sheets … as they are in the southern Sierra Nevada?

For those who take the high roads especially and can look out on the flat, tranquil desert to the east, can you imagine what is happening beneath those thousands of feet of Plio-Pleistocene sediments and metasediments that are being intruded by basaltic dikes and sills — by magma that is coming up from the mantle and forming gabbro in the lower-middle crust, quite possibly even while you're standing there!

We won't go that far east, but if the air is clear enough you will be able to see the Chocolate Mountains, just on the other side of the Salton Sea and the San Andreas fault. You know now that those mountains are made, in part, of 1.8 billion year old rocks that formed the coastline near Yuma when that part of Rodinia split up 600 million years ago and parts of exotic continents like Antarctica left North America and drifted out to sea! The other rocks in the Chocolate Mountains have a history that is almost as amazing. These slabs of Cretaceous, greenschist-facies seafloor sediments and pieces of oceanic crust used to lie offshore under the Pacific Ocean, but were then subducted under the Peninsular Ranges batholith and inserted into that old crust. Outcrops of this greenschist, now scattered along both sides of the San Andreas fault, are what gave geologists the first clues about what kind of fault the San Andreas really is!

So, on that note, the authors hope that you will have a very good time on this field trip, and that by asking questions—of yourselves and others, you will learn a lot! :>)

REFERENCES CITED

Blakey, R. C. and Ranney, W., 2008, Ancient Landscapes of the Colorado Plateau. Grand Canyon Association, 156 p.

Condie, K. C., 2003, Supercontinents, superplumes and continental growth: the Neoproterozoic record, in Yoshida, M., Windley, B. F. and Dasoupta, S., eds., Proterozoic East Gondwana Supercontinent Assembly and Breakup: Geological Society, London, Special Publications 206, p. 1–21.

Dickinson, W.R., 2009, Anatomy and global context of the North American Cordillera, in Kay, S.M., Ramos, V.A., and Dickinson, W.R., eds., Backbone of the Americas: Shallow Subduction, Plateau Uplift, and Ridge and Terrane Collision: Geological Society of America Memoir 204, p. 1–29. doi: 10.1130/2009.1204(01)

Elders, W. A. and Sass, J. H., 1988, The Salton Sea scientific drilling project: Journal of Geophysical Research, Solid Earth, v. 93, no. B11, p. 12,953–12,968.

Fuis, G.S., and Mooney, W.D., 1990, Lithospheric structure and tectonics from seismic-refraction and other data, in Wallace, R.E., ed., The San Andreas fault system, California: U.S.G.S. Professional Paper 1515, p. 207–236.

Grotzinger, J., and Jordan, T., 2010, Understanding Earth: 6th ed., W. H. Freeman, New York, 654 p.

Irwin, W.P., 1990, Geology and plate-tectonic development, in Wallace, R.E., ed., The San Andreas fault system, California: U.S.G.S. Professional Paper 1515, p. 61–80.

Kimbrough, D. L., Abbott, P. L., Balch, D. C., Bartling, S. H., Grove, M., Mahoney, J. B., and Donohue, R. F., 2014, Upper Jurassic Peñasquitos Formation—Forearc basin western wall rock of the Peninsular Ranges batholith, in Morton, D. M., and Miller, F. K., eds., Peninsular Ranges Batholith, Baja California and Southern California: Geological Society of America Memoir 211, p. 625–643.

Langenheim, V.E., Jachens, R.C., and Aiken, C., 2014, Geophysical framework of the Peninsular Ranges batholith—Implications for tectonic evolution and neotectonics, in Morton, D.M., and Miller, F.K., eds., Peninsular Ranges Batholith, Baja California and Southern California: Geological Society of America Memoir 211, p. 1–20

Mace, N.W., 1981, A paleomagnetic study of the Miocene Alverson Volcanics of the Coyote Mountains, western Salton Trough, California: a Master's Thesis at San Diego State University, 142 p.

Marshall, M., Mace, N., Erskine, B., and Jansen, E., 2005 and 2006, What regional and local paleomagnetic data tell us about the tectonics of the Coyote Mountains, southwest Imperial Valley, southern California, in Bloom, D.M. and P. Farquharson, eds., Geology and History of southeastern San Diego County, California: San Diego Association of Geologists field trip guidebook, p. 55–62.

Marshall, Monte, 2012, The Unappreciated Role that the Amalgamation and Break-up of Supercontinents has Played in the Geologic Evolution of the Western Margin of North America, in Snyder, S., Miller-Hicks, B., and Miller, J., eds., Palms to pines: geological and historical excursions through the Palm Springs Region, Riverside County, California: San Diego Association of Geologists and South Coast Geological Society field trip guidebook, p. 48–72.

McElhinny, M.W. and McFadden, P.L., 2000, Paleomagnetism—Continents and Oceans: International Geophysics Series, v. 73, Academic Press, New York, 386 p.

Miller, R.B., and Snoke, A.W., 2009, The utility of crustal cross sections in the analysis of orogenic processes in contrasting tectonic settings, in Miller, R.B., and Snoke, A.W., eds., Crustal cross sections from the western North American cordillera and elsewhere: Implications for tectonic and petrologic processes: GSA Special Paper 456, p. 1–38.

Moore, James G., 2000, Exploring the Highest Sierra. Stanford University Press. 427 p.

Walawender, Michael J., 2000, The peninsular ranges: a geological guide to San Diego's back country. Kendall/Hunt Publishing Company, 114 p.

Overview of the Geology of the southern Coyote Mountains, Salton Trough, Imperial County, Southern California

GEORGE MORGAN
georgemorgan@cox.net

J. R. MORGAN

ABSTRACT

THE OLDEST UNITS IN THE COYOTE MOUNTAINS are in the metamorphic section that is made of marbles, amphibolites, schists, cherts and quartzites. These units represent Paleozoic(?) age miogeoclinal to flysch sediments that were deposited in a passive continental margin in the state of Sonora, Mexico. Starting before or during the Jurassic, the passive margin in Sonora became an active continental margin and the sediments were intruded, deformed, and metamorphosed. Some of igneous units associated with the active margin are: diorite, gabbro, dikes of various compositions, foliated tonalite and the White Cross Gneiss. All of these units are Jurassic (~162 Ma). The deformation has completely destroyed most of the original sedimentary relationships between the marbles, schist, quartzites, cherts, and amphibolites. The metamorphics have been left with steep foliation and tight folds.

After an interval, when no Cretaceous to Oligocene rocks were preserved in the Coyote Mountains, Basin and Range extension took place in Sonora, Mexico. In the Coyote Mountains the sedimentary and volcanic rocks that record this activity were deposited in extensional basins. These rocks include: 1) White and Red Fanglomerates (Early Miocene), with the Red Fanglomerate interfingering with the Alverson Formation; 2) Alverson Formation(~17 Ma) (Early Miocene) consisting of basalt and andesite flows, dikes, plugs, tuffs, lahars, and pyroclastics; basalt flows in the upper Alverson section (younger than ~16.9 Ma) interfinger with Imperial Group sediments; 3) marine and non-marine rocks of the Viejo Formation, a proposed new formation of the Imperial Group (Early [?] Miocene to Pliocene).

During the deposition of the Imperial Group sediments the in Coyote Mountains, the tectonics changed from Basin and Range extension to right-lateral transtension associated with the San Andreas fault. During Imperial Group deposition, the Coyote Mountains started to move to the northwest and came under the depositional influence of the Colorado River. The widespread Deguyños Formation, the upper part of the Imperial Group, and the non-marine, widespread Palm Spring Group (Pliocene to Pleistocene) sediments in the Coyote Mountains record this change in tectonics and sedimentation.

The Coyote Mountains are now located in the western part of the Salton Trough, dominated by right-lateral faulting and are being differentially uplifted.

The types of faults found in the Coyote Mountains are: active right-lateral faults (Elsinore and related faults), active left-lateral faults (Painted Gorge Wash and related faults), detachment faults, high-angle normal faults, and thrust faults.

INTRODUCTION

The Coyote Mountains are situated on the western side of the Salton Trough in southern California (Fig. 1) west of the new plate boundary. The trough is bounded on the west by the Peninsular Ranges, on the north by the Transverse Ranges, on the east by the Basin and Range Province, and to the south by the Gulf of California. Prior to 5.5 Ma, the Coyote Mountains were located approximately 155 miles (250 km) to the southeast in the state of Sonora, Mexico (Winker and Kidwell, 1986).

We will make reference to the Coyote Mountains in many places in this paper. The Coyote Mountains we see today are the result of recent, less than 1 million years ago, uplift. There are many times in the past that the Coyote Mountains were a basin and completely covered by volcanics, sediments, or both. There are other times when they had enough relief that they could shed sediment. We refer to this area at these different times as the "Coyote Mountains."

Presented below is our work on an ongoing project to map the Coyote Mountains at 1:2,000 scale. As our work continues some of the ideas presented here will evolve.

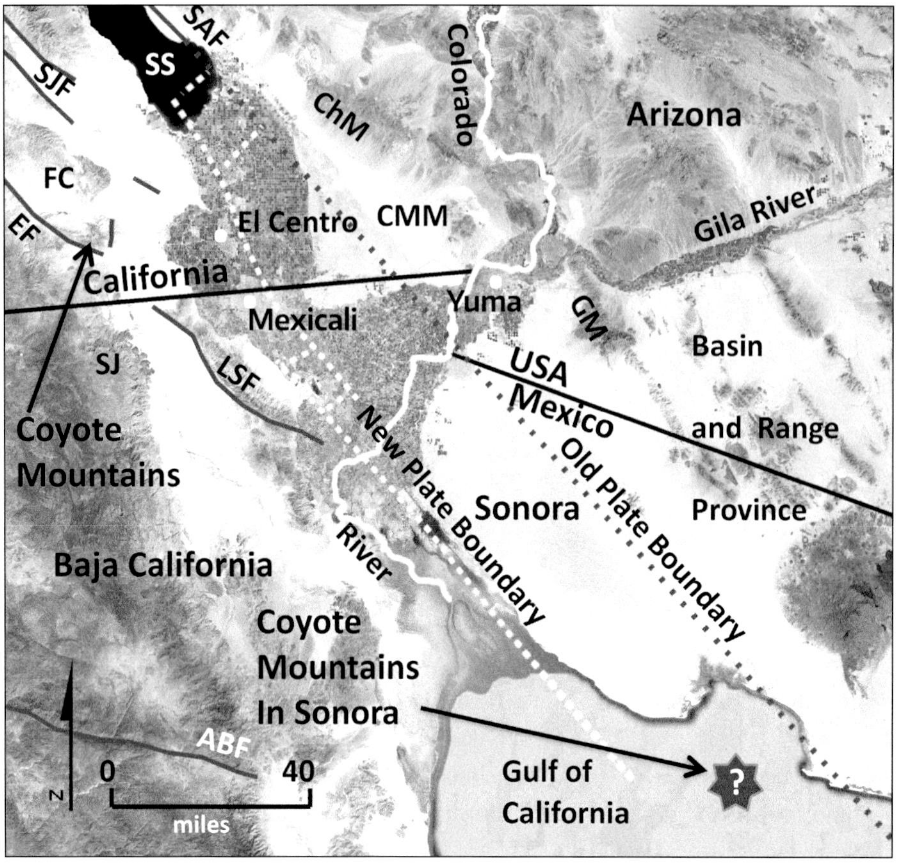

Figure 1. **ABF**-Agua Blanca Fault, **CMM**-Cargo Muchacho Mountains, **ChM**-Chocolate Mountains, **EF**-Elsinore Fault,**FC-** Fish Creek Mountains, **GM**-Gila Mountains, **LS**-Laguna Salada, **LSF**-Laguna Salada Fault, **SAF**-San Andreas Fault, **SJF**- San Jacinto Fault, **SJ**- Sierra Juarez, **SS**-Salton Sea. Black star is the paleogeographic loction of the Coyote Mountains 5.5 Ma (Winker and Kidwell, 1996).

PREVIOUS WORK

Many workers have studied the Coyote Mountains, but only Dibblee (1954 and 2003), Christensen (1956), and Winker and Kidwell (1996) have mapped large areas of the Mountains (Figure 3). Morton (1977) used both Dibblee's (1954) and Christensen's work to compile a part of his report on the geology and mineral resources of Imperial County.

METAMORPHIC AND IGNEOUS ROCKS

Metamorphic rocks in the Coyote Mountains were assigned a Paleozoic age by Dibblee (1954) and are composed of marbles, schists, quartzites, cherts, and amphibolites. These rocks are thought to represent a passive continental margin with miogeoclinal to flysch deposits from the North American craton. Ordovician conodonts are reported from some of the marbles (Miller and Dockum, 1983), but our sampling of the marbles has not yielded any conodonts. In our search for conodonts, we have found scolecodonts, which are jaw parts of annelids, a common fossil-producing segmented worm. Figure 7 in the Field Guide, shows the conodont location and our scolecodont location. Blaettler (1991), using thin-section petrography, found that the amphibolites in the southern Coyote Mountains were at one time basalts. One of these amphibolite outcrops Blaettler sampled contains relic pillow structures that we believe represent sea-floor basalts.

Mapping indicates that at some time before and during the Jurassic, the Coyote Mountains were involved in an active continental margin, which deformed and metamorphosed the pre-Jurassic sediments. We sampled nine outcrops of gneisses and plutonic rocks in the Coyote Mountains for zircons. The dates produced from these samples range from 160.2 to 164.1 Ma. These Jurassic ages date the magmatic activity of the active margin. Schmidt and Paterson (2002) have dated a Jurassic magmatic event on the eastern side of Baja California, south of the Agua Blanca fault in the Sierra San Pedro Martir with a 164 Ma U-Pb orthogneiss. We have not found any Cretaceous igneous rocks in the Coyote Mountains.

Sometime before the Eocene, deformation and metamorphism ended and the Coyote Mountains were part of an uplifted block with steep foliation, tight folds and a metamorphic grade of lower amphibolite with pervasive andalusite. The deformation has completely destroyed most of the original sedimentary relationships among the marbles, schist, quartzites, cherts, and amphibolites. We have been unable to establish which parts of the metamorphics are stratigraphically up. No Cretaceous to Oligocene rocks are found in the Coyote Mountains.

During the Eocene several river systems flowed across the region from the Sonora, Mexico, to the Pacific Ocean (Link and Abbott, 1991). The only evidence in the Salton Trough for these rivers are the exotic silicified meta-volcanics clasts that are found reworked in sediments under the Alverson Formation in the Coyote Mountains and in similar sediments throughout the Salton Trough. During this interval, erosion exposed the Coyote Mountains.

After the Eocene, extensional tectonics produced the Basin and Range Province in Sonora (Nourse et al., 1994).

SEDIMENTARY AND VOLCANIC ROCKS

Metamorphic and igneous rocks in the Coyote Mountains are unconformably overlain by sedimentary and volcanic rocks that were deposited in extensional basins when the Coyote Mountains were in Sonora, Mexico (Fig. 1). The sedimentary and volcanic rocks are divided into: 1) Early Miocene, the lower-White and the upper-Red Fanglomerates with the Red Fanglomerate inter-

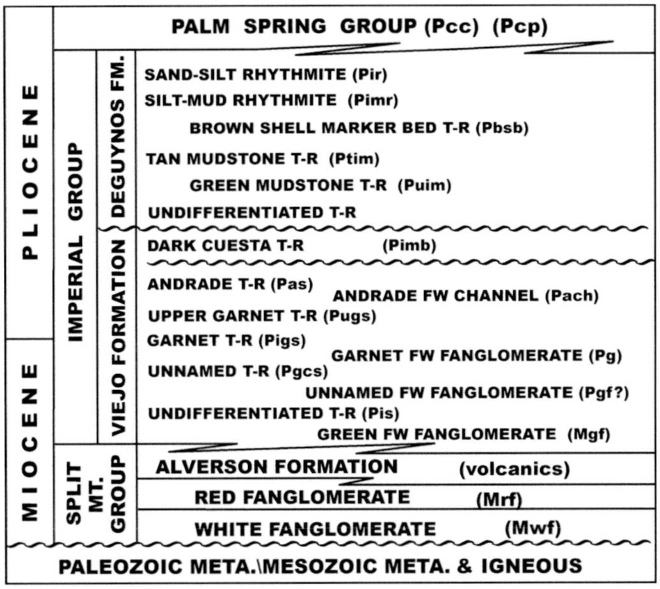

Figure 2. This simplified working columnar section for only the Coyote Mountains reflects the work we have done. We have done little work on the Palm Spring Group. **FW**- fresh water, **Meta.**-metamorphic. All Transgression-Regression (**T-R**) sequences are marine. Letters in parentheses are map designations.

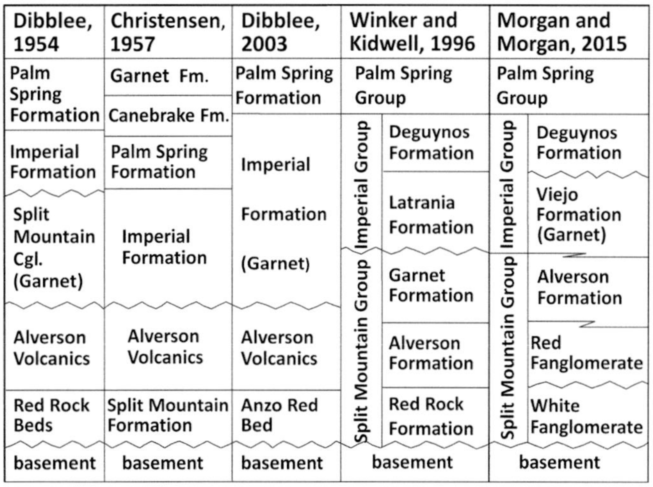

Figure 3. Edited columnar sections from Dibblee (1954, 2003), Christensen (1957), and Winker and Kidwell (1996) for units they used in the Coyote Mountains. Our working columnar section only for the Coyote Mountains is indicated by "Morgan and Morgan." Only relative age of units shown. "(Garnet)" designates units that contain Garnet Formation rocks.

fingering with Alverson Formation; 2) Early Miocene Alverson Formation (17 Ma) consisting of basalt and andesite flows, dikes, plugs and domes; with tuffs, lahars, and pyroclastics; basalt flows in the upper Alverson interfinger with lower Viejo Formation marine sediments (Figures 2 and 4); 3) Early Miocene to Pliocene non-marine and marine sediments of the Viejo Formation, a lower part of the Imperial Group (Figure 2). The Viejo Formation is a proposed new formation of the lower Imperial Group, which is discussed in detail later in this paper.

The Pliocene Deguyños Formation (Winker and Kidwell, 1996), the upper part of the Imperial Group, and the Pliocene to Pleistocene non-marine Palm Spring Group (Figure 2) were not deposited in extensional basins in Sonora, Mexico. Both units are widely deposited in a transtensional tectonic environment related to the San Andreas fault zone. Both of these units have been influenced by the Colorado River (Winker and Kidwell, 1996). At one time the Deguyños and Palm Spring Group probably completely covered the Coyote Mountains.

Our mapping shows that subsequent faulting, differential uplift, and erosion have left only remnants of the sedimentary and volcanic units that at one time covered the Coyote Mountains.

White and Red Fanglomerates of the Coyote Mountains (Early Miocene)

Dibblee (1954, 2003), Christensen (1957), and Winker and Kidwell (1996) named various fanglomerates below the Alverson Formation in the Coyote Mountains by unit names: Red Rock, Anza Beds, Garnet Formation, Split Mountain Conglomerate, and Split Mountain Group (Fig. 3). These units have not been defined using type sections, which has led to correlation problems. These unit names were also used for similar sediments found elsewhere in the Salton Trough. We avoid using some of these names in the Coyote Mountains.

We use "White Fanglomerate" and "Red Fanglomerate" instead of "Red Rock Formation" (Figure 2 and 3). The clasts in the White and Red Fanglomerates are, for the most part, derived from local metamorphic and plutonic rocks of the Coyote Mountains. The lower White Fanglomerate does contain a few exotic clasts of silicified metavolcanics. Volcanic clasts of the overlying Alverson Formation are found in an upper Red Fanglomerate that interfingers with the Alverson. The Red Fanglomerate in the Ship Rock area contains sand dune deposits. These fanglomerates were likely deposited in basins associated with Basin and Range extension when the Coyote Mountains were in Sonora. In the Coyote Mountains the White and Red fanglomerates are found in scattered, isolated and to some extent, fault-bounded outcrops of limited extent.

There are similar fanglomerate units under Miocene basalt flows in the Fish Creek Mountains (Dibblee, 1954), Davies Valley area, Superstition Mountains, Volcanic Hills (Fourt, 1979), and the in the Sierra Cucapah, Mexico. The fanglomerates of these areas are locally derived and some do contain exotic clasts that came from the area of Sonora (Link and Abbott, 1991). These fanglomerate units also appear to be isolated, somewhat fault bounded, and of limited extent, but are widely dispersed in the western Salton Trough. The similar fanglomerates and the Coyote Mountains fanglomerates represent small parts of individual basins in the Basin and Range province when the Coyote Mountains were in the Sonora, Mexico. Each of these similar fanglomerates were influenced not only by extension, but by right-lateral faulting and rotation (Marshall et al., 2006; and Mace, 1981). We believe that correlating among these similar fanglomerates would be very difficult.

Alverson Formation (Early Miocene)

Dibblee (1954) named the Alverson Volcanics for the volcanic rocks he observed in Alverson Canyon. (Ruisaard, 1979) renamed the unit the Alverson Formation. The Alverson Formation volcanic rocks erupted through plugs and were deposited on the earlier rocks as cinder cones, flows, lahars (?), tuff beds, and pyroclastics (Morgan and Morgan, 2014). The volcanism is dated at 17 Ma (zircon U-Pb), (Morgan et al., 2012) from lower tuffs in Alverson Canyon and Painted Gorge. In Alverson Canyon, a basalt below the dated tuff was dated at 20.1 ± 1.6 Ma using K-Ar by Ruisaard (1979) (Figure 4). We have concerns about the accuracy of K-Ar dates that Ruisaard published. The compositions of the Alverson flows and plugs vary from basalt (Ruisaard, 1979) to andesitic basalt (Dibblee, 1954), to hornblende andesite and rhyolites.

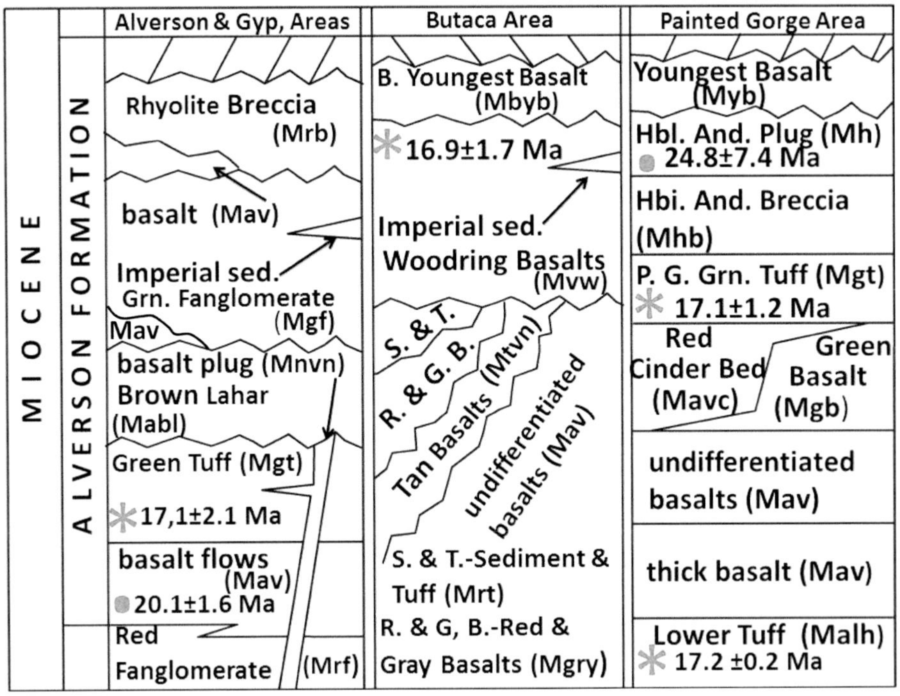

Figure 4. Simplified working columnar sections for the Alverson Formation in the Coyote Mountain areas of Alverson-gypsum, Butaca Canyon, and Painted Gorge. No units have been correlated among the areas. The grey asterisks are zircon age dates. The zircon date in Woodring's Basalts is detrital. The two small grey squares are K-Ar dates from Ruisaard (1979). We consider Ruisaard's K-Ar dates suspect. Letters in parentheses are map designations.

There are several volcanic plugs and cinder cones in the Coyote Mountains. Most of the plugs and cinder cones are in the Painted Gorge area. Basalts are found on a landslide under Carrizo Mountain, highest part of the Coyote Mountains, and indicate that the volcanics may have completely covered the mountains. The Alverson Formation is likely part of the volcanism associated with Basin and Range extension, when the Coyote Mountains were in Sonora. The upper(?) Alverson Formation interfingers with Viejo marine sediments in the areas of Gypsum Canyon and Butaca Canyon, southern Coyote Mountains (Figures 4 and 9).

In Figure 4, three columnar sections are presented for the Alverson Formation: Alverson and Gypsum canyons, Butaca Canyon, and Painted Gorge areas. We have not found a volcanic unit that correlates among these areas, but mapping is not complete. The Alverson and Gypsum area has the most compete stratigraphic section, but we have not finished mapping volcanics in upper Alverson Canyon. Butaca area appears to have the thickest section, but only the upper part of the section is mapped. We have a good grasp of the upper part of the section in Painted Gorge, but the number of plugs and flows are complicating the middle of this section.

The plugs and flows in the Painted Gorge area likely represent a single magma chamber at depth. Some of the plugs are close to each other and may represent a single volcanic event. Most of the plugs and flows are basaltic in composition. There are two (?) hornblende-andesitic plugs (Mh) that are stratigraphically above most of the basalt flows in the Painted Gorge area (Figure 4). These andesite plugs likely represent a change in the chemistry of the magma chamber from mafic rich to silica rich. Stratigraphically above the andesite plugs are several basalt flows of limited extent. These are the youngest flows (Myb) in the Painted Gorge area. These younger basalt flows likely represent recharge of the magma chamber with mafic material.

The Imperial Group (Early (?) Miocene to Pliocene)

In the Coyote Mountains, the use of stratigraphy and formation names from the Split Mountain–Vallecito–Fish Creek Mountains area have caused correlation problems. The Split Mountain–Vallecito–Fish Creek stratigraphy (Winker and Kidwell, 1996) describes a shallow subsiding basin north of the Coyote Mountains. Our stratigraphy describes a subsiding Coyote Mountains near sea level. The different depositional environments produce somewhat dissimilar sedimentary units and sequences, i.e., non-marine sediments and transgressional-regressional sequences (T-R) of the Coyote Mountains are not recognized in the Split Mountain-Vallecito-Fish Creek Mountains area.

Hanna (1926) first used the term "Imperial"; Woodring (1931) defined the Imperial as including all sedimentary units in the Salton Trough that contain marine fossils. Winker and Kidwell (1996) promoted the Imperial Formation to Group (Figures 2 and 3). The Imperial Group in the Salton Trough is composed of marine sediments, but the Imperial Group (Viejo Formation) found in the Coyote Mountains consists of marine and freshwater sediments (Figures 2 and 8). Most of the Viejo marine sediments were deposited near sea level as transgressional and regressional (T-R) sequences (at least eight T-R sequences are recognized) and represent the arrival of the Gulf of California when the Coyote Mountains were in Sonora, Mexico.

Viejo Marine Transgressional and Regressional Sequences in the Imperial Group

A T-R sequence is the result of rising sea level with deposition followed by a falling sea level, erosion, and an unconformity. A typical T-R sequence produces sediments with a recognizable, predictable order. The first and lowest is a conglomerate (high energy) formed at sea-level by waves cutting into the underlying rocks. As sea-level rises the second sediment is a sandstone (lower energy) deposited below wave base. With a continuing rise in sea-level the third sediment deposited is a mudstone representing a low energy environment. At times sea level continues to rise and limestone is deposited. This succession of sediments represents the transgression of a T-R sequence. If sea-level starts to fall a reverse order may be deposited starting with limestone, then mudstone followed by sandstone and finally a conglomerate capped by an unconformity. This later succession of sediments represents the regression part of a T-R sequence. There are instances when the erosional part of the regression is great enough to erode the units that were deposited leaving only a part of the transgression.

The large number of T-R sequences (8+) in the Coyote Mountains likely resulted from sea level changes. The T-R sequences have been influenced by tectonic activity. The T-R sequences in the Viejo do show the recognizable predictable sediments of a typical T-R sequence (Figures 2, 5, and 8).

Figure 5. A typical Viejo T-R sequence on the eastern side of upper Alverson Canyon. Day pack and hat in shade for scale. The basal conglomerate is composed almost entirely of rounded volcanic clasts because the buttress is Alverson basalt. The sandstone is grayer than usual because of the basalt lithic fragments. Fossils are found in both the conglomerate and sandstone. Mudstone has been lost to erosion.

First and Lowest Sedimentary part of a T-R Sequence

The first and lowest sediments are high-energy sediments that are the results of vigorous wave action at sea level. The Viejo basal conglomerates and coarse sandstones are the result of this high-energy environment. The high-energy conglomerates and sandstones are cemented with calcium carbonate ($CaCO_3$) and form resistant outcrops. Conglomerates of the oldest T-R sequences in the Coyote Mountains only contain breccia clasts of the Alverson. At this time the Coyote Mountains basement rocks were completely covered by volcanic rocks. This is another indication that the Alverson completely covered the Coyote Mountains.

Conglomerates of younger T-R sequences contain rounded clasts of Alverson and basement rocks. The youngest Viejo T-R sequence is the Dark Cuesta member (Pimb), one of the few marker beds identified. The lowest part of the marker bed is a coarse, quartz-rich arkosic sandstone with abundant coarse granitic fragments; well-rounded, resistant, small quartz-rich cobbles; and the occasional fossil. We believe that the small cobbles have been reworked from earlier Viejo conglomerates. A dark desert-varnish commonly forms on the surface of the marker bed because of the abundant granitic fragments. The marker bed ranges in thickness from 20 cm to 4 meters. No schist, marble, and only the rare volcanic clast are found in the marker bed. This clast assemblage indicates that the Coyote Mountains were not shedding any basement rocks. We believe that this bed is the first Imperial unit to completely cover the Coyote Mountains. The clast composition of the Viejo conglomerates is controlled in a few places by the underlying rocks. The conglomerates of the T-R sequences record the erosion of the Coyote Mountains to low relief. Most of the basal conglomerates are clast supported but some are matrix supported. The basal sandstones are made of coarse calcite grains, fossil fragments, quartz grains, and lithic fragments. The thickness of the conglomerates can be dependent upon the underlying topography. Some of the sandstones could be considered calcarenites. Small pebbles and cobbles are common in the basal sandstones. Fossils are common in both the conglomerates and sandstones. The basal conglomerates are commonly the most distinctive part of the T-R sequence and in some places the only part preserved.

Commonly associated with these high-energy conglomerates and sandstones are pholad holes (Figure 6 and 7). Pholad are small bivalves that make holes for themselves in marbles and sometimes even basalts. The holes are about 2.5 cm wide and do not last long without a sedimentary protective cover. If you find pholad holes start looking for Viejo Formation marine sediments. Pholad holes by themselves are an indication that the marine Viejo was present. Pholads are filter feeders and thrive in high energy environments. The buttress unconformities where marine Viejo conglomerates and sandstones form against the marbles and volcanics are good places to find pholad holes or as we sometimes call them, "bullet holes."

Figure 6. JR on a marble surface covered with Pholads (bullet) holes on the east side of the Coyote Mountains.

Figure 7. Close up of the marble surface.

Second Sedimentary part of T-R Sequences

The next sediments in the T-R sequence are deposited in a less-energetic environment that would be below the rising sea-level and just below wave base. Sandstones and siltstones are the result of this less-energetic environment. The calcium carbonate-cemented sandstones commonly form cliffs and buttress unconformities with the metamorphics and volcanics. The siltstones are also cemented but are not cliff forming. The sandstones and siltstones of different T-R sequences are surprisingly similar: a result of similar environments working on the same rocks to produce similar looking sedimentary rocks. In some places the basal conglomerate below the sandstone is less than 30 cm thick. The sandstones are light yellow brown (tan), cemented, coarse to medium fragments of calcite, shells, and quartz, with lithic fragments and the occasional pebble or cobble made of basement rock, volcanic rock, or both. Some of the sandstones could be considered calcarenite. The sandstones commonly contain abundant fossils and worm burrows

Third and Uppermost Sedimentary Part of a T-R Sequences

The last and uppermost sediments in the sequence result from a low-energy environment. The olive-green massive mudstones and light yellow brown (tan) siltstones are the result of this low-energy environment. The mudstones and siltstone are cemented with calcium carbonate, but are commonly missing because they are easily eroded by the regressions and later erosion. The mudstones are also cemented with gypsum that results in the weathered light yellow brown (tan) "popcorn" surface of the mudstones. The mudstones are more numerous than the siltstones in the Viejo Formation. Few macrofossils are seen in the mudstone.

Fresh-Water Sedimentary part of the Viejo Formation within the Imperial Group

The freshwater sediments of the Viejo Formation consist of fanglomerates, sandstones, and freshwater limestones that are outside the T-R sequences. Most of the non-marine sediments are fanglomerates that are, for the most part, matrix supported, poorly sorted, light green to light brown to light yellow brown (tan), and are associated with Basin and Range extension in Sonora. Outside of the T-R sequences, marine sediments do interfinger with one fanglomerate, the Green Fanglomerate (Mgf). Both the marine T-R sequences and the fresh water sediments were deposited in extensional basins near sea level.

Fresh-Water Garnet Member of the Viejo Formation

The Garnet was named a formation by Christensen (1957) for the fanglomerates in the upper central part of the southern Coyote Mountains and believed the unit to be Quaternary in age and above the Canebrake Formation (Figure 3). The placement of the Garnet Formation in a columnar section depends strongly on who was mapping and when they mapped. Christensen never designated a type section for the Garnet. The clast count Christensen took for the Garnet was from a "prospect trench" that was on a marble lag gravel that we believe may not even be Garnet.

Dibblee (1954) mapped the Garnet as Split Mountain Conglomerate. In later mapping, Dibblee (2003) combined the Garnet and the other fanglomerates above the Alverson with the Imperial (Figure 3). Winker and Kidwell (1996) expanded the Garnet to include all fanglomerates in the Coyote Mountains stratigraphically above the Alverson Formation and below the Imperial Group. They then put the Garnet in the Split Mountain Group, but indicated it had marine toes (?). Our mapping indicates that the Garnet as Christensen described and the other similar fanglomerates above the Alverson in the Coyote Mountains interfinger with the marine T-R sequences of the Viejo Formation (Figures 2 and 8). Because Christensen's Garnet Formation interfingers with marine T-R sequences of the Viejo Formation, the Garnet will have to demoted to a member of the Viejo Formation.

The Garnet member of the Viejo is composed of light red-brown to light brown, weathering to light brown-gray, matrix supported fanglomerates and light brown, arkosic, coarse grained sandstones. The Garnet is marble poor, but the surface is marble rich. The marble cobbles are able to heal themselves, while the clasts of the other rock types cannot heal themselves after they have been weathered. Water chemically attacks the cobbles of the Garnet but when the water evaporates because of the desert environment the calcium carbonate heals the marble. The marble lag gravels (mc) form in the same manner. Christensen believed that the Garnet did not contain any volcanic clasts, but we have found some facies of the Garnet do contain volcanic clasts.

Fresh-water Green Fanglomerate of the Viejo Formation

The fresh-water Green Fanglomerate is the prominent cliff forming fanglomerate at the lower end of Alverson (Fossil) Canyon (Figure 2). Dibblee (1954) and Bell (1980) considered the Green Fanglomerate to be a part of the Imperial. Winker and Kidwell (1996) combined the Green Fanglomerate with the Garnet and the other unnamed fanglomerates under the Imperial Group and above the Alverson Formation. They then put this group of fanglomerates with the Alverson and the Red and White Fanglomerates in the Split Mountain Group. We have designated the Green Fanglomerate as a fresh-water member of the Viejo Formation.

A 300m wide belt of outcrops of the Green Fanglomerate on the southern side of the Coyote Mountains can be followed from the eastern side of Ocotillo Canyon to under the cuestas in the Gypsum Canyon area. The color ranges from light yellow brown (tan), to light green-brown, to light green. The Green Fanglomerate is rich in volcanic clasts but does contain clasts of schist, quartzites, and granitic rock. The cemented cobble beds are matrix supported with arkosic, coarse sandstone, commonly graded.

The clasts found in the marine conglomerates and freshwater fanglomerates are locally derived.

The presence of T-R sequences and the trace fossil *Gastrochaenolites* (made by pholads) on wave-cut benches (Watkins, 1990) indicate that the Coyote Mountains were near sea level during deposition of the Viejo Formation. Bell (1980) studied the fossils of Imperial sediments (our Viejo) in the southern Coyote Mountains and put the depth of deposition from 0 to 30 meters.

The non-marine Viejo represents deposition in an extensional basin near sea level. A rise in sea level could result in seawater invading this extensional-basin (transgression) and depositing marine sediments. A fall in sea level would result in the partial erosion of these marine sediments, i.e., a regression forming a complete T-R sequence. After a fall of sea level, a local base-level within the extensional basin may have allowed for the deposition of non-marine sediments. A rise in sea level would start a new T-R sequence. A slowly subsiding Coyote Mountains would ensure that the T-R sequences would be preserved. In this way, it is possible to produce marine transgressions and regressions that interfinger with non-marine sediments.

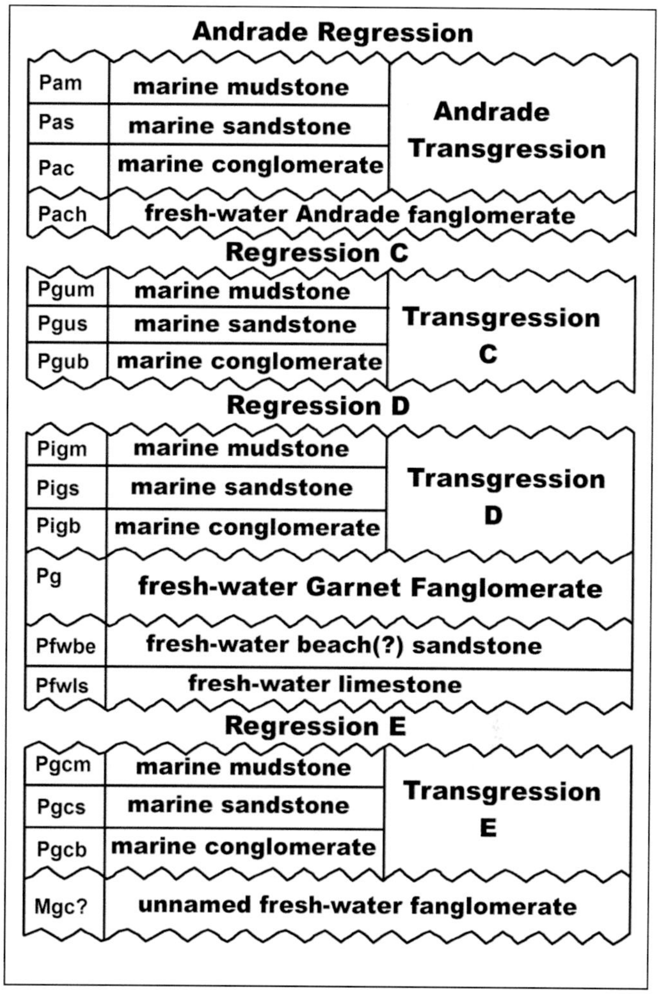

Figure 8. Working stratigraphic section of the Viejo Formation in the central area of the southern Coyote Mountains. This section is part of the Viejo Formation in which correlation of units is well constrained. Younger and older Viejo sediments recognized in the Coyote Mountains have not been correlated with this section. Wavy lines are unconformities.

During the deposition of the Viejo Formation, the fluctuations in sea levels have, in part, controlled the number of recorded T-R sequences and on a smaller scale the interfingering of marine and non-marine sediments. The angular unconformities between the T-R sequences suggest tectonic activity during the deposition of the Viejo Formation. There is lithological evidence that the slowly subsiding Coyote Mountains were completely buried by sediments of the upper Viejo Dark Cuesta T-R sequence (Figure 2).

Woodring's Outcrops

Because of the relationship between the Imperial and the Alverson has been mapped as an angular unconformity, the Imperial has been considered younger than the Alverson. Woodring (1931) observed that "sandstones carrying marine fossils lie between and interfinger with basalt" in the southeastern

Coyote Mountains. The rediscovery of Woodring's outcrops near Butaca Canyon (Morgan and Morgan, 2014) calls into question part of the lower Imperial being younger than the Alverson. In these outcrops Viejo marine sediments appear to interfinger with Alverson basalt flows (Figure 9). This is the only place in the Coyote Mountains where this relationship is seen. We have a detrital zircon date of 16.9±1.7 Ma from a tuffaceous sedimentary rock in the interfingering outcrop. This date indicates that the interfingering Viejo is younger than 16.9±1.7 Ma.

Woodring's Outcrops may be the only place where the interfingering of Imperial sediments with the Alverson flows are observed, but there are other lines of evidence in other areas that support this relationship. In several places in Alverson Canyon, Viejo marine sediments interfinger with the Viejo fresh water Green Fanglomerate (Mgf). Bell (1980) reported finding sand dollar fossils in "the middle of" a conglomerate in Chuckawalla Canyon. We believe Bell's conglomerate to be our Green Fanglomerate. The Green Fanglomerate can be mapped eastward to the Gypsum Canyon area. In the Gypsum Canyon area the Green Fanglomerate is unconformably capped by basalt and rhyolite breccia (Mrb) of the Alverson Formation. These relationships support Woodring's observations (Figure 4).

Deguyños Formation of the Imperial Group

Winker and Kidwell (1996) named the Deguyños Formation for the upper Imperial Group in the Split Mountain-Vallecito-Fish Creek Mountains basin north of the Coyote Mountains. They show the Deguyños Formation in the flats surrounding the Coyote Mountains. The Deguyños Formation is characterized by olive-green mudstone [weathering to light-yellow brown (tan)], light yellow brown (tan) siltstones, and brown sandstones with wave-base shell beds. We have seen no conglomerates or fanglomerates in the formation in the Coyote Mountains. Within the siltstones are rhythmites of sand-silt and silt-mud. Winker and Kidwell (1996) concluded that the Deguyños represents a Salton Trough-wide marine basin into which the Colorado River flowed. The rhythmites represent a part of the stratigraphic section that show Colorado River influence (Winker and Kidwell, 1996). There are angular unconformities in the Deguyños, which we believe represent T-R sequences without the high energy sediments (conglomerates and coarse sandstones). There is one marker bed (Pbsb) that we have mapped in the Deguyños that is found on the eastern side of the Coyote Mountains and on the eastern side of Dolomite Ridge southeast of Wind Cave area. The upper part of the Viejo and all of the Deguyños probably covered the Coyote Mountains, but we have not found any Deguyños in the Coyote Mountains.

Figure 9. Woodring's southern outcrop. The fossils in the Imperial sediments are broken and not in place. We have identified fossil turritellas, sand dollars, sea urchin spines, oysters, barnacles, and clams. There are no baked zones associated with the volcanic flow.

Palm Spring Group (Pliocene to Pleistocene)

The Palm Spring Formation (Woodring, 1931) was promoted to Group by Winker and Kidwell (1996). The Palm Spring Group interfingers with the youngest beds of the Imperial Group (Winker and Kidwell, 1996). At one time the Palm Spring Group likely covered the Coyote Mountains completely. We have not found any Palm Spring units in the Coyote Mountains. Dibblee (1954) mapped most of the Palm Spring sediments in the flats around and away from the Coyote Mountains. In the southern Coyote Mountains there are two outcrops of the Palm Spring Group that have been faulted against the basement rocks. In the northwestern Coyote Mountains (Dolomite Ridge) the Palm Spring Group sediments are close to basement rock but appear to be separated from basement rocks by faults. The Group's lithologies include friable tan sands and silts, and pink clays derived in part from the Colorado River; and locally derived conglomerates (Winker and Kidwell, 1996). The Palm Spring Group is a widespread unit in the Salton Trough. We have done little mapping in the Palm Spring Group.

Viejo Formation, a new formation for the Imperial Group

Winker and Kidwell (1996) use the Andrade Member of the Latrania Formation to describe the lower part of Imperial Group sediments in the Coyote Mountains. They recognize only a single transgression in the Coyote Mountains and the Andrade represents that transgression. The Latrania is well defined as the lower part of the Vallecito Basin without any T-R sequences or high-energy freshwater sediments. The paleo-environment of the Latrania Formation found in the Vallecitos-Split Mountain-Fish Creek Mountains basin is different than the paleo-environment for the lower Imperial in the Coyote Mountains. Stratigraphic relationships suggest that the Viejo may be older than Latrania. Therefore, we decided to avoid using the term "Latrania" in the Coyote Mountains, despite its first use for lower coarser sedimentary beds in the Coyote Mountains by Hanna (1926).

Winker and Kidwell's (1996) Andrade Member limestone (calcarenite) is one of the most recognized Imperial Group units in the Coyote Mountains. It is a part of a marine T-R sequence in the Viejo (Figure 8) and we use the high energy Andrade to define this T-R sequence.

There are differences between the Viejo Formation sediments in the Coyote Mountains and the Deguyños Formation sediments found on the flats around the Coyote Mountains. The Viejo sediments are T-R sequences mixed with non-marine sediments. Except for the mudstones these marine and non-marine sediments represent high-energy environments near sea level. The Viejo appears to have no Colorado River influences and Viejo marine sediments interfinger with the Alverson Formation. Previous workers mapping the Imperial Group in areas outside the Coyote Mountains do not report T-R sequences, except for Johnson et al. (2003) who reported a T-R sequence near El Mangle, Baja California Sur, Mexico. We have not found any unit that we can correlate between Viejo and Deguyños. There are T-R sequences in the Deguyños Formation, but there are no non-marine or marine high-energy sedimentary units in these T-R sequences. The sediments of the Deguyños represent a Colorado River delta that was deposited in a shallow-marine basin (Winker and Kidwell, 1996). Rhythmic units are found in the Deguyños. The shell beds in the mudstones may represent high-energy wave-base sediments or may represent the high-energy part of a T-R sequence. There is one unit (Pbsb) of the Deguyños that laps onto the Coyote Mountains and this unit unconformably overlies a T-R unit of Viejo Formation. The Deguyños is widespread in the Western Salton Trough. Viejo sediments are found in extensional basins of limited extent only in the Coyote Mountains. We believe that the Viejo was deposited when the Coyote Mountains were in Sonora, Mexico.

The characteristics of the Viejo beds are poor matches to the descriptions of either the Deguyños or Latrania Formations. We propose a new division of the Imperial Group: the Viejo ("old") Formation (Figure 2).

The Coyote Mountains in Sonora, Mexico

Winker and Kidwell's (1986) 5.5 Ma paleogeographic reconstruction based on fault slip rates places the Coyote Mountains in Sonora, Mexico. Extensional tectonics started in Sonora after the Eocene. Nourse et al. (1994) reports 25 to 18 Ma deformation related to Basin and Range extension in Sonora, Mexico. The early Miocene to Pliocene sedimentary and volcanic units of the Coyote Mountains represent this extensional province. The White and Red Fanglomerates and Alverson volcanics likely filled basins that resulted from extensional tectonics. The fanglomerates, sands, freshwater limestones, and the marine sediments of the Viejo Formation were also deposited in these extensional basins. The Alverson Formation represents the volcanic part of the extension. McDowell et al. (1997) mapped similar aged sediments and volcanics in extensional basins in Sonora, Mexico. McDowell et al. (1997) considered the volcanics associated with the sediments they mapped to be the volcanic parts of Basin and Range extension. McDowell et al. did not report any marine sediments in their extensional basins. The angular unconformities within these units attest to the tectonic activity taking place in Sonora.

Sometime during or after the deposition of the Viejo Formation, the Coyote Mountains started moving north as a result of San Andreas transtensional tectonics. Winker and Kidwell (1986) suggest 5.5 Ma for initial northwestward translation of the Coyote Mountains. We do not know if this date is accurate; however, we observe that when the Deguyños was deposited, the Coyote Mountains had moved north far enough for the Deguyños to be influenced by the Colorado River. Transtension was accompanied by concurrent spreading center activity, so the basin was likely smaller than today's Salton Trough. At some point in the early Pliocene the Colorado River started to build a delta into the basin. As the Coyote Mountains move north the extensional tectonics of the West Salton detachment (Axen and Fletcher, 1998) also began to influence the mountains.

Late Cenozoic Deformation in the Coyote Mountains

Late Cenozoic deformation within the Coyote Mountains has produced detachment, thrust, and strike-slip faults. The detachments are second order and belong to the West Salton detachment fault (Axen and Fletcher, 1998). Two second-order detachments are mapped: one on the eastern side of the Coyote Mountains (Morgan and Morgan, 2010), and the other is on the western side (Morgan and Morgan, 2013). The main West Salton detachment fault is not exposed at the surface, but we infer its presence below the Coyote Mountains. The second-order detachment faults exposed in the Coyote Mountains displace Imperial Group sediments.

About one million years ago, right-lateral (with minor left-lateral), high-angle faults started to dominate the western Salton Trough and the Coyote Mountains. These younger faults cut the detachments and ended their movement (Steely et al., 2008). The right-lateral and left-lateral faults are active today, and the Coyote Mountains are presently being uplifted.

The Elsinore fault is the main active high-angle right-lateral fault in the Coyote Mountains (Dibblee, 1954). It strikes northwest and forms the southwest side of the Coyote Mountains.

The Painted Gorge Wash fault is an active left-lateral high-angle fault that is conjugate to the Elsinore fault (Morgan and Morgan, 2005). It is mapped for 9.5 km, strikes northeast and is southeast of the Coyote Mountains.

Christensen (1957) mapped the Painted Gorge fault and it is an active right-lateral fault that cuts Terrace 2 on the northeast side of the Coyote Mountains. There are other right-lateral faults in the Coyote Mountains that do not show recent activity.

In some cases, pre-existing high-angle faults and second-order detachment faults appear to have been reactivated as right-lateral faults (Morgan and Morgan, 2013).

In the Alverson Canyon area (Fossil Canyon), thrust faulting is associated with right-lateral faulting. The thrusting was first recognized by Ann Bykerk-Kauffman and her students (personal communication). Thrusting is northwest verging. This is the only area in the Coyote Mountains where thrust faulting is recognized.

Soft-sediment deformation took place in the Imperial Group during uplift of the Coyote Mountains. Evidence of this soft-sediment deformation can be seen in the morphology of the Viejo Andrade calcareous beds as they cling to the Coyote Mountains (Figure 10). It is unrealistic to believe that the Andrade was deposited on a 25 degree slope. On the east side of the Coyote Mountains, this soft-sediment deformation also raised the gently-dipping detachment surfaces to higher-angle surfaces (Morgan and Morgan, 2010).

Additional evidence of the episodic uplift of the Coyote Mountains is seen in at least 5 alluvial-fan terraces found in the mountains and on the surrounding flats (Morgan and Morgan, 2013). In Alverson Canyon and Painted Gorge, Terrace 4 is covered by Terrace 3 indicating at least one episode of subsidence occurred during the uplift of the Coyote Mountains. The uplift also explains some of the stream piracy that has and is taking place around the Coyote Mountains. Landslides and unstable slopes in the basement rocks attest to quick and uneven uplift. Rapid erosion of the overlying soft sediments during uplift has further destabilized the slopes.

A few of the geomorphic forms found in the Coyote Mountains are: an outlier (klippe?) made from the detachment faults, mullion structures on faults, hoodoos in tuff beds, horst and graben structures, stripped dip-slopes, cuestas, dissected fans, homoclinal ridges, and slot canyons.

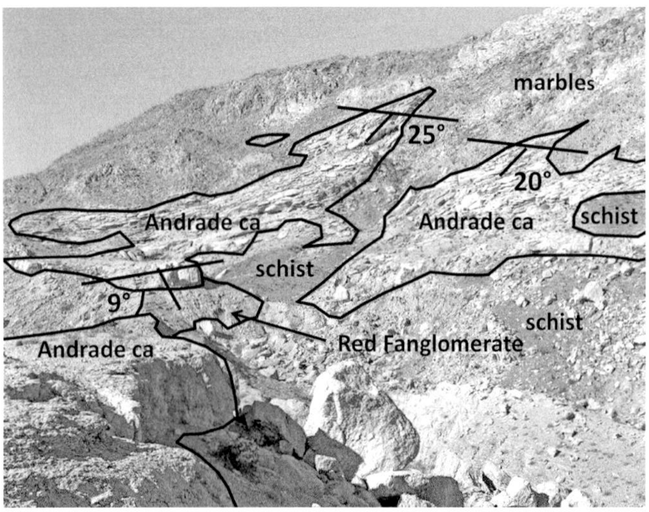

Figure 10. Looking northwest at the south side of Marble Mountain. The Viejo Andrade (Pas), a marine calcareous sandstone, is close to flat lying in the foreground but has a steeper dip up the hill. Strike and dip symbols are shown in black with dip values in black.

ACKNOWLEDGMENTS

Thanks to Dave Bloom for his editing of this paper and the following:, Bill Elliott, Tony Carrasco, Mike Hart, Ann Bykerk-Kauffman, Eric Frost, Stephen Schellenberg, Monte Marshall, Gordon Gastil, Gary Axen, Randy Wagner, Larry Busch, Charles Winker, Jerry Treiman, Jon Spencer, John Prall, and Jim Senn.

REFERENCES

Axen, G. J., and Fletcher, J. M., 1998, Late Miocene-Pliocene extensional faulting, northern Gulf of California, Mexico, and Salton Trough, California: International Geological Review, V 40, p. 217–244.

Bell, Patricia J., 1980, Environment of Deposition, Pliocene Imperial Formation Southeast, Coyote Mountains, Imperial County, California: M.S. thesis, San Diego State University.

Blaettler, Karen Griffith, 1991, A petrographic study of an amphibolite outcrop in the Coyote Mountains, Imperial County, California, San Diego State University, Under Graduate Research Report, V. 53.

Christensen, A. D., 1957, Part of the geology of the Coyote Mountains area, Imperial County, California: Unpublished M.S. thesis, University of California at Los Angeles, 188 p.

Dibblee, T. W., 1954, Geology of the Imperial Valley region, California, in Jahns, R. H., ed., Geology of southern California: California Division Mines Bulletin, v. 170, p. 21–28.

Dibblee, T. W., Jr., 2003, Geology of the Coyote Mountains, Imperial and San Diego Counties, California, in Murbach, M.L. and Hart, M. W., eds., 2013, Geology of the Elsinore Fault, San Diego Region. San Diego Association of Geologists and South Coast Geological Society.

Fourt, R., 1979, Post-batholithic Geology of the Volcanic Hills and Vicinity, San Diego County, California, San Diego State University, Master Thesis.

Johnson, M. E., Backus, D.H. and Ledesma-Vázquez, J., 2003, Offset of Pliocene ramp facies at El Mangle by El Coloradito fault, Baja California Sur: Implications for transtensional tectonics, in Johnson, S, E., Paterson, S. R., Fletcher, J. M., Girty, G. H., Kimbrough, D. L. and Martin-Barajas, A., Tectonic Evolution of Northwestern Mexico and the Southwestern USA, Geological Society of America Special Paper 374, p.407–420.

Link, M. H., and Abbott, P. L., 1991, Eocene sedimentary history, San Diego, California: overview and Field trip stops,in Abbott, P. L., and May, J. A., eds., 1991, Eocene geological history San Diego Region, Pacific section SEPM, Vol. 68, p.1–26.

Marshall, M., Mace, N., Erskine, B., and Jamsen, E., 2006, What regional and local paleomagnetic data tells us about the tectonics of the Coyote Mountains, southwestern Imperial Valley, southern California, in Bloom, D. M. and Farquharson, P. T., eds., Geology and history of southeastern San Diego, California, p. 55–62.

McDowell, F. W., 1997, Interrelationship, of sedimentary and volcanic deposits associated with Tertiary extension in Sonora, Mexico GSA Bulletin, v. 109,no. 10; p. 1349–1360.

Miller, R, H., and Dockum, M. S., 1983, Ordovician conodonts from Metamorphosed carbonates of the Salton Trough, California: Geology, v. 11, p. 410–412.

Morgan, G. J., and Morgan, J. R., 2005, The Painted Gorge Wash Fault: recent left-lateral faulting, east side Coyote Mountains, Imperial County, California: Bloom, D. M., ed. Geology and history of southeastern San Diego County, California, San Diego Association of Geologists, p.189–204.

Morgan, G. J., and Morgan, J, R., 2010, Post Pliocene detachment on the eastern side of the Coyote Mountains, Salton Trough, southern California:

GSA, Abstract and Programs, v. 42, n. 4, p. 94,

Morgan, G. J., and Morgan, J, R., 2013, Recent mapping in the Coyote Mountains, western Salton Trough, Imperial County, southern California, USA: GSA, Abstract and Programs, v. 45, n. 6, p. 61.

Morgan, G. J., 2013, Evidence of fresh-water limestone (Bouse Formational equivalent?) in the transgressional-regressional, marine and non-marine Imperial Group sediments, Coyote Mountains, Imperial County, Salton Trough, southern Californian, GSA Abstracts with programs, v, 45, n. 6, p. 19.

Morgan, J. R., Morgan, G. J. and Pecha, M., 2012, U/Pb dating of tuffs from the Alverson Volcanics in the Fossil Canyon and Painted Gorge areas of the Coyote Mountains, western Salton Trough, California.

Morgan, G. J., and Morgan, J, R., 2014, A 2014 update of an Overview of the Geology and Geological Problems of the Coyote Mountains, Salton Trough, Imperial County, southern California. In: Coast to Cactus, Geology and Tectonics San Diego to Salton Trough, California, Morton, J. B., Editor, p. 135–148.

Morton, P. K., 1977, Geology and mineral resources of Imperial County, California: California Division of Mining and Geology, County Report No.7.

Nourse, J. A., Anderson. T. A., and Silver, L. T., 1994, Tertiary metamorphic core complexes in Sonora, northwest Mexico: Tectonics, v.13, p. 1161–1182

Schmidt, K. L., and Paterson, S. R., 2002, A doubly vergent fan structure in the Peninsular Range batholith: Transpression or local complex flow around a continental margin buttress?: Tectonics, v. 21, no. 5, DOI 10.1029/200ITC001353.

Ruisaard, C. I., 1979, Stratigraphy of the Miocene Alverson Formation, Lmperial County, California: Masters of Science Thesis, San Diego State University, 125p.

Steely, A. N., Janecke, S. U., Dorsey, R. J., and Axen. G. J., 2009, Early Pleistocene initiation of the San Felipe fault zone, SW Salton Trough, during reorganization of the San Andreas fault system: Geological Society of America Bulletin, v. 121, p. 663–687, doi. 10.1130/B26239.1

Watkins, R., 1990, Paleoecology of a Pliocene rocky shoreline, Salton Trough Region, California: Society for Sedimentary Geology, Palaios, Research Letters, V. 5, p. 167–175

Winker, C. D., and Kidwell, S. M., 1986, Paleocurrent evidence for lateral displacement of the Pliocene Colorado River delta by the San Andreas fault system, southeastern California: Geology, V14, p. 788–791.

Winker, C. D., and Kidwell, S. M., 1996, Stratigraphy of a marine rift basin: Neogene of the western Salton Trough, California, in Abbott, P. L., and Cooper, J. D., eds., Field Conference Guidebook and Volume for the Annual Convention: San Diego, California, American Association of Petroleum Geologists, Pacific Section, p. 295–336.

Woodring, W. P., 1931, Distribution and age of the Tertiary Deposits of the Colorado Desert: Carnegie Institution of Washington Publication 148, p. 1–25

First report of early Miocene land mammals from the Red Rock Formation, southern Coyote Mountains, Imperial County, California

THOMAS A. DEMÉRÉ
Department of Paleontology, San Diego Natural History Museum
BRIDGET BORCE
Department of Paleontology, San Diego Natural History Museum
Department of Biology, San Diego State University

ABSTRACT

PALEONTOLOGICAL FIELDWORK CONDUCTED IN the Coyote Mountains since 1985 has resulted in the recovery of significant new fossils from Neogene marine and nonmarine strata. Of particular note is the discovery and recovery of land mammal fossils from the Red Rock Formation, the stratigraphically lowest and oldest Neogene rock unit exposed in the region. The Red Rock Formation is thought to predate the initiation of strong extensional tectonics in the Salton Trough. The recovered fossils include fragmentary mandibular, dental, and postcranial remains of a small to medium-sized camelid, tentatively assigned to the protolabine taxon, *Protolabis*. This group of stem camelids has an early Miocene (late Arikareean North American Land Mammal Age) evolutionary origin, which serves to constrain the maximum geologic age of the Red Rock Formation. Although fragmentary, the camelid fossils recovered from the Red Rock Formation serve as an indication of the paleontological potential of this rock unit.

INTRODUCTION

The Coyote Mountains, located in the western part of the Salton Trough (Figure 1), have drawn the attention of geologists and paleontologists since the mid-19th century, when W.P. Blake (1857) and T.A. Conrad (1857) first reported fossils from the area. Subsequent early paleontological work in the mountains was reported by Orcutt (1890), Vaughn (1917), Kew (1914), Nomland (1916), and Hanna (1926). Since Hanna (1926) interest in the paleontology of the Coyote Mountains has occurred sporadically, with pulses of work in the 1930s (e.g., Woodring, 1931; Grant and Hertlein, 1938), 1960s (e.g., Mitchell, 1961), 1970s (e.g., Stump, 1972; Foster, 1979), 1980s (e.g., Miller and Dockum, 1983; Bell-Countryman, 1984; Quinn and Cronin, 1984; Kidwell, 1988), 1990s (e.g., Deméré, 1993; Watkins, 1990a, 1990b, 1992; Winker and Kidwell, 1996), and 2000s (e.g., Deméré, 2006; Deméré and Rugh, 2006; Redman et al., 2007; Rugh, 2014). The vast majority of this published research has focused on the extensive marine invertebrate fossil assemblages preserved in strata of the Imperial Group. These upper Miocene and lower Pliocene strata and their contained fossil record preserve evidence critical to understanding such major regional events as the formation of the proto-Gulf of California, the initial uplift of the Colorado Plateau, and the birth of the Colorado River Delta. Imperial Group strata, however, are not the only sedimentary rocks exposed in the Coyote Mountains that preserve evidence of important geologic events. Gradationally overlying Imperial Group strata are rocks of the Palm Spring Group, which document both the late Pliocene and Pleistocene migration of the ancestral Colorado River Delta across the head of the proto-gulf to form the below-sea level Salton Basin, as well as the formation of the first in a series of large, ephemeral freshwater lakes that have characterized portions of the Salton Basin since the late Pliocene.

In the Coyote Mountains, Imperial Group strata either interfinger with and gradationally overlie Miocene fanglomerates of the Split Mountain Group (Garnet Formation of Winker and Kidwell, 1999) or nonconformably overlie volcanic rocks of the middle to lower Miocene Alverson Formation and/or plutonic and metamorphic rocks of the Mesozoic and Paleozoic crystalline basement complex (Dibblee, 2003). Good exposures of the former stratigraphic relationship occur in the Fossil Canyon area near Ocotillo, while exposures of the latter stratigraphic relationship occur in the Painted Gorge area near Plaster City (Figure 1). Locally, a reddish sandstone and conglomerate unit underlies and in places interfingers with the alkaline to tholeiitic basalts of the Alverson Formation (Ruisaard, 1979; Kerr and Kidwell, 1991). These reddish sandstones and conglomerates have a complex stratigraphic nomenclatural history and have been referred to the Split Mountain Formation (Ruisaard, 1979; Bell, 1980), or the Anza Formation (Kerr, 1982; Todd, 2004), or the Red Rock Formation (Winker and

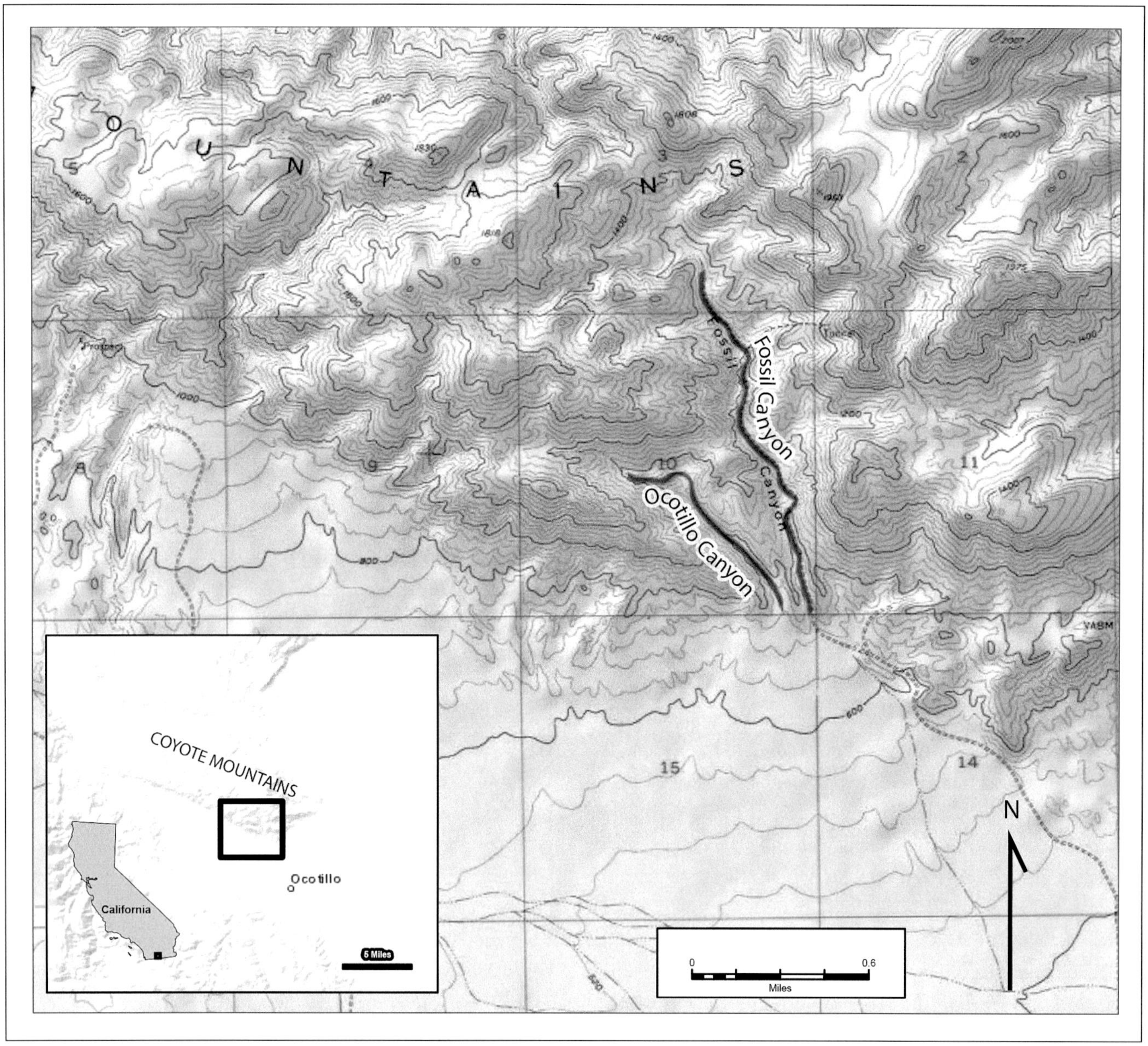

Figure 1. Index map of the southeastern Coyote Mountains showing the location of Fossil Canyon and Ocotillo Canyon. Base map: Carrizo Mountain, CA USGS 7.5′ topographic quadrangle.

Kidwell, 1996). The latter formational name is used in this report following the work of Winker (1987), Winker and Kidwell (1996), and Dorsey et al. (2011).

The Red Rock Formation was originally named for exposures of lower Miocene non-marine strata in Red Rock Canyon in the Fish Creek Mountains. In the Coyote Mountains, the Red Rock Formation is only well-exposed in Ocotillo Canyon, just west of Fossil Canyon, and underlies and interfingers with volcanic flow rocks of the Alverson Formation (Ruisaard, 1979; Kerr and Kidwell, 1991). The superpositional relationship of the Red Rock Formation relative to the Alverson Formation indicates that the former is no younger than 14–22 Ma—the age of the Alverson Formation (Dorsey et al., 2007). As an early Miocene rock unit, the Red Rock Formation predates the onset of structurally controlled basin formation and strong extensional tectonics in the Salton Trough (Dorsey et al., 2011). This factor suggests that more intense study of the Red Rock Formation has the potential to yield new evidence about geologic conditions and events during this poorly known chapter in the geologic history of the region.

Kerr (1984) and Kerr and Kidwell (1991) discussed the sedimentology and facies relationships of strata of the Red Rock Formation as exposed in the type area in the Fish Creek Mountains and concluded that they accumulated in a braided stream depositional system with longitudinal bars, channel lags, migrating sand waves, and seif sand dunes. To date, no fossils have been reported from strata of the Red Rock Formation and as a consequence nothing is known of the organisms and ecosystems that existed in this early Miocene braided stream setting. In an effort to correct this paleontological deficiency, we report for the first time land mammal fossils from the Red Rock Formation.

GEOLOGY AND STRATIGRAPHY

The geology of the Coyote Mountains is petrologically and structurally complex. The current elevation of the range is largely the result of strike-slip faulting along the Elsinore Fault Zone, which runs NW along the southern flank of the mountains (Todd, 2004). A series of NNW and NE striking high-angle normal faults also occur in the area, where they have produced tilted fault blocks and local horsts and grabens. The oldest rocks exposed in the Coyote Mountains are rocks of the crystalline basement complex, which includes Paleozoic metasedimentary rocks and Mesozoic plutonic rocks. The metasedimentary rocks are mostly marble and quartz-mica schist, with local outcrops of amphibolite, gneiss, and quartzite (Ruisaard, 1979; Dibblee, 2003). Miller and Dockum (1983) reported the occurrence of early Ordovician conodonts from low-grade marbles exposed on the SE flank of Carrizo Mountain, in the eastern part of the mountain range. Plutonic rocks in the Coyote Mountains consist of intrusions of quartz diorite, granodiorite, and pegmatite dikes. These rocks likely share a common history with Mesozoic plutonic rocks of the Peninsular Range Batholith. Cenozoic sedimentary rocks rest noncomformably on the crystalline basement complex and include the lower Miocene Red Rock Formation, the lower to middle Miocene Alverson Formation, the upper Miocene Split Mountain Group, the Mio-Pliocene Imperial Group, and the Plio-Pleistocene Palm Spring Group (Winker and Kidwell, 1996). Locally, elevated Pleistocene terrace deposits unconformably overlie the Neogene rocks and create a stepped series of dissected pediment surfaces.

OCOTILLO CANYON

Along the southern margin of the Coyote Mountains are a series of fault controlled canyons that provide good exposures of the Neogene stratigraphic section (Figure 1). Certainly the best known and best studied of these canyons is Fossil Canyon, also called Alverson Canyon or Shell Canyon. A prominent NNW striking high angle normal fault runs along the eastern slopes of Fossil Canyon and in places juxtaposes crystalline basement rocks against NE dipping Neogene marine strata of the Imperial Group. A smaller canyon located to the west of Fossil Canyon and called Ocotillo Canyon, also is associated with a NNW striking, high angle normal fault that in this case exposes and offsets NE dipping Neogene nonmarine strata and volcanic flow rocks of the Red Rock Formation and Alverson Formation, respectively. The Red Rock Formation rests directly on crystalline basement rocks and represents the beginning of significant Neogene deposition in the Coyote Mountains area. As mapped and described by Ruisaard (1979), the Red Rock Formation in Ocotillo Canyon consists of approximately 140 meters of grayish-orange pink to moderate red, fine- to coarse-grained, arkosic sandstones and interbedded pebble to cobble conglomerates. The conglomerates typically contain clasts derived from the crystalline basement complex. However, near the major bend in Ocotillo Canyon, a stratigraphic sequence consisting of the Red Rock Formation and Alverson Formation is exposed in the steep eastern wall of the canyon. In this area, the approximate strike of bedding in the Red Rock Formation is N35°W, with an approximate dip of 20°NE. Lithologies include reddish-tan, fine-grained, laminated sandstone, reddish poorly sorted, fine- to coarse-grained arkosic sandstone with interbedded pebble to cobble conglomerate. Clasts within the conglomerates include gray feldspar, granitic rock fragments, and clear, white translucent and gray quartz. At the top of the section is a well-developed baked zone at the contact between the Red Rock Formation and the basaltic flow rocks of the Alverson Formation.

Paleontology

Artiodactyla

Camelidae

cf. *Protolabis* sp.

Referred specimens: SDNHM 140180 (Figure 2); partial left dentary with dp3-4 and unerupted m1. SDNHM 140352; associated left calcaneum and astragalus

Locality and Stratigraphic Horizon: SDNHM Locality 5961, Ocotillo Canyon, Coyote Mountains, Imperial County, California, USA. Collected November 2002 and May 2009 from the upper part of the Red Rock Formation, approximately 20 feet stratigraphically below basalt in the Alverson Formation.

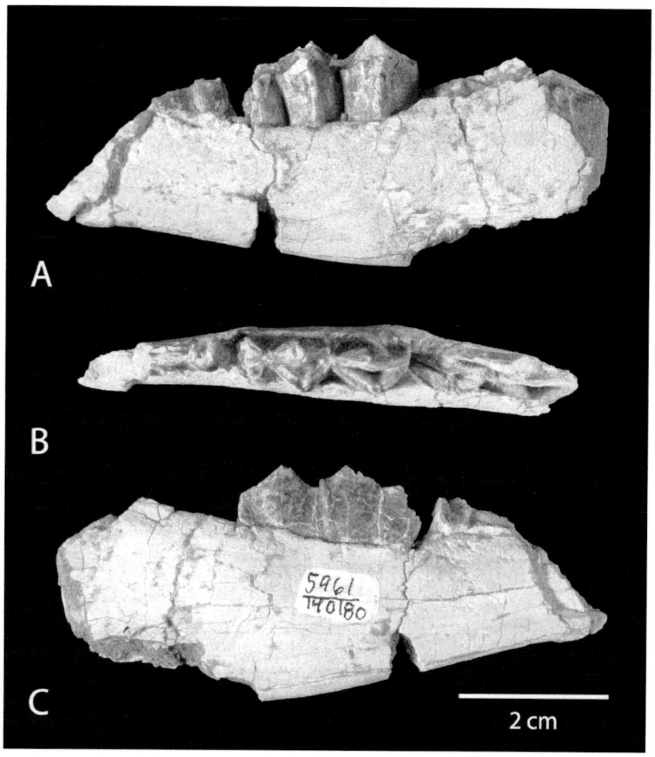

Figure 2. cf. ***Protolabis*** sp., partial left dentary with dp3-4 and unerupted m1 (SDNHM 140180) recovered from SDNHM Locality 5961. A. labial view; B. occlusal view; C. lingual view.

Description

Dentary: The partial left dentary (SDNHM 140180) is represented by the central portion of the horizontal ramus containing a partial dp3, a complete dp4, and an unerupted m1 (Figure 2). The preserved ramus measures 73.10 mm in greatest length, and is 18.06 mm in height and 10.14 mm in width as measured at the level of dp4. The anterior and posterior breaks are sharp and unabraded. The dentary is relatively small and gracile and reflects a rather small subadult individual.

dp3: The dp3 is fragmentary and preserves a nearly complete posterior crest with well-worn hypoconid and broken entoconid (Figure 2). The posterior fossettid is enlarged due to wear. The posterolingual corner of the crown is developed into a weak entostylid. The posterior border of the crown has a small facet, suggesting that the dp4 was in direct contact with the dp3. The anterior crest of the crown is largely missing, although the labial enamel surface of the primary cusp is intact. The crown measures 11.14 mm in length and the posterior lobe measures 4.74 in maximum width.

dp4: The dp4 is tri-lobed, distinctly seleodont, and except for damage to the anterior cusp is complete (Figure 2). The lingual surface is marked by a prominent metastylid and weakly developed entostylid. The entoconid has a very weak median ridge. There is no median ridge developed on the lingual surface of the metaconid. There are no anterior or posterior cingula present. With greater wear, the median fossettid likely would have reached the lingual surface behind the prominent metastylid. The broken anterior lobe preserves a small anterior fossettid, but the labial and lingual anterocondis are both missing due to breakage. A small piece of enamel adhering to the posterior margin of dp3 is likely a remnant of the anterior margin of the anterior lobe. As preserved, the dp4 measures 23.45 mm in length, although the actual length of the tooth was likely closer to 26 mm. The maximum width of the posterior lobe is 7.08 mm, while the median lobe measures 7.01 mm and the anterior lobe 5.61 mm.

m1: The m1 is unerupted and unworn (Figure 2). The crown is visible within the alveolar crypt and is narrowly elongate and bilobed, with the inner and outer cusps separated by deeply inset fossettids. Based on the relative positions of the metaconid and entoconid, there is likely a distinct metastylid. There is no evidence of an entostylid, although the posterior margin of the crown appears to be slightly broken. There is no evidence of median ridges on the metaconid or entoconid. However, this may be a function of how the tooth is preserved in its alveolar crypt. The maximum length of the crown is 24.70 mm, while the estimated width of both the anterior and posterior cusps is ~6 mm.

Calcaneum: The left calcaneum (SDNHM 140352) is damaged (due to weathering) and is missing the proximal portion of the sustentaculum, as well as the distal terminus of the calcaneal tuber. The dorsal margin of the calcaneal tuber has lost the cortical bone layer. As preserved, the calcaneum measures 45.58 mm in greatest length. The calcaneal tuber measures 35.14 mm in greatest length and 7.25 mm in minimum width.

Astragalus: The left astragalus is very fragmentary and is missing the lateral margin and major portions of the distal trochlea. The maximum length as preserved measures 20.72 mm and the maximum preserved width is 13.30 mm. The fossil preserves enough features to confirm the characteristic artiodactyl "double pulley" morphology.

Taxonomy

Unfortunately the presence of a deciduous dentition in SDNHM 140180 makes it difficult to definitively diagnose the dentary to the species level, or for that matter to a taxonomic level below the family Camelidae. However, several features taken together suggest that the specimen is likely a protolabine camelid and provisionally a taxon of *Protolabis*. Length and width measurements of dp3 and dp4, as well as the length/width ratio of dp4 fall within the morphometric range of *Protolabis* (Honey and Taylor, 1978). In addition, the general morphology of dp3 and dp4 in SDNHM 140180 is compatible with features described for deciduous teeth of *Protolabis* by Loring and Wood (1969) and Honey and Taylor (1978), including the presence of a prominent metastylid and an anterolingually placed median fossettid on dp4. In addition, Honey (2007) has recently suggested that the two, coeval protolabine taxa *Protolabis* and *Michenia* might actually represent sexually dimorphic pairs of a single camelid genus of small to medium-sized ungulate browsers.

DISCUSSION

The camelid fossils described in this study represent the first identifiable land mammal fossils known from the Red Rock Formation. Although the fossils are fragmentary and do not preserve sufficient anatomical detail to allow for taxonomic identification to the species level, the fossils do document the occurrence of land mammal fossils in the Red Rock Formation and indicate the likelihood that additional well-preserved vertebrate fossils will be discovered with increased paleontological prospecting. The eventual recovery of additional, identifiable vertebrate fossils will provide new information critical to better understanding the biochronology, biostratigraphy, and paleoecology of the Red Rock Formation. This information can, in turn, be used to better understand the early Miocene paleoenvironments and paleoclimate of the Salton Trough.

Camelids have had a long geological history in North America and, in fact, the entire clade had its evolutionary first appearance on this continent during the middle Eocene (Uintan NALMA; 46.2–42 Ma). It is noteworthy that fossils of one of the earliest stem camelids, *Poebrodon californicus*, have been recovered from middle Eocene strata of the Santiago Formation as exposed in Carlsbad and Oceanside (Golz, 1976). Camelids remained a North American endemic lineage until the late Miocene (Turolian ELMA; 9.0–5.3 Ma), when a northern dispersal event brought the first camelids to Eurasia (Honey et al., 1998). The period of greatest camelid diversity in North America occurred during the early through middle Miocene (late Arikareean to Barstovian NALMAs; 19.2–13.6 Ma). In some early and middle Miocene mammalian faunas, camelids were the most common large herbivores, where they were represented by a diversity of small to medium-sized animals (Honey et al., 1998). Camelids remained a conspicuous element of North American land mammal assemblages through the late Miocene (Hemphillian NALMA; 10.3–4.9 Ma), with a decrease in diversity during the Pliocene (Blancan NALMA; 4.9–1.8 Ma) until their extirpation from North American at the end of the Pleistocene (Rancholabrean NALMA, 240–10 ka).

The earliest protolabine camelids are known from the early Miocene (late Arikareean NALMA; 23–19 Ma) of the great basin and central plains (Honey et al., 1998) and the lineage extended through the middle Miocene into the late Miocene (Clarendonian NALMA; 13.6–10.3 Ma). Protolabines were relatively long-legged (compared to miolabines) camelids with moderately hypsodont cheek teeth, and are considered to have been ground-level feeders (browsers) in contrast to the tree-browsing and giraffe-like, later diverging aepycamelines (Janis, 1982). The relatively diverse protolabine lineage of stem camelids occupied the woodland savannah habitats that dominated certain regions of North America during the early Miocene (Janis, 1982). Given the taxonomic diversity and broad biogeographic range (i.e., central California, Great Basin, and Great Plains) of protolabines, it is not surprising that fossil remains of this camelid lineage have been found in the Coyote Mountains.

The recovery of camelid fossils from the Red Rock Formation has biochronological implications and suggests that this rock unit possibly is no older than the late late Arikareean NALMA (i.e., no older than ~19 Ma). This tentative age correlation, in turn, has implications for constraining the timing for initiation of strong extensional tectonics in the Salton Trough. However, it needs to be emphasized that this biochronological evidence is weak and in need of testing by the collection of additional vertebrate fossil remains and possible radiometric dating.

SUMMARY

Paleontological fieldwork in the southern part of the Coyote Mountains has resulted in the recovery, for the first time, of land mammal fossils from Neogene fluvial deposits of the Red Rock Formation. The fossils represent a small- to medium-sized camelid that is tentatively referred to the protolabine taxon, *Protolabis*. Protolabine camelids were conspicuous members of woodland savannah mammalian assemblages of the early Miocene of central California, the Great Basin, and the Great Plains. The recovery of land mammal fossil remains in the Red Rock Formation has the potential to yield important new information about the early Miocene paleoenvironment, paleoclimate, and paleoecology of the Salton Trough.

ACKNOWLEDGEMENTS

The fossils reported in this study were collected by dedicated field paleontologists presently and formerly employed in the Department of Paleontology at the San Diego Natural History Museum, especially Richard A. Cerutti, Patrick L. Sena, and the late Stephen L. Walsh. Field work and fossil collection activities were conducted under BLM paleontological investigation permit CA-10-00-009P. It is important to emphasize that vertebrate fossils cannot be collected in the Coyote Mountains (a federally designated wilderness area) without an approved paleontological investigation permit from the BLM El Centro field office.

BIBLIOGRAPHY

Bell, P.J. 1980. Environments of Deposition, Pliocene Imperial Formation, Southeast Coyote Mountains, Imperial County, California. Department of Geological Sciences, San Diego State University, unpublished Master's thesis, 80 p.

Bell-Countryman, P. 1984. Environments of deposition, Pliocene Imperial Formation southern Coyote Mountains, Imperial County, California. In, C.A. Rigsby (ed.), The Imperial Basin—Tectonics, Sedimentation, and Thermal Aspects, Pacific Section SEPM: 45–70.

Blake, W.P. 1857. Geological report. U.S. Pacific Railroad Exploration. U.S. 33rd Congress, 2nd session, Senate Executive Document 78 and House Executive Document 91, vol. 5, 390 p.

Conrad, T.A. 1857. Report of Mr. T. A. Conrad on the fossil shells collected in California by W. P. Blake, geologist of the expedition under the command of Lieutenant R. S. Williamson, U.S. Topographical Engineers, 1853. In, W.P. Blake, Geological report. U.S. Pacific Railroad Exploration. U.S. 33rd Congress, 2nd session, Senate Executive Document 78 and House Executive Document 91, vol. 5, Appendix 2: 325–326.

Deméré, T. A. 1993. Fossil mammals from the Imperial Formation (upper Miocene-lower Pliocene), Coyote Mountains, Imperial County, California. In, R.E. Reynolds and J. Reynolds (eds.), Ashes, Faults, and Basins. San Bernardino County Museum Association Special Publication 93-1: 82–85.

Deméré, T.A. 2006. The Imperial Sea: Marine Geology and Paleontology. In, G.T. Jefferson and L. Lindsay (eds.), Fossil Treasures of the Anza-Borrego Desert. Sunbelt Publications, San Diego, chapter 2: 29–41.

Deméré, T.A., and N.S. Rugh. 2006. Invertebrates of the Imperial Valley. In, G.T. Jefferson and L. Lindsay (eds.), Fossil Treasures of the Anza-Borrego Desert, Sunbelt Publications, San Diego, California, chapter 3: 43–70.

Dibblee, T.W. 2003, Geology of the Coyote Mountains, Imperial and San Diego Counties, San Diego. In, M.L. Murbach and M.W. Hart (eds.), Geology of the Elsinore Fault Zone, San Diego Region: San Diego Association of Geologists, San Diego, South Coast Geological Society, Santa Ana, California, p.117–123.

Dorsey, R.J., A. Fluette, K. McDougall, B.A. Housen, S.U. Janecke, G.J. Axen, and C.R. Shirvell. 2007. Chronology of Miocene-Pliocene deposits at Split

Mountain Gorge, southern California: A record of regional tectonics and Colorado River evolution. Geology 35: 57–60.

Dorsey, R.J., B.A. Housen, S.U. Janecke, C.M. Fanning, and A.L.F. Spears. 2011. Stratigraphic record of basin development within the San Andreas fault system: Late Cenozoic Fish Creek-Vallecito basin, southern California. Geological Society of America Bulletin 123: 771–793.

Foster, A.B. 1979. Environmental variation in a fossil scleractinian coral. Lethaia. 12:245–264.

Golz, D.J. 1976. Eocene Artiodactyla of southern California. Natural History Museum of Los Angeles County Science Bulletin 26:1–85.

Grant, U.S. and L.G. Hertlein. 1938. The West American Cenozoic Echinoidea. Publications of the University of California at Los Angeles in Mathematical and Physical Sciences 2: 1–225.

Hanna, G.D. 1926. Paleontology of Coyote Mountains, Imperial County, California. California Academy of Sciences Proceedings, series 4, 14: 51–186.

Honey, J.G. 2007. The Camelidae. In, D.R. Prothero and S. Foss (eds.), The Evolution of Artiodactyls. Johns Hopkins University Press, p. 177–188.

Honey, J.G. and B.E. Taylor. 1978. A generic revision of the Protolabidini (Mammalia: Camelidae), with a description of two new protolabidines. American Museum of Natural History Bulletin 161:367–426.

Honey, J.G., J.A. Harrison, D.R. Prothero, and M.S. Stevens. 1998. Camelidae. In, C.M. Janis, K.M. Scott, and L.L. Jacobs (eds.), Tertiary Mammals of North America. Cambridge University Press, Cambridge and New York, p. 439–462.

Janis C. 1982. Evolution of horned ungulates: ecology and paleoecology. Biological reviews 57: 261–318.

Kerr, D.R. 1982. Early Neogene Continental Sedimentation, Western Salton Trough, California. Department of Geological Sciences, San Diego State University, unpublished Master's thesis, 150 p.

Kerr, D.R. 1984. Early Neogene continental sedimentation in the Vallecito and Fish Creek Mountains, western Salton Trough, California. Sedimentary Geology 38: 217–246.

Kerr, D.R., and S.M. Kidwell. 1991, Late Cenozoic sedimentation and tectonics, western Salton Trough, California. In, M.J. Walawender and B.B. Hanan (eds.), Geological Excursions in Southern California and Mexico: San Diego, California, Department of Geological Sciences, San Diego State University, p. 397–416.

Kew, W.S.W. 1914. Tertiary echinoids of the Carrizo Creek region in the Colorado Desert. University of California Publications in Geological Sciences Bulletin 8: 39–60.

Kidwell, S.M. 1988. Taphonomic comparison of passive and active continental margins: Neogene shell beds of the Atlantic Coastal Plain and northern Gulf of California. Palaeogeography, Palaeoclimatology, Palaeoecology 63: 201–223.

Loring, S.H. and A.E. Wood. 1969. Deciduous premolars of some North American Tertiary camels (family Camelidae). Journal of Paleontology 43:1199–1209.

Miller, R. and M. Dockum. 1983. Ordovician conodonts from metamorphosed carbonates of the Salton Trough, California. Geology 11: 410–412.

Mitchell, E.D., Jr. 1961. A new walrus from the Imperial Pliocene of southern California: with notes on odobenid and otariid humeri. Los Angeles County Museum Contributions in Science 44: 1–28.

Nomland, J.O. 1916. Corals from the Cretaceous and Tertiary of California and Oregon. University of California Publications in Geological Sciences 9: 59–76.

Orcutt, C.R. 1890. Geology of the Colorado Desert. California Mining Bureau Report 10: 899–919.

Quinn, H.A., and T.M. Cronin. 1984, Micropaleontology and depositional environments of the Imperial and Palm Springs Formations, Imperial Valley, California. In, C.A. Rigsby (ed.), The Imperial Basin—Tectonics, Sedimentation, and Thermal Aspects: Pacific Section, Society for Sedimentary Geology (SEPM), Los Angeles, p. 71–85.

Redman, CM., L.R. Leighton, S.A. Schellenberg, C.N. Gale, J.L. Nielsen, D.L.Dressler, and M.K. Klinger. 2007. Influence of spatiotemporal scale on the interpretation of paleocommunity structure: lateral variation in the Imperial Formation of California. Palaios 22: 630–641.

Rugh, N.S. 2014. Imperial Group invertebrate fossils Part 2: the Kidwell Collection. In, R.E. Reynolds (ed.), Not a Drop Left to Drink, California State University Desert Studies Center, 2014 Desert Symposium. Pp. 138–143.

Ruisaard, C.I. 1979. Stratigraphy of the Miocene Alverson Fomation, Imperial County, California. Department of Geological Sciences, San Diego State University, unpublished Master's thesis, 136 p.

Stump, T.E. 1972. Stratigraphy and paleontology of the Imperial Formation in the western Colorado Desert. Department of Geological Sciences, San Diego State University, unpublished Master's thesis, 128 pp.

Todd, V.R. 2004. Preliminary geologic map of the El Cajon 30' x 60' quadrangle, southern California. USGS Open-File Report 2004-1361.

Vaughan, T.W. 1917. The reef-coral fauna of Carrizo Creek, Imperial County, and its significance. United States Geological Survey Professional Paper 98: 355–386.

Watkins, R. 1990a. Paleoecology of a Pliocene rocky shoreline, Salton trough region, California. Palaios 5: 167–175.

Watkins, R. 1990b. Pliocene channel deposits of oyster shells in the Salton Trough region, California. Palaeogeography, Palaeoclimatology, Palaeoecology 79: 249–262.

Watkins, R. 1992. Sedimentology and paleoecology of Pliocene shallow marine conglomerates, Salton Trough region, California. Palaeogeography, Palaeoclimatology, Palaeoecology 95: 319–333.

Winker, C.D. 1987. Neogene stratigraphy of the Fish Creek-Vallecito section, Southern California: implications for early history of the northern Gulf of California and Colorado Delta. Department of Geosciences, University of Arizona, unpublished Ph.D. dissertation, 494 pp.

Winker, C.D. and S.M. Kidwell. 1996. Stratigraphy of a marine rift basin: Neogene of the western Salton Trough, California. In, P. L. Abbott and J. D. Cooper (eds.), Field Conference Guide 1996. Pacific Section AAPG GB 73, Pacific Section SEPM book 80: 295–336.

Woodring, W.P. 1931. Distribution and age of the Tertiary deposits of the Colorado Desert. Carnegie Institute of Washington Publication 418: 1–25.

Geomorphic analysis of late Quaternary activity of the Earthquake Valley and Hot Springs–San Felipe fault zones, and their role in slip transfer between the northern Elsinore and southern San Jacinto faults

ERIK M. GORDON, THOMAS K. ROCKWELL, GARY H. GIRTY, CHRIS GOETZ

ABSTRACT

WE STUDIED THE RELATIVE TECTONIC GEOmorphic expression of faults within and between the Elsinore and San Jacinto fault zones in southern California to test recent models on the evolution of the southern California fault system. Prior work characterized the Hot Springs/San Felipe fault zone as the primary player in the transfer of slip between the northern Elsinore fault to the southern San Jacinto fault, and also inferred that this fault zone formed during the past 1.1–1.3 Ma. From analysis of geomorphic expression based on field studies, aerial photography and GoogleEarth imagery, these faults are essentially transparent in the modern geomorphology and display little to no evidence of Quaternary motion. The Lancaster fault, which is one of the primary northern elements of this fault system, is demonstrated to have been active during the Pliocene and inactive for at least the past several hundred thousand years. The San Felipe fault does have some expression of Quaternary activity, albeit at a very low rate. In contrast, the Earthquake Valley fault is very well expressed in the modern landscape with numerous examples of offset and deflected drainages, sags, pressure ridges, linear hill-side scarps and benches, consistent with recent assessments of a 2–3 mm/yr slip rate. We also identify and document numerous minor faults throughout the Vallecito and Fish Creek mountains that show evidence of latest Quaternary motion in the form of scarps in alluvium and bedrock, sidehill benches, and sag features. Based on the location and orientation of these small faults, we infer that the Vallecito-Fish Creek Mountain uplift is a consequence of the transfer of slip between the Earthquake Valley and southern San Jacinto fault zones, similar to the Palomar-Agua Tibia Mountain uplift that we attribute to the transfer of slip between the northern Elsinore and Earthquake Valley faults. From these new observations, we conclude that 1) the Hot Springs/San Felipe fault zone likely originated in the Pliocene and is mostly inactive in the modern tectonic regime; 2) the Earthquake Valley fault is the primary fault that transfers displacement from the northern Elsinore fault to the southern San Jacinto fault; and 3) Palomar Mountain and the Vallecito-Fish Creek Mountains are large pressure ridge systems that are a result of left-restraining steps between the Elsinore and Earthquake Valley faults and the Earthquake Valley and southern San Jacinto faults, respectively.

INTRODUCTION

Significant work has been done to understand the timing and character of tectonic evolution of the San Andreas fault system in onshore southern California from the primary and possibly only master fault strand in the Pliocene to the ~80 km wide fault zone observed today (Sharp, 1967; Powell and Weldon, 1992; Johnson et al., 1983; Morton and Matti, 1993; Magistrale and Rockwell 1996; Fialko, 2006; Steely et al., 2009; Blisniuk et al., 2010; Blisniuk et al., 2013; Lutz et al., 2006; Kirby et al., 2007; Belgarde, 2007; Janecke et al., 2010; Dorsey et al., 2012). The relationship between and evolution of the Elsinore and San Jacinto faults has been a particular topic of debate, which has important implications for understanding the modern tectonic system in southern California. In particular, there are two current issues that require further work: 1) the timing of initiation of the San Jacinto and Elsinore fault zones in southern California; and 2) the distribution of slip rate on and among these faults.

The age of initiation of the San Jacinto has been estimated as early as 1.8–2.5 Ma (Magistrale and Rockwell, 1996). In contrast, other workers place this age at 1.1–1.3 Ma based on the presence of an unconformity and anticline in the San Felipe Hills and San Felipe-Borrego Basin, respectively (Steely et al., 2009; Lutz et al., 2006; Kirby et al., 2007; Belgarde and Janecke, 2007; Janecke et al., 2010; Dorsey et al., 2012). Recent work indicates a slip rate of 12.1+3.4/−2.6 mm/yr on the San Jacinto for the past 0.5 Ma (Blisniuk et al., 2013), implying a much higher slip rate of ~30 mm/yr in the first half of its history if the fault is only on the order of 1.2 Ma and has an estimated total displacement of ~25 km (Sharp, 1967;

Morton and Matti, 1993; Rockwell et al., 2014). This would indicate a substantial change in rate sometime in the middle Pleistocene. If a constant slip rate for the San Jacinto fault is assumed using the 12.1+3.4/−2.6 mm/yr value from Blisniuk (2013), an age of 1.6–2.6 Ma is required to account for the ~25 km of displacement, which is much closer to that estimated by Magistrale and Rockwell (1996).

Magistrale and Rockwell (1996) place an age of initiation for the Elsinore fault coeval to the San Jacinto fault at 1.8–2.5 Ma based on total offset and late Quaternary slip rate. According to the model proposed by Magistrale and Rockwell (1996), slip was constrained prior to 0.9 Ma to the trace of the Elsinore fault north of Palomar Mountain, south of which slip was concentrated on the Earthquake Valley fault between Palomar Mountain southeast through the Vallecito-Fish Creek Basin. South of the basin, slip was again localized along the southern Elsinore fault in the Coyote Mountains. The authors postulated that it was not until after 0.9 Ma that the modern trace of the Elsinore fault linking the already-existent northern and southern sections developed, paralleling the Earthquake Valley fault. In comparison, a different model has been presented based on the much later proposed initiation of the Elsinore and Earthquake Valley faults at 1.2 Ma during a tectonic reorganization, as evidenced by an unconformity in the San Felipe Hills (Steely et al., 2009; Lutz et al., 2006; Kirby et al., 2007; Balgarde and Janecke, 2008; Janecke et al., 2010; Dorsey et al., 2012). This model also incorporates the Hot Springs/San Felipe fault (HSSF) zone as an active member in the overall southern California San Andreas fault system.

Complexity in understanding the age of initiation of both the Elsinore and San Jacinto faults is compounded by slip gradients along these faults. Previous work shows that the Elsinore fault has a change in slip rate between the northern and southern segments, decreasing from ~5 mm/yr in the north (Millman and Rockwell, 1986; Vaughan and Rockwell, 1986; Rockwell et al., 2000) to 2–3 mm/yr in the south (Magistrale and Rockwell, 1996; Fletcher et al., 2011). Evidence has also been developed that indicates a slip gradient along the San Jacinto fault zone, showing an increase of 6 mm/yr from the northern to southern sections of the fault zone (Fialko, 2006). Individual strands also display slip rate gradients, as is clear with the Clark strand which exhibits a maximum rate near Anza of 12-13 mm/yr (Blisniuk et al., 2013, Rockwell et al., 1990) decreasing southeastward to about 9 mm/yr at Rockhouse Canyon (Blisniuk et al., 2010) and to 1.5 mm/yr near the southern surface termination of the fault near the southern Santa Rosa Mountains (Blisniuk et al., 2010).

Studies have been conducted which offer an explanation for the varied distribution of slip between these two faults, using the Hot Springs/ San Felipe fault zone (HSSF) as a mechanism to actively transfer slip from the northern Elsinore fault southeast to the southern San Jacinto fault (Steely et al., 2009; Lutz et al., 2006; Kirby et al., 2007; Belgarde and Janecke, 2007; Janecke et al., 2010; Dorsey et al., 2012). Evidence for this is based on an inferred ~5.8 km right lateral separation of the Miocene-Pliocene aged West Salton detachment fault (WSDF) south of Yaqui Ridge (Steely et al., 2009, Dorsey et al., 2012, Axen and Fletcher, 1998). However, in our work we show that the HSSF, which includes the Lancaster, Agua Caliente, Hot Springs, and San Felipe faults, exhibits little or no displacement in the late Quaternary, which implies that the HSSF is not the modern primary system accounting for transfer of slip between these two faults. Furthermore, slip on the Lancaster and Aguanga faults, which comprise the northern portion of this zone, produced the formation of a pull-apart basin in the Pliocene, indicating that they are considerably older elements of the southern SAF system. In this paper we provide new observations on the activity of specific faults in the HSSF zone as well as present newly-identified faults with Holocene displacement in the Vallecito and Fish Creek mountains, which we interpret as being a transpressive pressure ridge system that accommodates transfer of slip from the Earthquake Valley fault to the southern San Jacinto fault zone. Based on these new observations, we propose that the HSSF zone was the primary system transferring slip in the late Pliocene and early Pleistocene, but has since become at most a minor member of this process, which is now dominated by the Earthquake Valley fault zone.

GEOLOGIC BACKGROUND

Regional Tectonics

South of the "Big Bend" in the San Andreas fault near Cajon Pass in the San Bernardino Mountains, the San Andreas fault transitions into a set of parallel faults forming a ~80 km wide fault system that encompasses the majority of southern California, collectively accounting for most (80%) of the ~49 mm/yr relative motion between the Pacific and North American plates (Sharp, 1967; Bennett et al., 1996; Magistrale and Rockwell, 1996; Fialko, 2006; Li and Liu, 2006; Li and Liu, 2007). From east to west, the primary faults of this zone include the San Andreas, San Jacinto, and Elsinore faults (Figure 1) (Sharp, 1967; Powell and Weldon, 1992; Magistrale and Rockwell, 1996). Slip rate distribution within this zone indicates that the San Andreas fault still accommodates most of the slip, with the San Jacinto fault a close second (Fialko, 2006; Magistrale and Rockwell, 1996; Becker et al., 2005; Blisniuk et al., 2013) and the Elsinore fault a distant third (Rockwell et al., 2000; Fletcher et al., 2011).

The San Jacinto fault comprises a 260 km-long SE trending zone of fault strands that transfers slip from the San Andreas fault near Cajon Pass in the San Bernardino Mountains to the Cerro Prieto and Imperial faults in northern Baja California (Sharp, 1967; Sanders and Magistrale, 1997). Multiple strands within the San Jacinto fault are related to each other by releasing and restraining bends and step-overs. The principal strands, from north to south, include

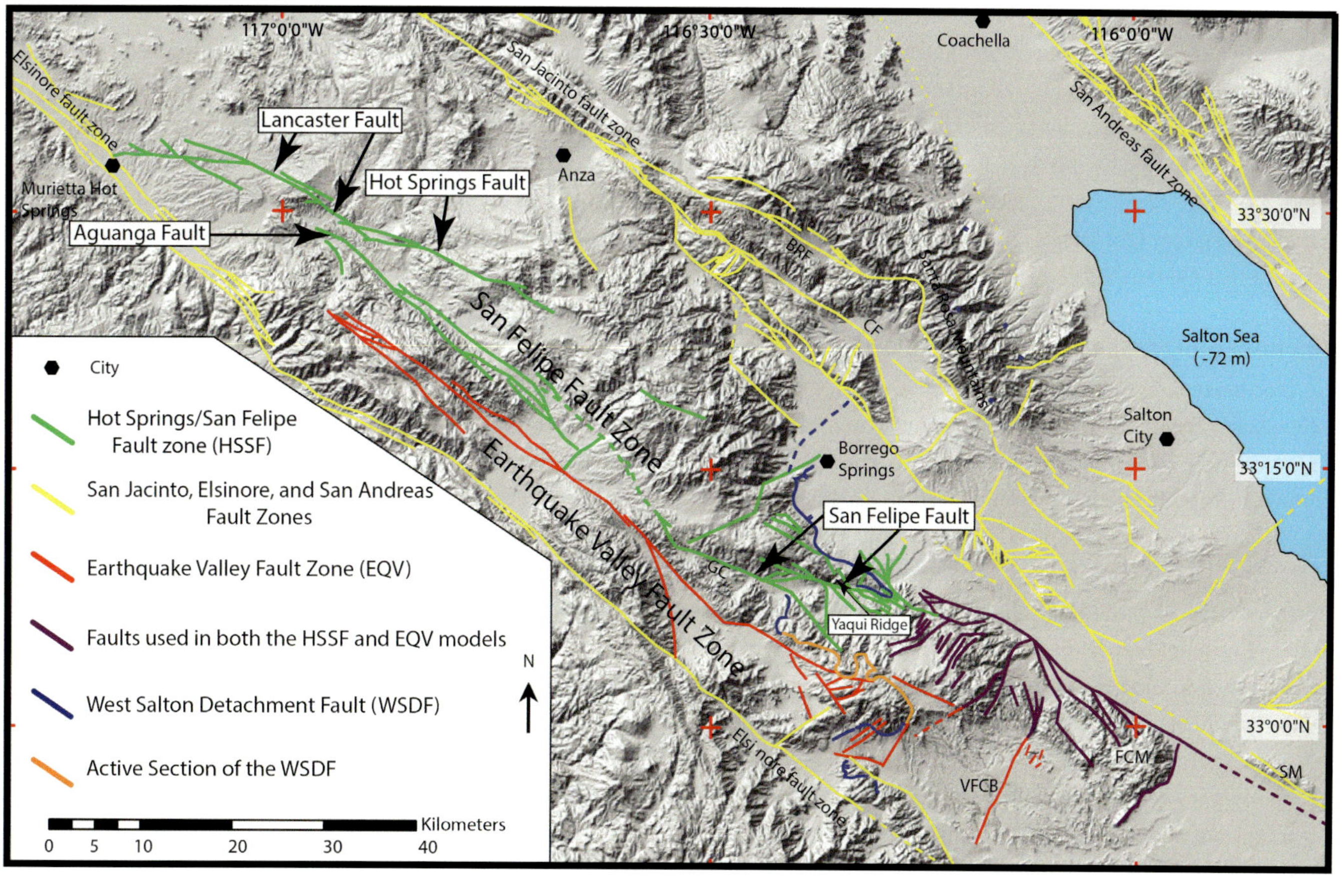

Figure 1. Diagram showing the overall seismic system of southern California, modified from Steely et al (2009). Faults in yellow compose the San Andreas, San Jacinto, and Elsinore fault zones. Faults in green form the bulk of the Hot Springs/San Felipe system (HSSF). Faults in red form the Earthquake Valley fault system (EQV). Faults in purple are used as part of the HSSF model used in Steely et al, 2009, but are reinterpreted as part of the EQV zone in this study. The fault in pink along the NE side of the Fish Creek Mountains (FCM) was a fault interpreted in Steely et al. (2009), which we were not able to confirm based on geomorphology.

the Claremont fault north of the Hemet step-over at Mystic Lake, the Casa Loma and Clark faults from Mystic Lake in the Hemet Valley southeast through Anza to the San Felipe Hills, the Coyote Creek fault, which extends from near Horse Canyon southeast through Coyote Mountain and Ocotillo Wells and steps across the Ocotillo Badlands to the Superstition Mountain fault, and the Superstition Hills fault, which lies east of and parallel to the Superstition Mountain fault (Sharp, 1967; Sanders and Magistrale, 1997; Dorsey and Roering, 2006). Cumulative displacement along the overall length of the San Jacinto fault has been estimated at ~25 km (Sharp, 1967; Morton and Matti, 1993; Rockwell et al, 2014).

The 230 km-long Elsinore fault zone trends SE from the Los Angeles Basin through the Peninsular Ranges batholith to the Coyote Mountains, located just north of the border with Mexico. The zone continues into Mexico as the Laguna Salada, Borrego, Paso Superior and Pescadores faults, which ruptured in 1892 and 2010 (Mueller and Rockwell, 1995; Fletcher et al., 2014). There are three NW striking segments that comprise the Elsinore fault zone, the northernmost of which is located north of 33.5°N in the Elsinore Trough. The central segment, which is double-stranded between 33.5°N and 33°N, consists of the main trace of the Elsinore fault and the parallel Earthquake Valley fault to the east (Figure 1) (Magistrale and Rockwell, 1996). The main trace of the Earthquake Valley fault extends SE from the eastern side of Palomar Mountain to Whale Peak just northwest of the Vallecito-Fish Creek Basin, and exhibits strong geomorphological expression including fault scarps across Holocene alluvium, offset and captured drainages, uplift due to localized transpression, sag depressions and side-hill benches and valleys (Jahns, 1954; Magistrale and Rockwell, 1996; Rockwell et al., 2013; this study). South of 33°N, the southern Coyote Mountains segment extends from the Tierra Blanca Mountains southeast to Ocotillo near Interstate 8, where deformation steps across Yuha Basin to the Laguna Salada fault in the southern part of Yuha Basin, continuing SE into Baja California as the Laguna Salada system (Kennedy, 1977; Hull and Nicholson, 1992; Magistrale and Rockwell, 1996).

Previous work indicates a total cumulative displacement of 10–15 km across the northern section of the Elsinore fault in the Elsinore Trough (Weber, 1977; Woyski et al., 1991; Hull and Nicholson, 1992) and only around 2.5 km near Granite Mountain at the southern end of the central segment (Lampe, 1988). The average slip rate on the central segment of the Elsinore fault is split between the main strand, estimated to be around 2.5–3 mm/yr (Magistrale and Rockwell, 1996; Vaughn et al., 1999), and the Earthquake Valley fault. Magistrale and Rockwell (1996) and Sundell (2009) infer a slip rate of 2–2.8 mm/yr for the Earthquake Valley fault based on the difference in total slip between the northern and southern sections of the Elsinore fault and offset of alluvial surfaces, respectively. More recently, Rockwell et al. (2013) determined a rate of about 2.5mm/yr for the Earthquake Valley fault in Warner basin based on a 1.9 km right-lateral offset of mid-Pleistocene valley fill deposits that contain the ~800 ka Bishop Tuff.

The Hot Springs/San Felipe Fault (HSSF) Zone

Northeast of the Earthquake Valley fault, the Hot Springs/San Felipe fault zone (HSSF) is a right-lateral fault zone branching off of the Elsinore fault northeast of the city of Murrieta, trending southeast through the Peninsular Ranges towards the Superstition Mountains segment of the San Jacinto fault (Figure 1) (Steely et al., 2009; Dorsey et al., 2012). The Lancaster fault, the principal northern element of the HSSF zone, interacted with the Aguanga fault to form a pull-apart basin in the Pliocene. These two faults collectively bounded the basin and juxtaposed Cretaceous crystalline basement against the basin-filling Pliocene-aged Temecula Arkose (Figure 1) (Pajak et al, 1996; Hutchison, 1987, Hull and Nicholson, 1992). Southeast of the Lancaster fault, the Hot Springs, Agua Caliente, and San Felipe faults comprise a continuous zone of SE-trending faults that extend to the NW corner of the Vallecito and Fish Creek mountains (Steely et al., 2009; Dorsey et al., 2012).

Cumulative dextral displacement across the HSSF have been estimated at ~5.8 km, based on an inferred right-lateral displacement of the West Salton detachment fault just south of Yaqui Ridge (Figure 1) (Steely et al, 2009). The West Salton detachment fault is a 20 degree NE-dipping normal fault which is believed to have been active prior to the onset of the tectonic reorganization of the San Andreas system, resulting from oblique rifting associated with the opening of the Gulf of California (Axen and Fletcher, 1998; Shirvell et al, 2009). Though the majority of the WSDF is now considered inactive, observations of a prominent fault scarp on an alluvial surface along the east side of Whale Peak indicates recent activity of at least this part of the fault (Steely et al., 2009; Dorsey et al., 2012).

NEW OBSERVATIONS

Hot Springs/San Felipe Fault Zone (HSSF)

We analyzed GoogleEarth imagery, aerial photography and LandSat imagery of the portion of the HSSF from the Lancaster fault near Vail Lake southeast to Yaqui Ridge, as mapped by Steely et al. (2009). Our specific goal was to identify and map evidence of Pleistocene and Holocene tectonic activity to assess this fault zone's current role in the southern SAF system. Two locations of specific interest were identified and assessed in the field along the Lancaster fault to establish late Quaternary activity: one site north of Vail Lake and another 9.5 km to the ESE of Vail Lake (Figures 2 and 3). The location north of Vail Lake shows the Lancaster fault juxtaposing the light-colored Temecula Arkose against Cretaceous gabbro (Figure 2). At this location, the fault is capped or sealed by a massive argillic horizon underlain by a stage IV K horizon, and the gabbro is also capped by a stage IV K horizon. The strength of the soil implies at least several hundred thousand years of inactivity, as Stage IV calcic horizons with well-formed laminar caps require a long period of formation (Machette, 1975). At the location 9.5 km to the ESE, the fault juxtaposes the Temecula Arkose against Cretaceous granitic rock, resulting in a strong contrast in the topographic formation of the numerous, deeply incised rills and drainages across the fault (Figure 3). Despite the resulting excellent exposures of the Lancaster fault in this area, we could find no deflected streams, scarps, or other indicators of active faulting (Figure 3), as are observed along nearby active faults. These observations demonstrate that the Lancaster fault has not been active for several hundred thousand years, or longer, and is not part of the modern fault system.

Southeast of the Lancaster fault, the main traces of the HSSF fault zone are the Hot Springs and San Felipe faults (Figure 1). The Hot Springs fault is primarily evident in GoogleEarth imagery as vegetation lineaments and erosional ravines along the fault in crystalline Mesozoic basement rock (Figures 4a and 4b). We could find little or no evidence of offset of the basement rocks in aerial photography or GoogleEarth imagery, although previous mapping north of Hot Springs Mountain reveals bedrock offset of up to ~0.75 km in Mesozoic granite/mafics (Figures 4c and 4d) (Rogers, 1965). Similarly, we could find no scarps, deflected or offset drainages, or other indicators of youthful activity along the Hot Springs fault, suggesting a level of activity that is similar to that of the Lancaster fault.

The Agua Caliente fault zone, which includes the San Felipe fault, generally shows a similar lack of expression. The fault is mapped in bedrock (Rogers, 1965) but does not exhibit any significant offset of basement contacts along most of its length. Along the entire length of this zone, only the San Felipe fault exhibits evidence for Quaternary activity in the area of Yaqui Ridge (Figure 5) near where highway S3 diverts off of Interstate 78. In this one area, the fault is expressed as a degraded scarp in crystalline rock along Yaqui Ridge but does not appear to offset Pleistocene alluvial deposits (Figure 5). This is the same area that Steely et al. (2009) suggest that the West Salton detachment fault is offset ~5.8 km by the San Felipe fault (Figure 1). We could find no evidence in offset basement rock to support such a large displacement and suggest an alternative hypothesis: The San Felipe fault in this one area acted as a tear in the detachment sheet. If correct, then the apparent 5.8 km of right-lateral separation of the detachment fault would not be expected to extend as displacement to the northwest into basement rock.

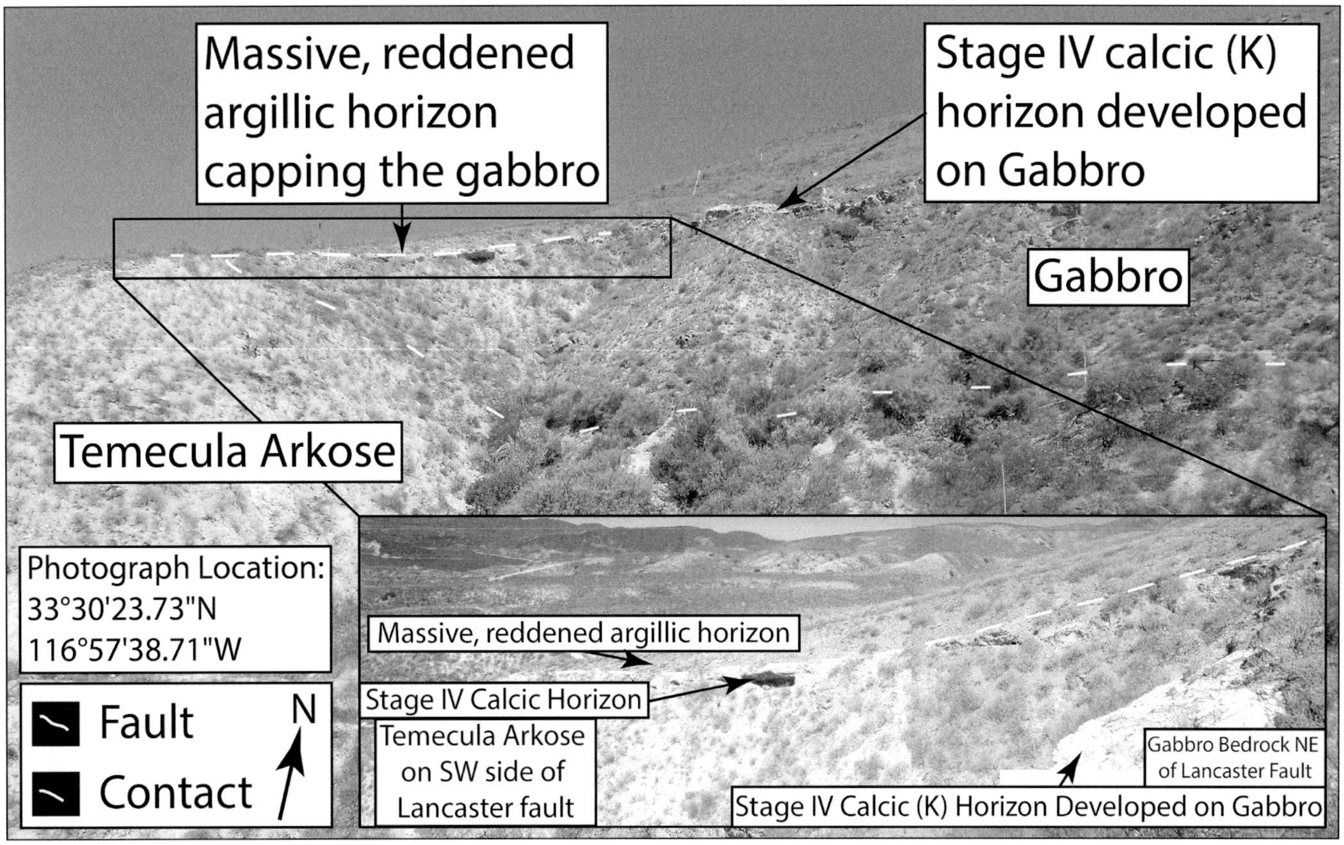

Figure 2. Section of the Lancaster fault north of Vail Lake separating the Pliocene-aged Temecula Arkose from basement gabbro. The fault here is capped by a massive argillic horizon, and the gabbro is capped by a stage IV calcic (K) horizon. A line of vegetation marks the fault leading towards the ridge.

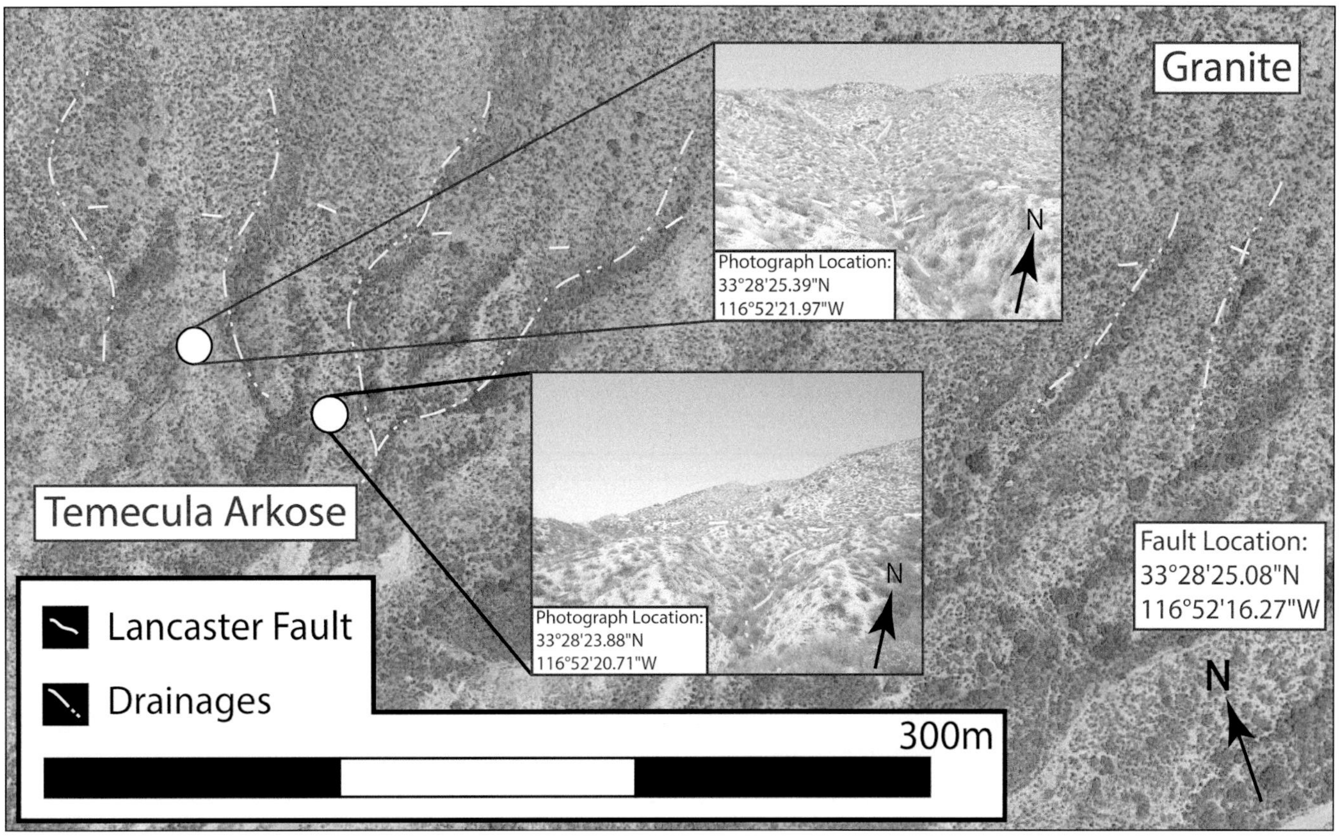

Figure 3. Photographs showing various locations along the Lancaster fault. Note the difference in overall channel incision across the fault as well as the absence of consistent channel offset.

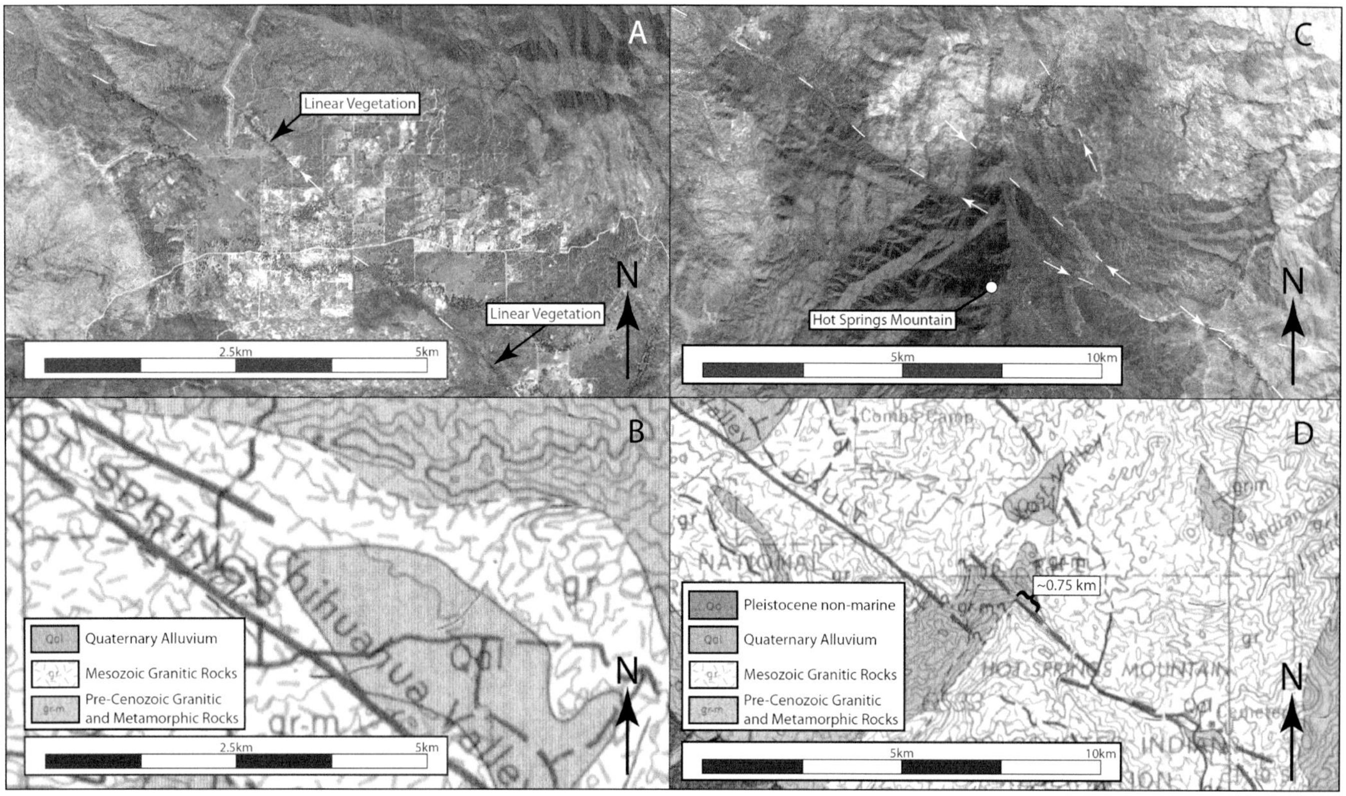

Figure 4. GoogleEarth images and corresponding geologic maps of parts of the Hot Springs fault. (A and B): The Hot Springs fault is observable due to linear vegetation and erosional ravines through Mesozoic granitic rock. (C and D): Images showing a location N/NE of Hot Springs Mountain where the Hot Springs fault causes an ~0.75 km offset in Pre-Cenozoic granitic and metamorphic rock. Geologic maps are adapted from Rogers (1965).

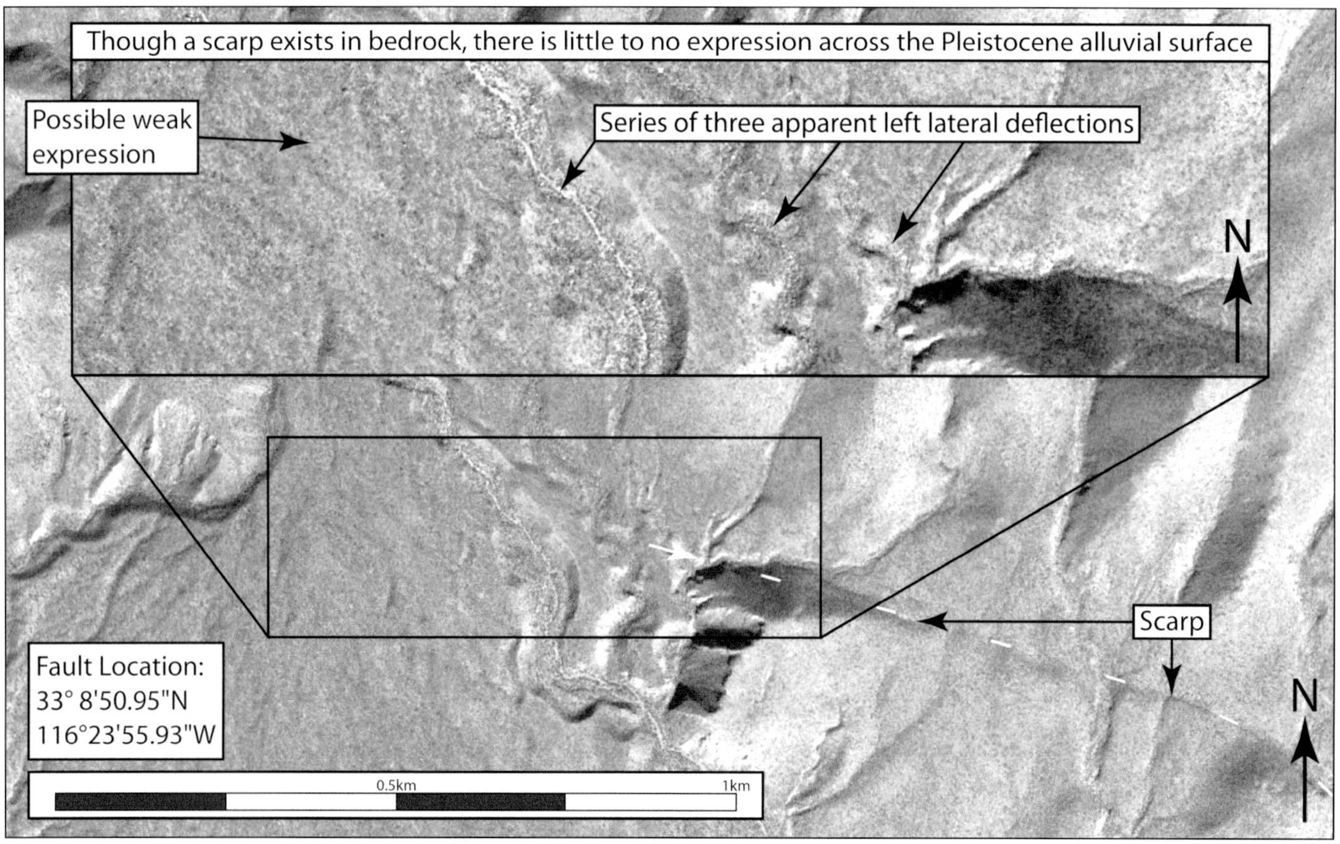

Figure 5. Fault scarp of the San Felipe fault at Yaqui Ridge. Though a scarp is identified in the older crystalline bedrock, little to no expression is observed in the Pleistocene alluvial surface. A series of three apparent linear left-lateral offsets are observed, but are counter to the right lateral nature of the San Felipe fault. We interpret these to not be fault related.

In marked contrast to all fault strands associated with the Lancaster, Hot Springs, Agua Caliente and San Felipe faults, the Earthquake Valley fault displays abundant evidence for youthful faulting, and exhibits evidence for a significant long-term slip rate. There are prominent freshened scarps and right lateral offset in channels cutting through young Quaternary alluvium. Two locations show localized uplift (pressure ridges) in late Quaternary deposits at small left steps along the fault east of Lake Henshaw in Warner Basin. Along the same stretch of fault, there are linear closed depressions (sags) and scarps in Holocene alluvium. At Big Lake in Warner Basin, which is also interpreted as a sag depression, the fault was trenched and shown to have ruptured at least three times in the past 2 ka (Rockwell et al., 2013). In this same area, sediments containing the Bishop Tuff are interpreted to have been offset 1.9 km, yielding a middle Pleistocene to present slip rate of about 2.5 mm/yr (Rockwell et al., 2013).

In Earthquake Valley, located about ~18 km to the southeast of Warner Basin, there are a series of offset drainages that show consistent right-lateral deflections (Figure 6), as well as beheaded drainages, scarps in Holocene alluvium, and ponded alluvium. As the Earthquake Valley fault extends to the southeast and approaches Whale Peak, where the main trace of the fault ends, the fault is visible as a series of linear pressure ridges cropping out of late Quaternary alluvium just NE of the community of Shelter Valley and Highway S2 (Figure 7). These pressure ridges, as well as the consistent right-lateral offset throughout the length of the Earthquake Valley fault, provide strong geomorphological evidence that the fault is presently active and accommodating significant late Quaternary slip.

Though the main trace of the Earthquake Valley fault disappears in the vicinity of Whale Peak, a large number of much smaller faults were identified in GoogleEarth imagery in and surrounding this area to the northeast (Figure 8, and electronic supplement). Despite their smaller size, many of these faults provide evidence for young faulting in the form of fresh scarps in late Quaternary alluvial deposits (Figure 9). Several faults are also expressed as uphill-facing scarps in the crystalline rock that comprises the mountains surrounding and including Whale Peak. One such scarp to the west of Whale Peak shows changes in the slope of the exposed scarp surface (a bevel), providing evidence for multiple, late Quaternary slip events (Figure 10).

Vallecito-Fish Creek Mountains

In the Vallecito and Fish Creek Mountain areas, we mapped numerous faults in GoogleEarth imagery that show evidence of late Pleistocene or Holocene activity (see electronic supplement), similar to those identified around Whale Peak (Figure 8). Evidence includes the presence of linear freshened surfaces and scarps that

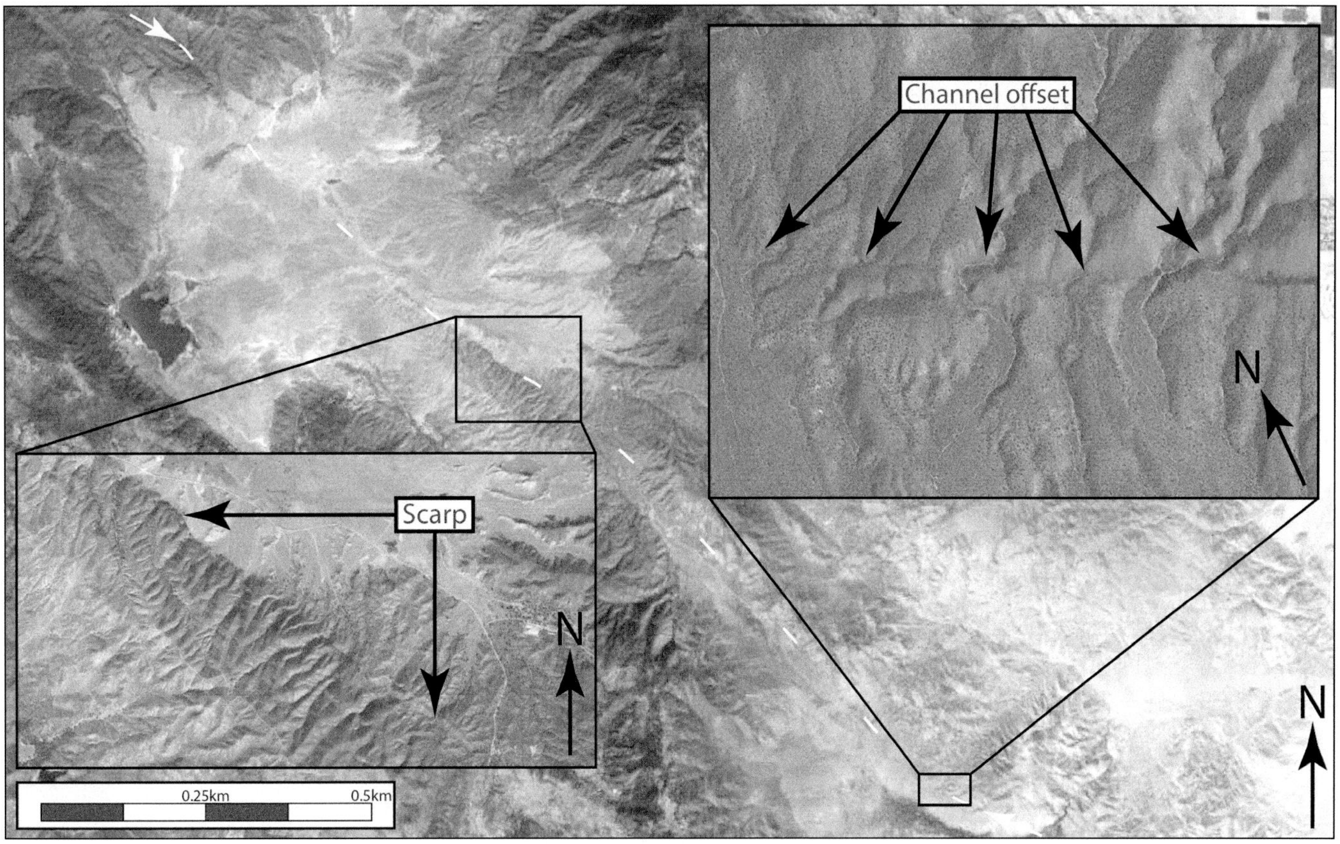

Figure 6. Section of the Earthquake Valley fault east of Lake Henshaw showing prolific scarps and consistent right-lateral channel offset indicating a young age for fault activity.

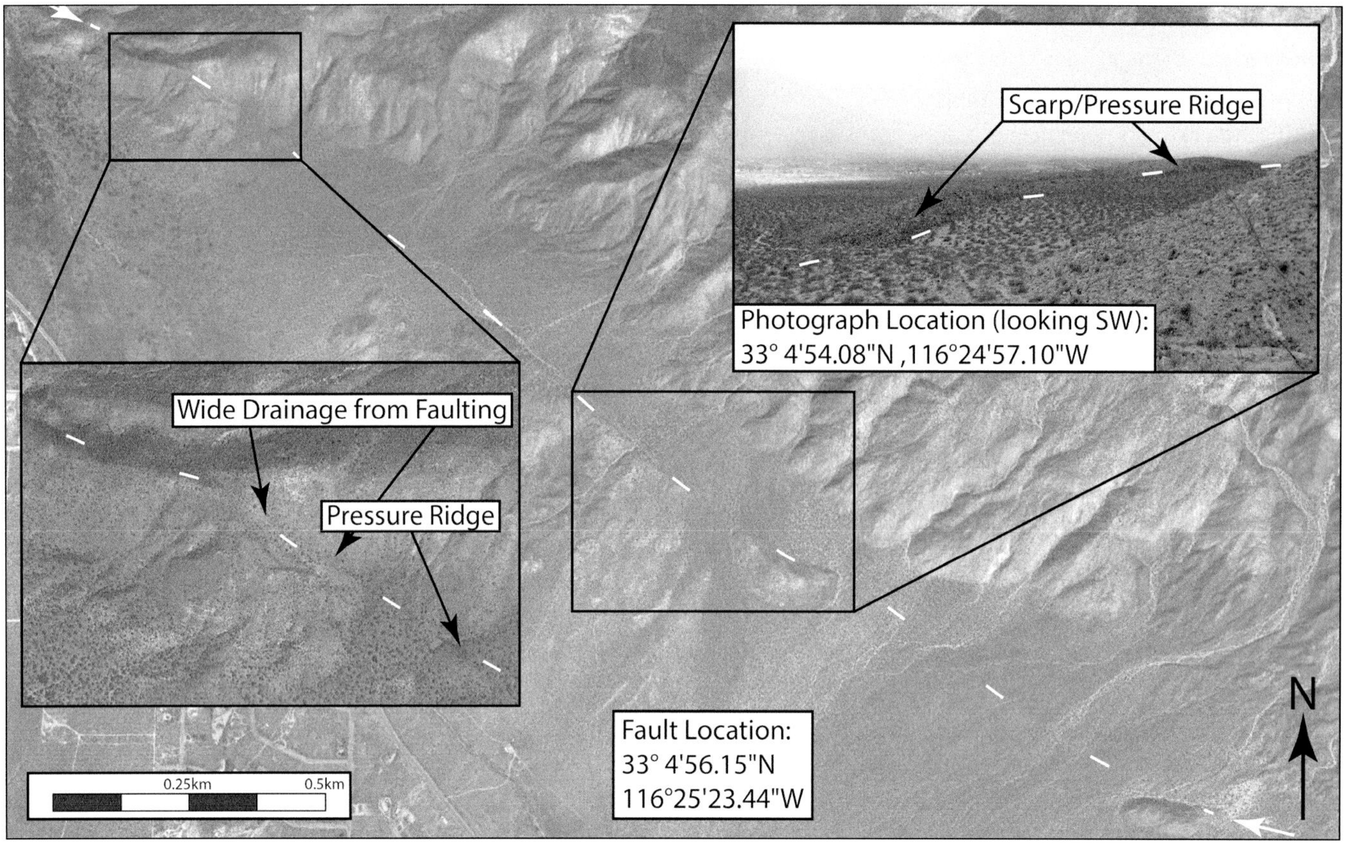

Figure 7. Figure showing the Earthquake Valley fault just NE of the city of Banner, trending towards Whale Peak to the SE. The fault is expressed as a series of pressure ridges cropping out of Quaternary alluvium.

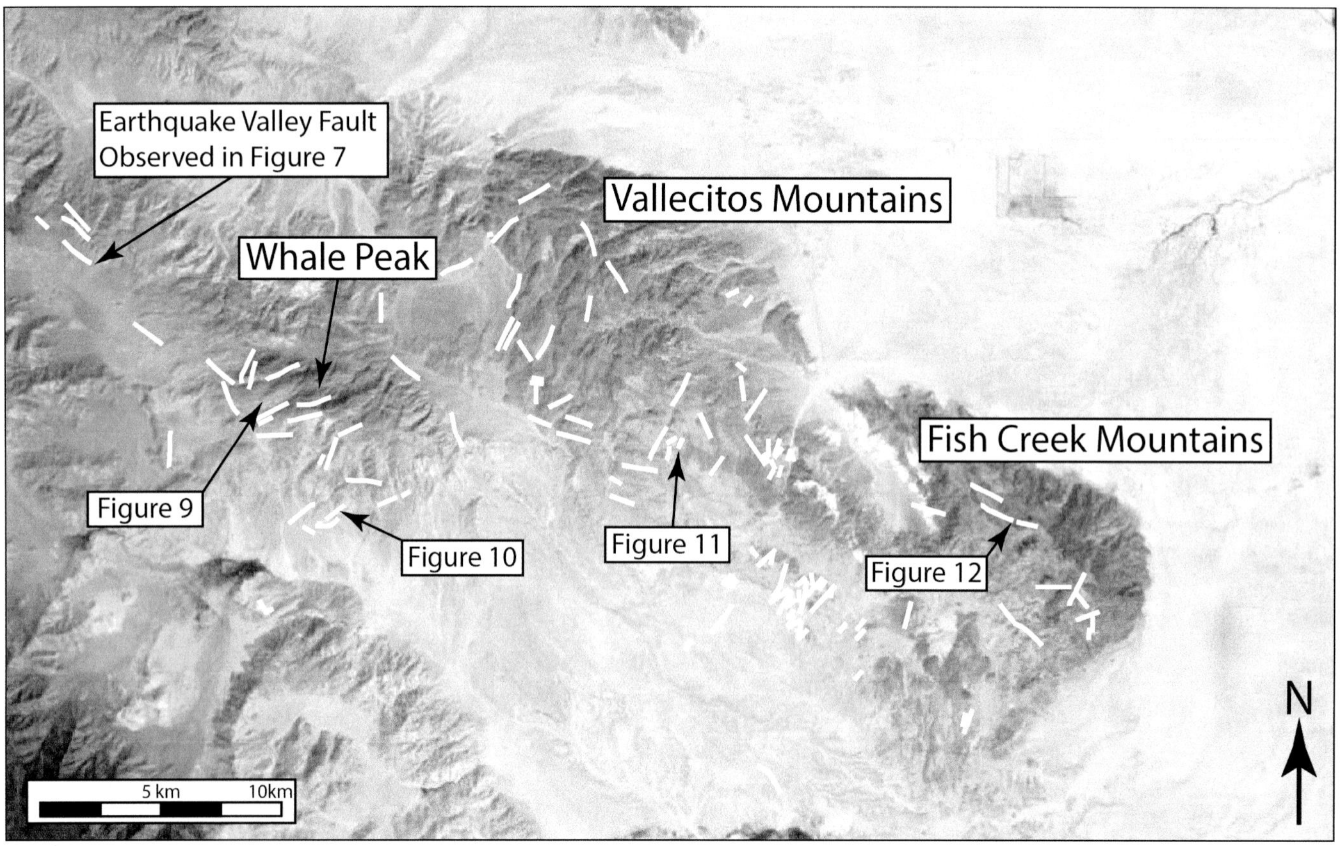

Figure 8. Faults identified within the area surrounding Whale Peak, the Vallecito Mountains, and Fish Creek Mountains which collectively act as a mechanism of transferring slip from the Elsinore to San Jacinto faults. Two fault trends are identified in the rose diagram shown in figure 13.

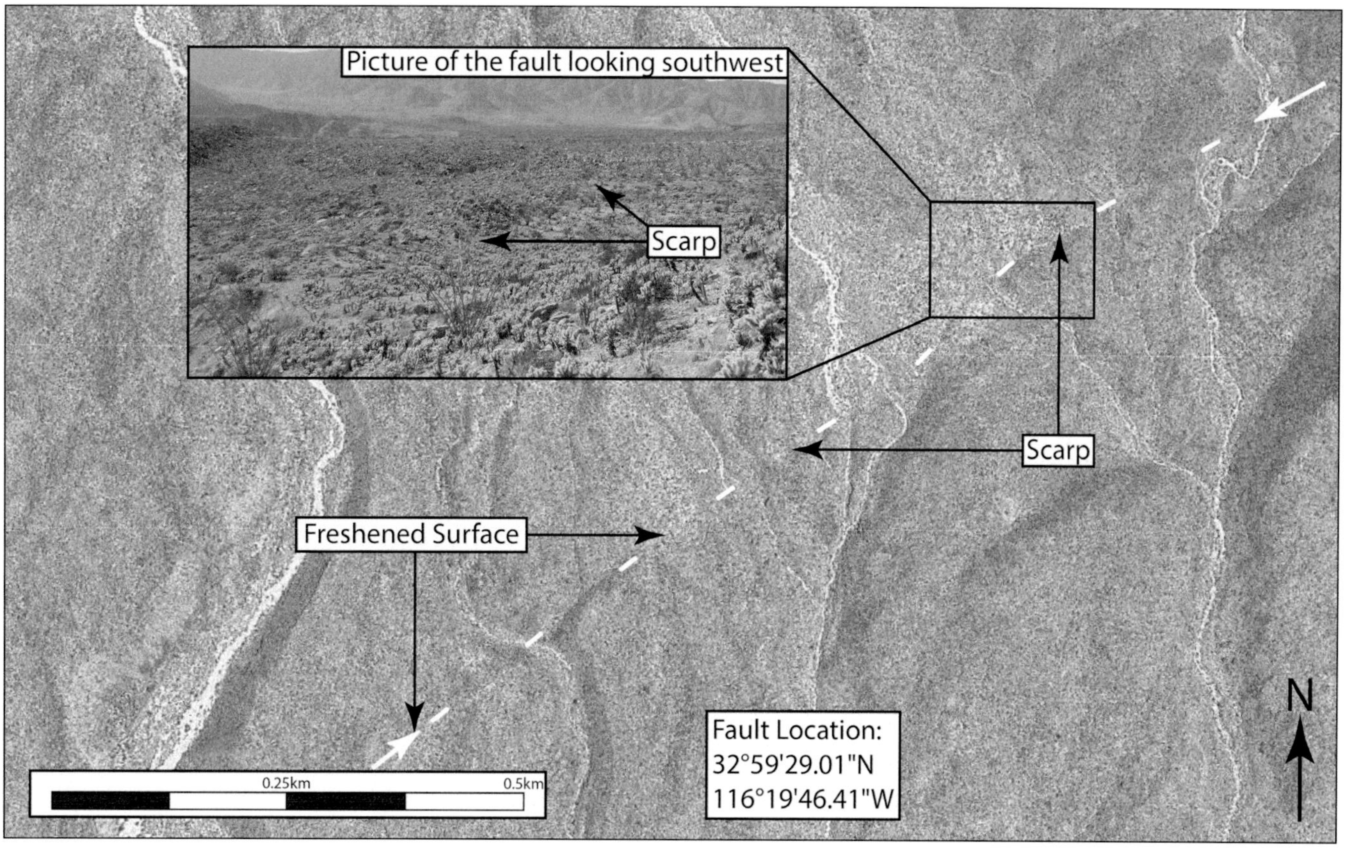

Figure 9. Fault scarp cutting across older alluvium. The photograph shows a point in which the scarp juxtaposes older alluvium against younger.

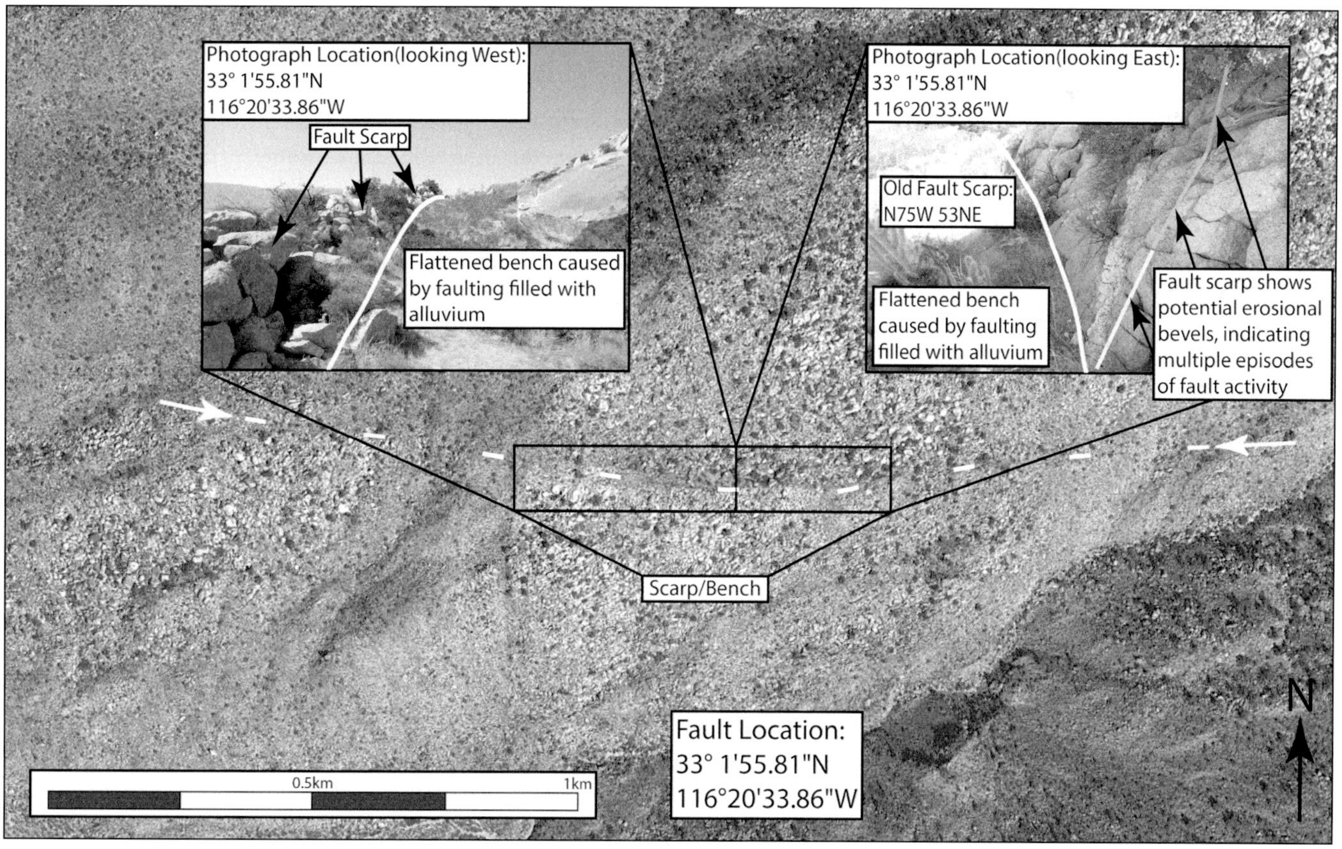

Figure 10. Uphill facing scarp west of Whale Peak capturing alluvium creating a flattened bench. Erosional bevels on the scarp face indicate multiple faulting events.

are interpreted as mole-tracks, offsets of rills and small drainages, scarps in alluvium and bedrock, and to a lesser degree linear vegetation in Quaternary alluvium and crystalline basement. Twenty three of these faults were selected and documented in the field to confirm their late Pleistocene or Holocene activity (see electronic supplement). Special attention was given to faults that cut through late Quaternary deposits, providing direct evidence of young activity (Figures 11 and 12). All recognized young faults from Whale Peak through the Vallecito and Fish Creek mountains in Figure 9 were photographed, and the photos were annotated and organized into geographical groups to produce a digital compendium that is included as an electronic supplement to this study.

Faults identified in the HSSF from the Lancaster fault southeast to Yaqui Ridge, as well as the mapped strands of the Elsinore and San Jacinto faults, and the faults identified in the Vallecito/Fish Creek mountains area in Figure 9, were subdivided into half kilometer segments and measured for strike. Strike measurements from the HSSF and Vallecito-Fish Creek Mountains were plotted in rose diagrams and compared to the strike of the principal strands of the Elsinore and San Jacinto zones, respectively (Figure 13). These rose diagrams show that the HSSF faults have an average strike of approximately N63W, which is 13-180 more westerly than the average strikes of the Elsinore fault (N50W) and San Jacinto fault (N45W) zones (Figure 13a). In comparison, faults within the Vallecito-Fish Creek Mountains have two orthogonal components with fault strikes that average N34E and N55W. The northwesterly component is similar to that of the Elsinore and San Jacinto faults and supports the idea that it is part of the modern tectonic regime. The northeasterly component is consistent as conjugate Reidel faults that together with the northwest-striking faults, form a conjugate set that act together to accommodate slip transfer and block rotation.

DISCUSSION

Age of Fault Initiation

The timing of initiation of the various elements of the San Jacinto and Elsinore faults remains a topic of debate. However, there is clear evidence that the Lancaster and Aguanga strands of the HSSF fault zone originated in the Pliocene, as the basin formed by the step between these two faults is filled with the Pliocene-aged Temecula Arkose (Pajak et al, 1996; Hutchison, 1987; Hull and Nicholson, 1992). Furthermore, the entire HSSF zone or group of faults is currently inactive or active at a very low rate, which indicates that they are likely older structures and no longer important players in the distribution of slip in the southern San Andreas fault system. This point is important from the perspective of the age of initiation of both the San Jacinto and Elsinore faults in southern California. The presence of a Pliocene element of the southern San Andreas fault system invalidates the assertion that onshore faults west of the San Andreas fault formed in the Quaternary during a major plate reorganization (Steely et al., 2009; Lutz et al., 2006; Kirby et al., 2007; Balgarde and Janecke, 2008; Janecke et al., 2010; Dorsey et al., 2012). The HSSF zone does have a more westerly trend than the currently active Elsinore and San Jacinto fault strands, but it also represents an earlier fault system that transferred slip from the northern Elsinore-Whittier fault zone to the southern San Jacinto fault zone, implying that both were in existence by late Pliocene to earliest Pleistocene time. Other fault strands within the Elsinore and San Jacinto fault zones may have evolved at later times, which explains why the active system of faults have more northwesterly strikes than fault strands of the HSSF fault zone.

The slip rate inferred for the San Felipe fault by Steely et al. (2009) was based on both an interpreted right lateral offset of the West Salton Detachment of 5.8 km and an inferred age of inception of 1.1-1.4 Ma. This age estimation of 1.1–1.4 Ma is inferred from a sequence stratigraphic reassessment of a conglomerate formation on the other side of Yaqui Ridge to the east (Lutz et al., 2006; Kirby et al., 2007; Steely et al., 2009), but no new dating has been done to establish the absolute age control of the conglomerate that is inferred to be offset. The inferred offset along with the interpreted mid-Pleistocene age of the offset yielded a rate of about 5 mm/yr for the life history of the San Felipe fault, which is clearly not the current rate, as is evident in the modern landscape. From our observations of fault-related geomorphology, the slip rate on the San Felipe fault is probably an order of magnitude lower that along the Earthquake Valley or Elsinore fault. As the San Felipe fault along Yaqui Ridge is also the only place in the HSSF where significant lateral offset is observed, and there is a lack of significant observable offset elsewhere along the HSSF, we believe this interpretation warrants further study.

Other elements of the HSSF zone appear to have been completely inactive in the late Quaternary, with the Lancaster fault apparently inactive for at least several hundred thousand years. Thus, the current rate that the HSSF fault zone accommodates in the modern tectonic environment is close to zero, which is consistent with the fault zone being an older element of the southern San Andreas fault system.

In contrast, the recent estimate of the late Quaternary slip rate for the Earthquake Valley fault is about 2.5 mm/yr, based on a 1.9 km offset of Warner Basin alluvial fill that contains the 800 ka-aged Bishop Tuff (Rockwell et al, 2013). Our study of the geomorphology along the Earthquake Valley fault supports the interpretation that this is an active and relatively fast-moving fault, as evidenced by the consistent right lateral offsets of rills and erosional drainages in Quaternary alluvium, abundant scarps in late Quaternary alluvium, the presence of pressure ridges and sags in late Pleistocene and Holocene alluvium (Figures 6 and 7). These observations indicate that the Earthquake Valley fault is actively accommodating slip from Warner Basin along the southeast flank of Palomar Mountain southeast to Whale Peak (Figure 1).

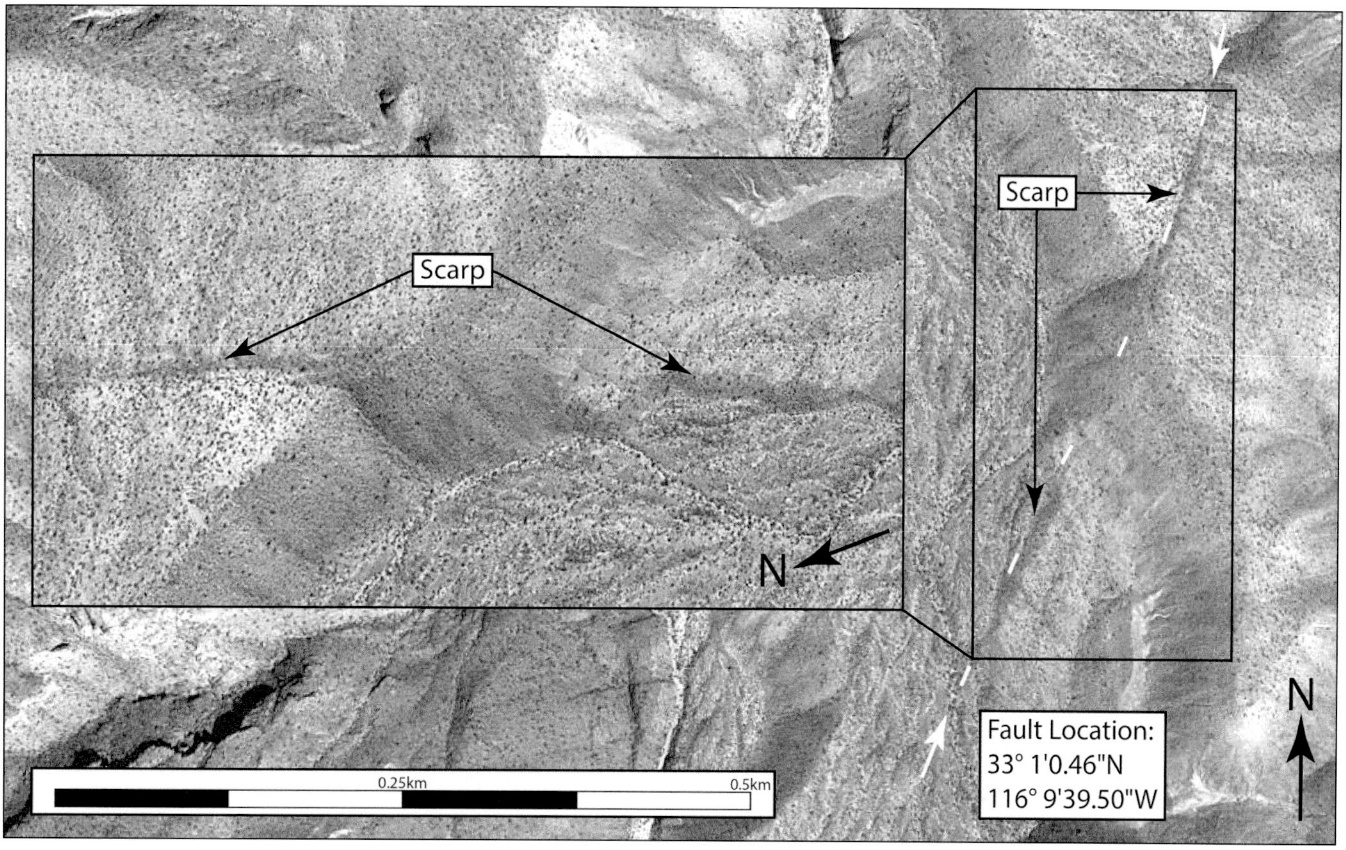

Figure 11. Example of a prominent scarp cutting through two alluvial surfaces. The younger, lower alluvial surface is further eroded by an active wash to the west.

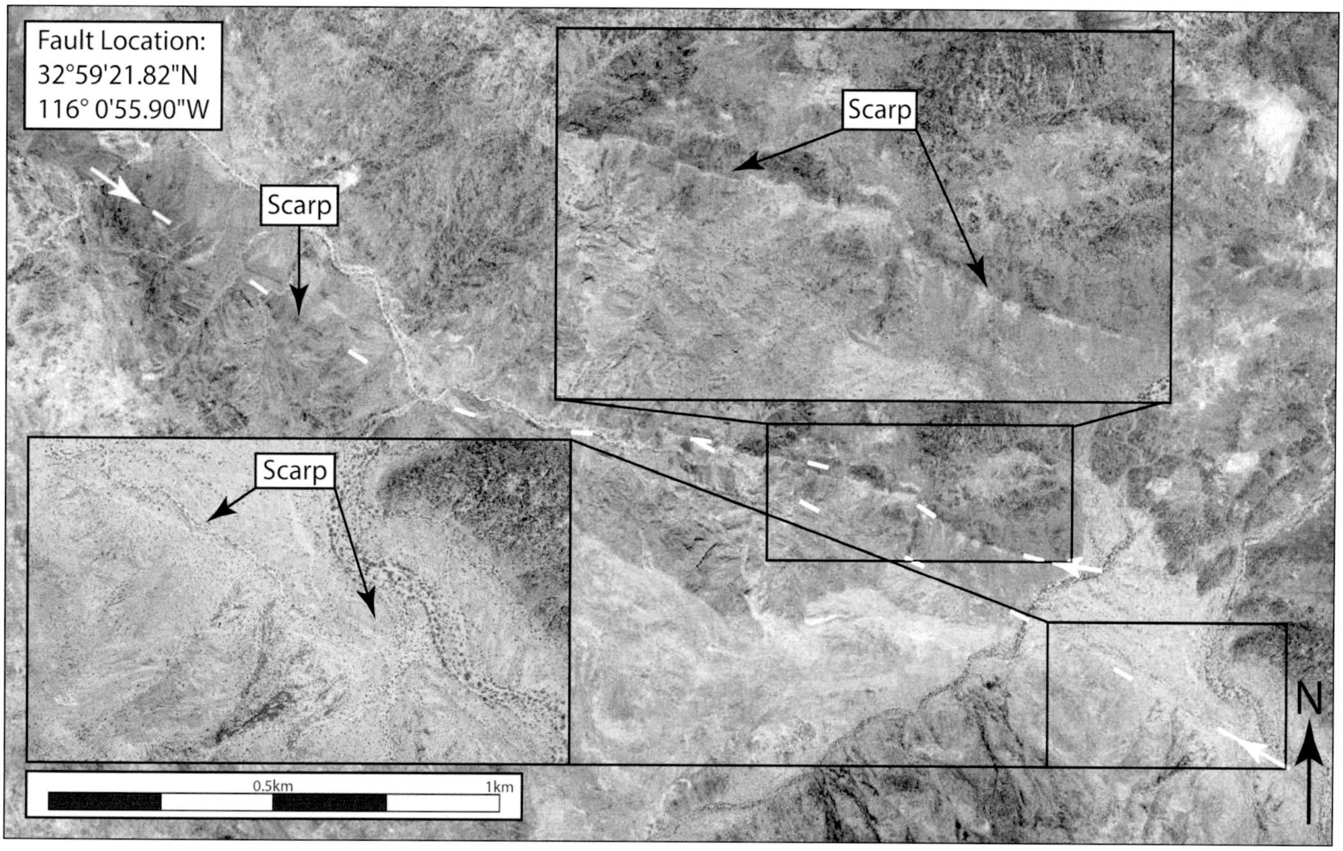

Figure 12. Example of a fault in the Vallecito Mountains showing a fault with scarps in both granitic rock, zoomed in above, as well as Holocene alluvium, zoomed in at the bottom left.

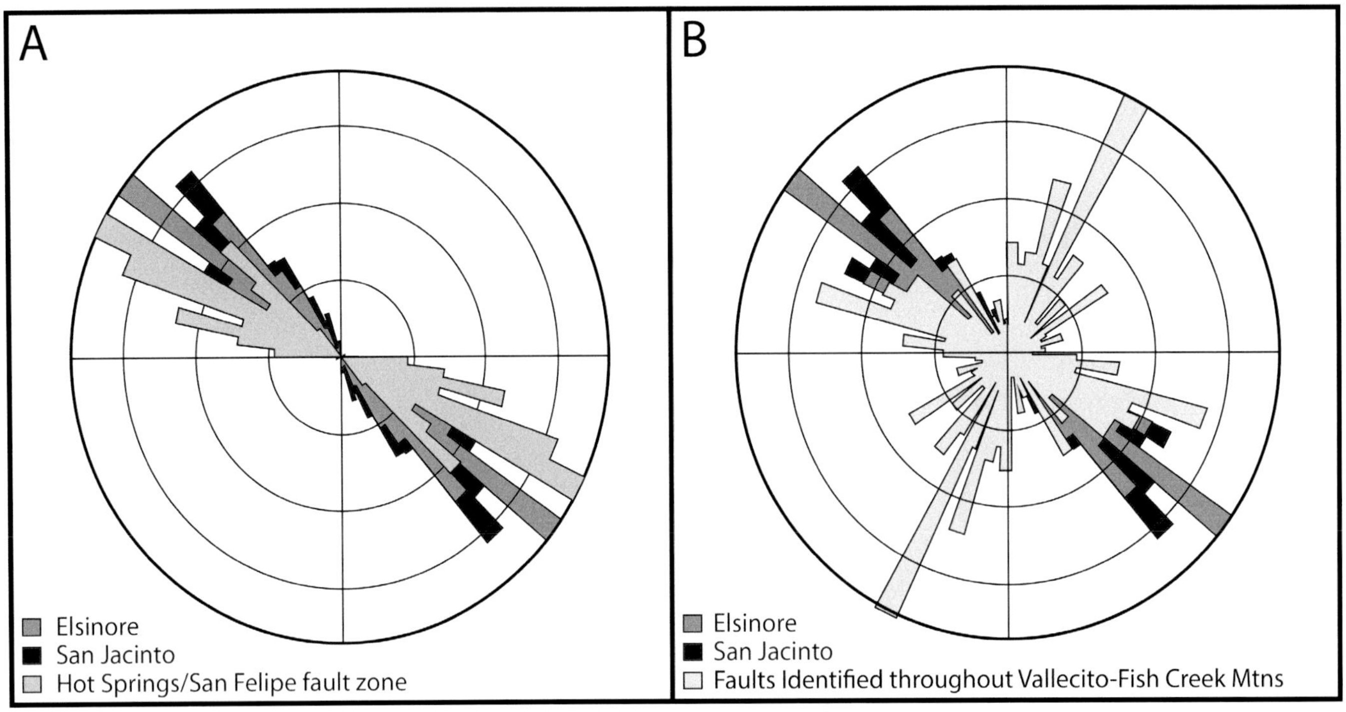

Figure 13. Rose diagrams comparing strikes of major fault systems in southern California. (A) The Hot Springs/San Felipe fault system shows a more western trend to the Elsinore and San Jacinto. (B) Faults identified between the Elsinore and the San Jacinto in the Vallecito/Fish Creek Mountains have two orthogonal components, one of which trends more westerly compared to the Hot Springs/San Felipe Mountains system, and another to the northeast.

The slip rate decrease along the Elsinore fault is inferred to occur southeast of Agua Tibia Mountain (Magistrale and Rockwell, 1996), as the slip rate along the northern Elsinore fault is estimated at about 5 mm/yr (Millman and Rockwell, 1986; Vaughan and Rockwell, 1986; Rockwell et al., 2000) whereas the total fault displacement and inferred slip rate along the central and southern Elsinore fault is less at about 2.7 km and 2–3 mm/yr, respectively (Lampe, 1977; Fletcher et al., 2011). This decrease occurs in the same region that the Earthquake Valley fault becomes expressed in the modern topography, which we infer to mean that the northern Elsinore slip rate is partitioned south of Agua Tibia and Palomar mountains between the Elsinore and Earthquake Valley faults, as first suggested by Magistrale and Rockwell (1996). This, in turn, explains the Palomar and Agua Tibia Mountain topographic uplift as a large pressure ridge system that results from the transfer of slip from the Elsinore to the Earthquake Valley fault.

In a similar fashion, the Earthquake Valley fault loses expression in the southeast near Whale Peak. To the east and northeast of the Earthquake Valley fault, we recognize a large number of minor faults that all express evidence of late Quaternary activity with the presence of scarps in alluvium or bedrock. These faults occur as a conjugate set, with a northeast-trending group with an average strike of about N34E, and a northwest-trending group with an average strike of about N55W, similar to that of the Elsinore fault. Many of the northeast-trending faults terminate in the Vallecito-Fish Creek Mountain area, and others to the south of the Fish Creek Mountains extend the entire distance between the Elsinore and San Jacinto faults. From this, we make two interpretations. First, the Vallecito–Fish Creek Mountains are the result of a restraining step through which the Earthquake Valley slip is transferred east to the southern San Jacinto fault zone. This is important because it explains the anomalously elevated topography of the mountains in a region that is generally interpreted to be subsiding, and because it also explains why geodetic observations suggest that the southern San Jacinto slip rate is higher than to the north (Fialko, 2006). The second is that the conjugate set of faults, along with the northeast-striking faults between the Elsinore and San Jacinto faults, is best explained as accommodating the broad transfer of slip from the southern Earthquake Valley and southern Elsinore faults to the southern San Jacinto fault through block rotation. This is also important in the interpretation of geodetic data, as deformation will not be seen as concentrated on a single fault zone, but would appear more distributed and virtually indistinguishable from the elastic deformation of the crust.

CONCLUSION

We use new geomorphological observations of evidence for fault activity along the Hot Springs/San Felipe and Earthquake Valley fault zones, and minor faults in the Vallecito-Fish Creek mountain area, to reassess the current activity of faults between the Elsinore and San Jacinto faults, and to better understand strain distribution on this complex fault system. Prior work suggests that the HSSF is the primary active fault zone that transfers slip from the northern

Elsinore fault to the southern San Jacinto fault, and has an inferred age of inception of 1.1–1.4 Ma based on a sequence stratigraphic reassessment of a conglomerate and an apparent ~5.8 km right lateral separation of the West Salton detachment fault. The notable lack of geomorphic indicators of young activity, such as scarps and pressure ridges through Quaternary alluvium, lateral offset of rills and erosional drainages, as well as the presence of very well developed soils capping or sealing the fault in the NE portion of the zone, instead indicate this fault zone has not been significantly active in the middle to late Quaternary. Furthermore, activity on the Lancaster and Aguanga strands of the HSSF fault zone during the Pliocene, resulting in deposition of the Temecula Arkose, demonstrates that the HSSF zone is much older than previously inferred. In turn, this may indicate that the ~5.8 km right-lateral separation of the West Salton detachment fault inferred for the San Felipe fault is not the result of Quaternary lateral offset, but rather, is a tear fault in the detachment sheet. The meager evidence for Quaternary activity of the San Felipe fault supports this alternative model.

In marked contrast, the Earthquake Valley fault expresses many geomorphological characteristics that indicate modern activity at a significant rate. The northern end of the Earthquake Valley fault corresponds to where displacement and slip rate on the Elsinore fault decreases along the southern flank of the Palomar and Agua-Tibia Mountain uplift. We therefore interpret that the uplift itself is a consequence of slip being transferred from the Elsinore to the Earthquake Valley fault, making the Palomar and Agua Tibia Mountains a large pressure ridge system.

Similarly, the Earthquake Valley fault loses expression in the vicinity of Whale Peak, northeast of which we identify and map a number of minor faults throughout the Vallecito-Fish Creek mountains. These minor faults collectively show two prominent sets; one which strikes similar to the Elsinore fault zone, and another that strikes to the northeast. From Whale Peak and to the south, faults with northeast strikes collectively extend the distance between the Elsinore fault to the San Jacinto fault. Together these observations indicate that not only are these minor faults related to slip transfer between the Elsinore and San Jacinto faults, but they also act as a conjugate set of faults that accommodate block rotation between the northwest-striking master faults. Further, we attribute the uplift of the Vallecito-Fish Creek mountains to also be the result of slip transfer between the Earthquake Valley fault and southern San Jacinto fault, and is therefore interpreted as a large pressure ridge system, similar to the Palomar-Agua Tibia Mountain uplift.

ACKNOWLEDGEMENTS

We thank the many landowners and the Bureau of Land Management for field access. We also thank Will Munson and Cheryl Martin for assistance with field work.

REFERENCES CITED

Axen, G.J., and Fletcher, J.M., 1998, Late Miocene-Pleistocene extensional faulting, Northern Gulf of California, Mexico and Salton Trough, California: International Geology Review, v. 40, p. 217–244.

Becker, T. W., Hardebeck, J. L., and Anderson, G., 2005, Constraints on fault slip rates of the southern California plate boundary from GPS velocity and stress inversions, Geophys. J. Int. 160, 634–650.

Bennett, R.A., Rodi, W., Reilinger, R.E., 1996, GPS constraints on fault slip rates in southern California and northern Baja, Mexico: Journal of Geophysical Research, v. 101, no. B10, p. 21,943–21,960.

Belgarde, B.E., 2007, Structural characterization of the three SE segments of the Clark fault, Salton Trough, California [M.S. thesis]: Utah State University, 4 plates, map scale 1:24,000, 216 p.

Blisniuk, K., Oskin, M., Mériaux, A.-S. , Rockwell, T., Finkel, R. C., Ryerson, F. J., 2013, Stable, rapid rate of slip since inception of the San Jacinto fault, California, Geophys. Res. Lett., 40, p. 4209–4213, doi:10.1002/grl.50819.

Blisniuk, K., Rockwell, T., Owen, L. A., Oskin, M., Lippincott, C., Caffee, M. W., Dortch, J., 2010, Late Quaternary slip rate gradient defined using high-resolution topography and ^{10}Be dating of offset landforms on the southern San Jacinto fault zone, California, J. Geophys. Res., 115, B08401, doi: 10.1029/2009JB006346.

Dorsey, R. J., Axen, G. J., Peryam, T. C., Kairouz, M. E., 2012, Initiation of the Southern Elsinore fault at ~1.2 Ma: Evidence from the Fish Creek–Vallecito Basin, southern California, Tectonics, 31, TC2006, doi:10.1029/2011TC003009.

Dorsey, R. J., Roering, J. J., 2006, Quaternary landscape evolution in the San Jacinto fault zone, Peninsular Ranges of Southern California: Transient response to strike-slip fault initiation, Geomorphology, v. 73, p. 16–32, doi: 10.1016/j.geomorph.2005.06.013.

Fialko, Y., 2006, Interseismic strain accumulation and the earthquake potential on the southern San Andreas fault system: Nature, v. 441, p. 968–971, doi: 10.1038/ nature04797.

Fletcher, K. E. K., Rockwell, T. K., Sharp, W. D., 2011, Late Quaternary slip rate of the southern Elsinore fault, Southern California: Dating offset alluvial fans via 230Th/U on pedogenic carbonate, J. Geophys. Res., 116, F02006, p. 1-11, doi: 10.1029/2010JF001701.

Hull, A. G. and Nicholson, C., 1992, Seismotectonics of the northern Elsinore fault zone, southern California, Bull. Seism. Soc. Am. 82, pp. 800–818.

Hutchison, J. H, 1987, Moles of the Scapanus Latimanus Group (Talpidae, Insectivora) from the Pliocene and Pleistocene of California, Contributions in Science, no. 386, pp. 1–15.

Jahns, R. H., 1954, Geology of the Peninsular Ranges province, southern California and Baja California, in Geology of Southern California, R. H. Jahns (Editor), Calif. Div. Mines Geol. Bull. 170, p. 29–52.

Janecke, S., Dorsey, B., Belgarde, B., 2010, Age and Structure of the San Jacinto and San Felipe fault zones, and their lifetime slip rates, Clifton, H. E., and Ingersoll, R. V., eds., Geologic excursions in California and Nevada: tectonics, stratigraphy, and hydrogeology: Pacific Section, SEPM (Society for Sedimentary Geology), Book 108, p. 233–271.

Johnson, N. M., Officer, C. B., Opdyke, N. D., Woodard, G. D., Zeitler, P. K., Lindsay, E. H., 1983, Rates of late Cenozoic tectonism in the Vallecito-Fish Creek basin, western Imperial Valley, California: Geology, v. 11, p. 664–667.

Kennedy, M. P., 1977, Recency and character of faulting along the Elsinore fault zone in southern Riverside County, California, Calif. Div. Mines Geol. Spec. Rep. 131, 12 pp.

Kirby, S.M., Janecke, S.U., Dorsey, R.J., Housen, B.A., McDougall, K., Langenheim, V., and Steely, A., 2007, Pleistocene Brawley and Ocotillo

Formations: Evidence for initial strike-slip deformation along the San Felipe and San Jacinto fault zones, California: The Journal of Geology, v. 115, no. 1, p. 43–62, doi: 10.1086/509248.

Lampe, C. M., 1988, Geology of the Granite Mountain area: Implications of the extent and style of deformation along the southeast portion of the Elsinore fault, MS thesis, 150 pp., San Diego State Univ., San Diego, Calif.

Li, Q., and Liu, M., 2006, Geometrical impact of the San Andreas fault on stress and seismicity in California, Geophys. Res. Lett., 33, L08302, doi:10.1029/2005GL025661.

Li, Q., and Liu, M., 2007, Initiation of the San Jacinto fault and its Interaction with the San Andreas fault: Insights from Geodynamic Modeling: Pure appl. geophys., doi: 10.1007/s00024-007-0262-z

Lutz, A.T., Dorsey, R.J., Housen, B.A., and Janecke, S.U., 2006, Stratigraphic record of Pleistocene faulting and basin evolution in the Borrego Badlands, San Jacinto fault zone, Southern California: Geological Society of America Bulletin, v. 118, no. 11, p. 1377–1397, doi: 10.1130/B25946.1

Machette, M. N., 1985, Calcic soils of the southwestern United States: Geological Society of America, Special paper 203, p. 1–21

Magistrale, H., Rockwell, T., 1996, The central and southern Elsinore fault zone, southern California: Bulletin of the Seismological Society of America, v. 861, p. 791–803.

Matti, J.C., Morton, D.M., 1993, Paleogeographic evolution of the San Andreas fault in southern California: A reconstruction based on a new cross fault correlation, *in* Powell, R.E., et al., eds., The San Andreas fault system: Displacement, palinspastic reconstruction, and geologic evolution: Geological Society of America Memoir 178, p. 107–159.

Millman, D. E. and T. K. Rockwell, 1986, Neotectonics of the Elsinore fault in Temscal Valley, California, in *Neotectonics and Faulting in Southern California,* prepared for the 82nd annual meeting of the Cordilleran section of the Geological Society of America, P. Ehlig (Compiler), 159–166.

Mueller, Karl J. and Rockwell, Thomas K., 1995, Late Quaternary activity of the Laguna Salada fault in northern Baja California, Mexico: *Geological Society of America Bulletin*, v. 107, no. 1, pp. 8–18.

Pajak, A. F., Scott, E., Bell, C. J., 1996, A review of the biostratigraphy of Pliocene and Pleistocene sediments in the Elsinore fault zone, Riverside County, California: PaleoBios, v. 17, no, 2–4, p. 28–49.

Powell, R. E., & Weldon II, R. J., 1992. Evolution of the San Andreas fault.: Annu. Rev. Earth Planet. Sci. v. 20, 431–468.

Rockwell, T., Akciz, S., and Gordon, E., 2013, Paleoseismology of the Aqua Tibia - Earthquake Valley fault, eastern strand of the Elsinore fault zone. Seismological Research Letters, v. 84, p. 332.

Rockwell, T., Bergmann, M., Kenney, M., 2000, Holocene slip rate of the Elsinore fault in Temecula Valley, Riverside County, California, in Geology and Enology of the Temecula Valley, edited by B. Birnbaum and K. Cato, pp. 105–118, San Diego Assoc. of Geol., San Diego, Calif.

Rockwell, T. K., Dawson, T. E., Ben-Horin, J. Y., Seitz, G., 2014, A 21-Event, 4,000-Year History of Surface Ruptures in the Anza Seismic Gap, San Jacinto fault, and Implications for Long-term Earthquake Production on a Major Plate Boundary fault, Pure Appl. Geophys., doi: 10.1007/s00024-014-0955-z

Rockwell, T., Loughman, C., Merifield, P., 1990, Late Quaternary rate of slip along the San Jacinto fault zone near Anza, Southern California: Journal of Geophysical Research, v. 95, p. 8593–8605, doi: 10.1029/ JB095iB06p08593.

Rogers, T. H., Compiler, 1965, Santa Ana Sheet, third printing, 1976, Geologic Map of California, Olaf P. Jenkins Edition.

Sanders, C., Magistrale, H., 1997, Segmentation of the northern San Jacinto fault zone, southern California. J. Geophys. Res. 102, B12, 27,453–27,467.

Sharp, R.V., 1967, San Jacinto fault zone in the Peninsular Ranges of southern California: Geological Society of America Bulletin, v. 78, p. 705–730, doi: 10.1130/0016-7606(1967)78[705:SJFZIT]2.0.CO;2.

Shirvell, C. R., Stockli, D. F., Axen, G. J., Grove, M., 2009, Miocene-Pliocene exhumation along the west Salton detachment fault, southern California, from (U-Th)/He thermochronometry of apatite and zircon, Tectonics, 28, TC2006, doi: 10.1029/2007TC002172.

Steely, A. N., S. U. Janecke, R. J. Dorsey, and G. J. Axen, 2009, Early Pleistocene initiation of the San Felipe fault zone, SW Salton Trough, during reorganization of the San Andreas fault system, Geol. Soc. Am. Bull., 121, 663–687, doi:10.1130/B26239.1.

Sundell, K., 2009, Late Quaternary Activity of the Earthquake Valley fault, Senior Thesis, University of California Davis.

Vaughan, P. and Rockwell , T., 1986, Alluvial stratigraphy and neotectonics of the Elsinore fault zone at Agua Tibia Mountain, southern California, in *Neotectonics and Faulting in Southern California,* prepared for the 82nd annual meeting of the Cordilleran section of the Geological Society of America, P. Ehlig (Compiler), 177–191.

Vaughan, P. R., Thorup, K. M., Rockwell, T. K., 1999, Paleoseismology of the Elsinore fault at Agua Tibia Mountain, Bulletin of the Seismological Society of America, v. 89, 6, pp. 1447–1457.

Weber, F. H., 1977, Seismic hazards related to geologic factors, Elsinore and Chino fault zones, northwestern Riverside county, California, Calif. Div. Mines GeoL Open-File Rept. 77-4, 96 pp.

Woyski, M. S., D. E. McKee, and L. P. Bisaha (1991). Evidence for 15 km right lateral displacement on the Elsinore fault zone, southern California, GSA Abstracts with Programs (Cordilleran section) 23, 110.

The Distribution of Pediments in the Coyote Mountains, Southern California

BELINDA J. STEVENS (M.A.)
Geography Department, California State University, Chico
DR. PAUL MELCON, *Editor*
Geography Department, California State University, Chico

INTRODUCTION

PEDIMENTS ARE GEOLOGICAL STRUCTURES THAT have been the subject of disputes and disagreements by geomorphologists since their first recognition and description in 1877 in the Henry Mountains of Utah (Dohrenwend, 1994, 321; Hadley 1967, 83; Oberlander 1974, 850; Tator 1952, 295).One issue open to debate is some researchers claim pediments are an erosional feature, while others regarded the structure to be a depositional form (Hadley 1967, 83–89).Pediments develop in many different climatic regimes and geomorphic settings, so it is not surprising that there are a variety of definitions describing the same structure that are similar in appearance (Dohrenwend 1994, 321–322; Oberlander 1974, 868). For my pediment research at the Coyote Mountains in southern California, I adopted John C. Dohrenwend's pediment definition because it focuses on pediments as an erosional feature without any specific lithology, implication of genesis, or processes:

A pediment is "a gently sloping erosional surface developed on bedrock or older unconsolidated deposits. The eroded surface may be subaerially exposed or covered by a continuous or discontinuous veneer of alluvial deposits, and the bedrock may include any lithologic type with any structural attitude" (1994, 323).

One problem stands alone concerning research on pediments in desert environments—the extraordinary difficulties presented by the sites (Cooke 1970, 26). Charles D. Winker and Susan M. Kidwell state that one of the struggles they faced in their field work in the Coyote Mountains was maneuvering through the range's very rugged terrain. This hardship caused access to be extremely limited, making geologic mapping arduous and consequently sparse (1996, 330). The inaccessibility and ruggedness of the terrain have deterred detailed pediment research in and around the range, which may reveal local interactions and regional correlations. My research involved mapping and classifying pediments in the Coyote Mountains, and examines factors in their morphogenesis (Stevens, Belinda Jean 2015).

PEDIMENTS IN GENERAL

Pediments are distributed in many different climatic regimes and geomorphic settings. Although they are most characteristic of arid and semi-arid environments (Denny 1967, 81; Dohrenwend 1994, 321; Oberlander 1974, 868; Royse and Barsch 1971, 3177; Tator 1952, 295; Twidale 1981, 423) they are not found in every desert basin (Cooke, Warren, and Goudie 1993, 191).In the western Basin and Range province, including the southwestern Arizona and southeastern California deserts, areas with smaller ranges (in width, height, relief, area, and volume) are characterized by pediments. Conversely, in areas occupied by larger ranges, alluvial fans are more common than pediments (Cooke, Warren, and Goudie 1993, 191).

Pediments are part of an open erosional and depositional system known as the Pediment Association (See Figure 1) (Cooke, Warren, and Goudie 1993, 192; Dohrenwend 1994, 324;

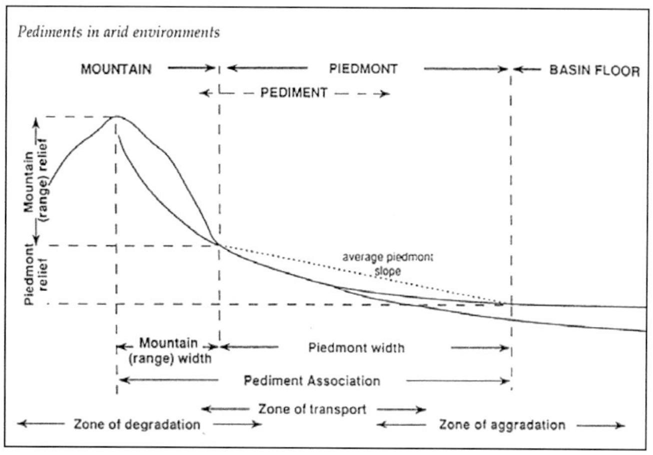

Figure 1. Pediment Association. The pediment association zone boundaries are determined by the rate of debris supply and removal. ***Source***: Data adapted from Dohrenwend, John C. 1994. Pediments in arid environments. In *Geomorphology of Desert Environments*, ed Athol D. Abrahams and Anthony J. Parsons, 324. London: Chapman & Hall.

Oberlander 1974, 851).The Pediment Association includes (1) the mountain area, a zone of degradation; (2) the piedmont area, zone of pediment formation and fluvial transport, and; (3) the basin floor area, zone of aggradation (Cooke 1970, 28; Dohrenwend 1994, 323–324). The boundaries between the three zones are determined by the rate of debris supply and removal. If the rate of debris supply increases or the rate of debris removal decreases, the zones tend to migrate upslope. Conversely, if the rate of debris supply decreases or the rate of removal increases, the zones tend to migrate downslope (Dohrenwend 1994, 342).

Pediments range in size from miniature structures measured in square feet to broad pediplains covering many square miles (Cooke, Warren, and Goudie 1993, 192; Rich 1935, 1023; Tator 1952, 298). Pediments are also referred to as "slopes of transportation," because alluvium is transferred across the pediment surface (Hadley 1967, 84; Twidale 1981, 423). General pediment surface profiles fall into three common shapes: concave or scooped-shape, convex or fan-shaped, or rectilinear (Cooke, Warren, and Goudie 1993, 192; Dohrenwend 1994, 327-328; Hadley 1967, 84; Tator 1952, 303; Twidale 1981, 425; Warnke 1969, 371). Pediment profiles can display one or a combination of all three general forms. The form is influenced by elements such as: (1) drainage area, (2) lithology, (3) precipitation, (4) stream discharge, (5) structural features, (6) size of regolith, and (7) soil cover (Hadley 1967, 84; Warnke 1969, 371).

A pediment slope is defined as the line joining the highest and lowest points of a pediment profile along the line of a pediment association length (Cooke, Warren, and Goudie 1993, 192). The slope has become an important portion of a pediment because it is more susceptible to analytical evaluation than other pediment properties; thus, leading to numerous studies of pediment slopes and hypotheses. For example, Ronald U. Cooke proposed "the slope of pediments associated with faults is significantly steeper than the slope of pediments not associated with faults" (1970, 33). His research in the western Mojave Desert confirmed that tectonic activity may be a factor in determining the general slope of pediments (Cooke, Warren, and Goudie 1993, 193). Pediment inclination measurements do not vary with different regions. Studies from South Australia report pediment gradients range from 1 to 6 degrees, with averages between 3 and 4 degrees (Bourne and Twidale 1998, 124). South Africa pediment gradients average about 5 degrees (Hadley 1967, 84), and the southwestern United States pediment gradient measurements range from 0.5 to 11 degrees (Cooke 1970, 28; Cooke, Warren, and Goudie 1993, 192; Dohrenwend 1994, 327; Hadley 1967, 84; Royse and Barsch 1971, 3177; Sharp 1940, 357). Pediment gradients appear to be steeper among more erosion resistant coarser rock types and gentler on less friable substrates. As well, profiles are steeper opposite intercanyon areas than below the canyon mouths of a mountain, and steeper along smaller streams than along larger ones due to lithological influences instead of factors concerning stream size (Tator 1952, 302).Two studies one from the Ruby-East Humboldt Range, Nevada, and the second from Mojave Desert and Sonoran Desert, Arizona report the causes of pediment gradient diversities included : (1) the amount and coarseness of debris, (2) permeability of underlying rocks, (3) periodicity of precipitation, (4) volume of water, (5) proximity to the mountain slope, and (6) the effects of tectonic activity for the differences in pediment slope gradients (Sharp 1940, 357, Mammerickx 1964, 427–431).

Pediments are characteristic of arid and semi-arid environments (Denny 1967, 81; Dohrenwend 1994, 321; Oberlander 1974, 868; Royse and Barsch 1971, 3177; Tator 1952, 295; Twidale 1981, 423), yet evidence is accumulating that pediments are ancient surfaces inherited from very different conditions than are currently operating (Cooke, Warren, and Goudie, 1993, 190; Oberlander 1974, 853). It is possible that some researchers are making cause-and-effect deductions from relations between a deductive process and a visible feature. There may be some confusion between historical pediment-forming agents and modifying processes of today (Dohrenwend 1994, 340; Cooke, Warren, and Goudie 1993, 189–190). According to Ben A. Tator, the proposed nature of the agents forming pediments varies with the researcher. He examined forty-three pediment studies conducted between 1880 and 1952 from the regions of Africa, Australia, Mexico, Mongolia, South America, and North America which suggest the following pedimentation processes: (1) basal sapping, (2) ephemeral stream erosion, (3) escarpment retreat, (4) lateral planation, (5) rainwash, (6) rills, (7) runoff, (8) sheetflood, (9) surface wash, (10) soil creep, (11) slope retreat, (12) wind, and (13) weathering to be dominant agent(s) in their studies (1952, 308-315). Some researchers believe more than one process has been active in pediment formation, either simultaneously or at different stages of a pedimentation cycle (Hadley 1967, 85; Sharp 1940, 356). However, as pediment research expands into diverse regions different pedimentation processes are going to dominate under diverse climatic, topographic, and geologic conditions (Dohrenwend 1994, 321–322; Sharp 1940, 362).

A critical factor of pediments preservation is its coverage (See Glossary, pediment sections) (Bourne and Twidale 1998, 127; Twidale 1981, 426–427). Researchers have reported various protective coverage types including: gravel, shale, limestone, and fossils (Miller 1950, 640). However, it is not uncommon for pediment surfaces to be bare without any specific type of coverage (Tator 1952, 306; Warnke 1969, 365). C.R. Twidale concludes "the development of protective factors such as mantle deposits and silcrete veneer, introduces a discontinuous element, a stable aspect, into an otherwise continuous and continuing erosional cycle" (1978, 1172).

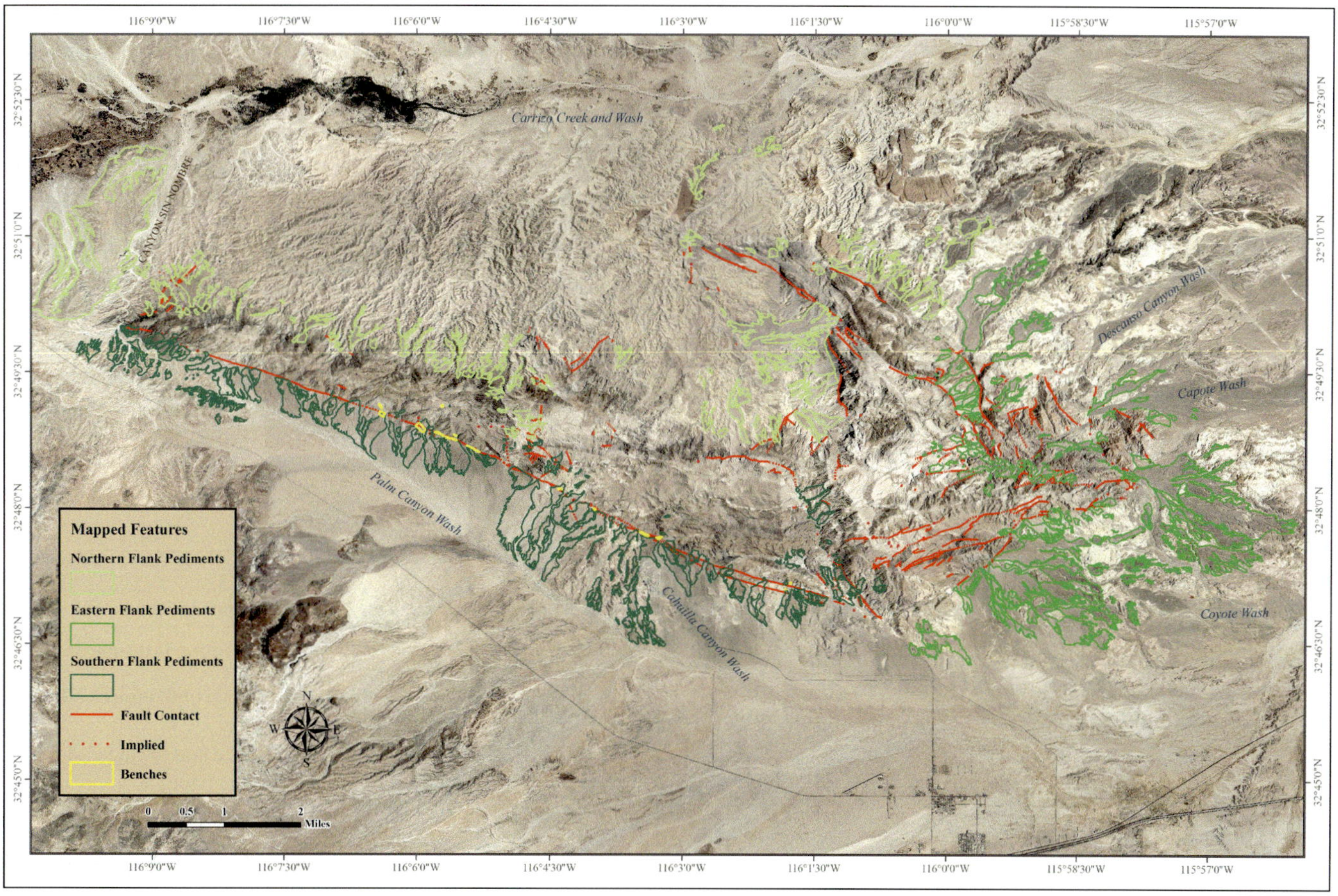

Figure 2. The Coyote Mountains. The Coyote Mountains are centered at 32° 48' 16.34" N, 116° 02' 46.74" W, and are approximately 13.5 miles long and 5 miles wide. The range is littered with minor faults, and bounded on the southern flank by the southern portion of the Elsinore Fault. For this study, the range is divided into southern, eastern, and northern flanks. Different shades of green represent the three flank's pediments. Fault contacts are represented in red. Structural benches are found to be along the Elsinore Fault's line and are represented in yellow. ***Source***: Stevens, Belinda 2013. ArcMap 10.2.

METHODOLOGY

My field work in the Coyote Mountains was conducted during the winters of 2012 and 2013, in segments lasting up to three weeks. All geological mapping was done manually on Google Earth Pro visible satellite imagery of the terrain. Ground photos and handwritten field notes were used to capture the aspects of specific locations. Travel within the Coyote Mountains was done mainly on foot through faulted canyons, eroded stream beds, and up the face of hills. On rare occasions, I was able to follow abandoned mining roads or game trails towards my destinations. Over 500 pediments, including partially active, inactive, and dissected pediments, structural benches, and fault traces mapped within the Coyote Mountains were transcribed into the ArcMap 10.2 computer application (Figure 2). The ArcMap layers included: San Diego and Imperial county imagery of the Coyote Mountains (USDA-FSA), 40 foot contour lines, degree of inclination (reclassified into twelve classes) (USGS-DEM), water resources (USGS-NHD), and manually digitized pediments, structural benches, and fault traces. Pediment profile slopes were produced using 3D Analyst tool within ArcMap, and attributes were transferred into Microsoft Office Excel to determine the degree of inclination for each pediment randomly selected. When creating the pediment profiles all efforts were used to include only the pediment from the piedmont junction to the end of the pediment toes (See Glossary, pediment sections).

The criteria for categorizing pediment groups were the pediment's association of their geographic location. The three pediment groups are: pediments located near faults (PNF), pediments located near streams, washes, or other water features (PNS), and peripediments (PP) (See Glossary) (Tator 1953, 51). Reference to a particular pediment may be made in accordance to the above initialism, the cardinal direction of the ranges, or a geographic location. For my study I divided the Coyote Mountains into three aspects: south, east, and north. The boundaries are not true cardinal directions, but instead are used for the general foundation of locating the position of pediments, and selected features (See Figure 2).Some fault traces were mapped by me, but the majority was mapped by Dr. Bykerk-Kauffman, Amy Gentry, and Cavan Ewing (both graduate students from California State University, Chico). Any error upon the placement of structural or fault accuracy is due to my lack of understanding, not of the data providers.

The following discussion examines the circumstances of some specific pediments in the Coyote Mountains by providing description and some suggestive interpretation.

GEOMORPHIC FEATURES OF THE SOUTHERN FLANK

Introduction

The Coyote Mountains' southern flank, as defined in this study, extends west from 32° 49′ 59.17″ North, 116° 09′ 21.55″ West, to the east 32° 46′ 35.03″ North, 116° 00′ 28.95″ West. For identification purposes, the southern flank pediments are color coded dark green (See Figure 2).

The southern flank is distinguishably different from the eastern and northern flank as the active right lateral strike-slip Elsinore Fault runs the full length of the southern flank, and trends southwest. Additionally, some minor faults are found to trend east and north. Several structural benches were mapped along the mountain front parallel to the path of the Elsinore Fault, and lay near the heads of some pediments. A few of the structural benches were devoid of any surface regolith, but had a ridge of regolith along the down side of mountain's slope. Other structural benches had surfaces of mixed gravel with granite cobbles ranging up to two inches in diameter (32° 48′ 51.15″ North, 116° 05′ 49.97″ West).

Most of the southern flank's drainage pattern flows into the southeast running Palm Canyon Wash. The remaining drainage runs into the northwest Canyon Sin Nombre Wash (See Figure 2). It is possible that the southern flank drainage system and faults have been formational and/or modifying agents of the southern flank's pediments.

Pediment Distribution and Characteristics

The distribution of pediments begins from the southwest (as defined in this study) end of the Coyote Mountains (32° 49′ 59.17″ North, 116° 09′ 21.55″ West) extending from the base of the mountain, and from a mesa landform across Palm Canyon Wash. The pediments adjoining the southern flank are bound at the head by the Elsinore Fault line and at the toes by the Palm Canyon Wash (See Glossary, pediment sections). The pediments bordering the mesa landform are bound at the head by the mesa's top planar surface and at the toes by the Palm Canyon Wash (32° 49′ 30.00″ North, 116° 09′ 02.69″ West). The pediments bordering the mesa landform slope north towards the southern flank, and mirror the same trending direction as the pediments extending from the mountain's base (See Figure 2). Several of the southern mesa pediments have a sparse protective surface of desert varnished gravels, which overlay the older strata of the Pliocene-Pleistocene non-marine Palm Spring Group (See Figure 3).

The southern flank pediments reflect the course of the Canyon Sin Nombre and Palm Canyon Wash drainage (32° 49′ 29.13″ North, 116° 08′ 19.48″ West). For approximately one and a quarter miles east of Canyon Sin Nombre, pediments extending from the base of the range trend southwest. Thereafter, pediments follow the Palm Canyon Wash drainage pattern and trend southeast (See Figure 2).

The general shapes of southern flank pediments are concave upwards, having higher slopes in the head, intermediate slopes in the body and a lower slope at the toes (See Glossary, pediment sections). Some pediments in plan view have a rectilinear shape and are possibly remnants of an extensive pediment that has been heavily dissected. Some pediments coverage consist primarily of basement debris that includes migmatite, schist, and gneiss, which gives the pediment surfaces a dark appearance (32° 49′ 48.43″ North, 116° 08′ 38.54″ West). Other pediments coverage includes basement debris consisting of dolomitic marble, biotite, schist, and metaquartzite, which gives pediment surfaces a lighter color appearance (Todd 2004, 7–8). Additional types of pediment coverage include: (1) a sparse covering of well-sorted gravel with desert varnished, sub-angular, granitic, schist, and quartz cobbles within a sand and silt matrix overlaying a C-Type clastic older

Time Units			Rock Units
Era	**Period**	**Epoch**	**Palm Spring Group (non-marine)**
Cenozoic	Quaternary	Pleistocene	• L-Type Clastics: sandstones and conglomerates • C-Type Clastics: fine-grain sandstones and mudstones
	Neogene	Pliocene	**Imperial Group (marine)**
			• C-Type Clastics: mudstones and fine-grain sandstones derived from the ancestor Colorado River
		Miocene	• L-Type Clastics: coarse sandstones and conglomerates
			Split Mountain Group (non-marine)
			• L-Type Clastics: sandstones, conglomerates, and sediments derived from the local environment • Volcanic Rocks
Mesozoic	Jurassic		**Basement**
			• Metaplutonic rocks
Paleozoic			• Metasedimentary rocks

Figure 3. Coyote Mountains' Stratigraphic Column. The Coyote Mountains' stratigraphic column reflects its historical activities of continental sedimentation, basin subsidence, marine incursion, and tectonic changes. *Source: Data adapted from Winkler, Charles D. and Susan M. Kidwell. 1996. "Stratigraphy of the Marine Rift Basin: Neogene of the Western Salton Trough, California." In* Field Conference Guide Pacific Section A.A.P.G. GB 73, Pacific Section S.E.P.M., *edited by Patrick L. Abbott and John D. Cooper, 80:299.*

strata, and (2) well-sorted gravel with marble clasts and reworked fossilized Imperial Group marine shells, overlying a C-Type clastic older strata. Some of the pediments coverage is dissected and the underlying strata are exposed. For example, one PNF pediment has a coverage of mixed boulders, cobbles, and gravel overlying a C-Type clastic stratum with layers of exposed tilted cemented coquina (32° 47′ 38.23″ North, 116 ° 03′ 36.98″ West) (See Figure 4).

Traces of the Elsinore Fault and numerous other fault paths are visible along the southern flank. These faults may have produced several structural benches, and may be influential in pediment formation and/or modification. One particular PNF pediment extends from the mountain base adjacent to the Elsinore Fault path. It does not exhibit a common pediment concave form, but instead appears to have a higher elevated head relative to adjacent pediments (32° 47′ 47.55″ North, 116° 03′ 40.96″ West). The pediment's elevated head extends as a hill into the middle of the pediment body (See glossary- pediment sections). Both the elevated head and the body have the same regolith coverage that includes boulders, and angular cobbles of granite, marble, and quartz clasts within a silt matrix. Approximately thirty percent of the regolith surface displays a film of desert varnish that overlay C-Type clastic older strata. An uplifted pediment may not be unusual as Jacqueline Mammerickx also reported faulting caused a pediment slope to be warped and uplifted in her Valyermo, California pediment research (1964, 430) (See Figure 5).

The Alverson and Fossil Canyons are two major canyons divided by a pediment draped ridge (32° 47′ 03.66″ North, 116° 01′ 15.82″ West) (See Figure 2). In the two canyons, the basement is overlaid by both the non-marine Split Mountain Group, and the marine Imperial Group sediment (See Figure 3). Both canyons show evidence of fault and stream activity and are littered with pediments along the cliffs, ridges, and near the canyons' heads. One PP pediment lays at the head of the Alverson Canyon with the remnants of a protective regolith coverage, which preserves the last segments of the eroded pediment's Split Mountain Group L-Type clastics older strata (32° 47′ 26.19″ North, 116° 01′ 54.18″ West) (See Figure 6).

Summary

In summary, the southern flank contains concave, convex, and rectilinear shape pediments, and are grouped into categorizes PNF, PNS, or PP according to the pediments association to the features. Evidence of dissection was seen upon pediments' surface, head, flanks, and toes. Prevailing evidence of formational and/or modifying agents includes: faults, mountain tributaries, rills, gullies, washes, lateral planation, modern day quarrying, and construction. The average southern flank pediment slope falls in the range of 3–8 degrees. In this study any pediments disturbed by local quarrying or construction are omitted.

Figure 4. Pediments with Coquina Layers. The tilted cemented coquina layers in this southern pediment created a wall that protected the pediment gravels from erosion. The exposed coquina layer measures approximately 318 feet (96 meters) in length (Google Earth). ***Source:*** Stevens, Belinda. 2013. Personal photograph.

Figure 5. Warped Pediment. The warped pediment profile displays an elevated head with possibly fault gouge. The pediment has a thicker coverage than the older strata. ***Source***: Stevens, Belinda. 2013. Personal photograph

Figure 6. Pediment with Dissected Covering. Remnants of a pediment surface suggest a pediment once extended across the head of the southern Alverson Canyon. Note the contrast of gentle pediment slope with the steeply dissected terrain. ***Source***: Stevens, Belinda. 2013. Personal photograph.

GEOMORPHIC FEATURES OF THE EASTERN FLANK

Introduction

The Coyote Mountains' eastern flank, as defined in this study, extends east from 32° 46′ 33.20″ North, 116° 00′ 26.75″ West, to the north 32° 50′ 04.19″ North, 116° 00′ 08.31″ West. The eastern pediments, for identification purposes are color coded a medium green (See Figure 2).

Two differences between the southern flank and the eastern flank are faults and drainages. Unlike the southern flank's major Elsinore Fault, the eastern flank is absent of a major fault line but has numerous minor faults. The majority of the eastern flank faults trend east by northeast and north by northwest. Different than the southern flank's two major drainages (Palm Canyon Wash and Canyon Sin Nombre Wash), the eastern flank has tributaries that form intricate networks of branching creeks and channels located between more erosion-resistant landform islands, and appear to be a compelling force in dissection and pediment modification. The eastern flank's drainage system generally flows east. Similar to the southern flank, the eastern faults and streams may possibly be formational and/or modifying agents of the eastern flank pediments.

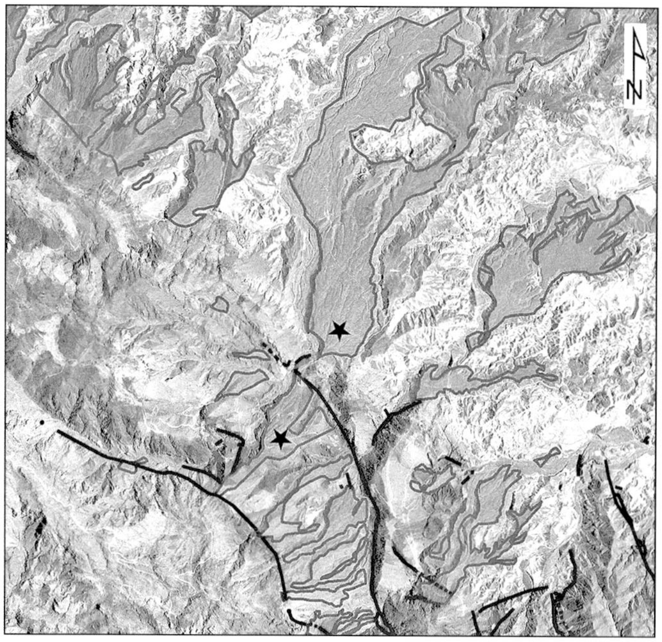

Figure 7. Pediment with a Dual Action Agent. Faults may have been a dual formational and a modifying agent in an eastern flank pediment. First in the formation of a large single pediment, and second in the modification of separating the same pediment landform by younger faults (In reference to the two pediments with black stars). ***Source***: Stevens, Belinda. 2013. ArcMap 10.2.

Pediment Distribution and Characteristics

The distribution of pediments within the eastern flank are found extending from the base of the mountain, draping from the tops of canyon cliffs, paralleling between faults, terracing along river banks, and spanning the length of washes. The eastern flank is littered with minor faults that have a dominant eastward strike south of the Painted Gorge Wash area. North of Painted Gorge, the faults transition and trend northwest parallel to the Elsinore Fault (See Figure 2).It is possible the eastern flank faults are both formational and modifying agents. For instance, in one eastern canyon several PNF pediments are bound at the head and toes by paralleling faults (32° 49′ 09.00″ North, 115° 59′ 46.79″ West). The faults cause the pediments to appear to be straight and uniform in length. However, one of the bounded pediments appears to have once been part of a larger pediment that is now truncated (See Figure 7, black stars).

Earlier fault activity may have been influential in the larger pediment formation and later modified by younger faults. The bound pediments coverage alternates between the Palm Spring C-Type and L-Type. Beneath the coverage, older strata sediment is from the Imperial Group's C-Type and L-Type, the Split Mountain Group's L-Type, and volcanic rock (See Figure 3).Yi-Fu Tuan remarks how a straight boundary between a pediment and a piedmont was influenced by faulting in the lower San Pedro Valley of Arizona (1962, 62-63).In the Painted Gorge area, faulting may also be a responsible agent in a pediment's underlying heterogeneous older strata (32° 48′ 32.90″ North, 115° 59′ 24.79″ West).

Draping from the top of a canyon cliff, a PNS pediment has a complex older strata that includes: (1) marble, part of the basement compound, (2) sandstone, part of the Split Mountain Group, and (3) volcanic clastics, also part of the Split Mountain Group—all overlaid by regolith (See Figure 8).

Unlike the southern flank's drainage system, the eastern flank drainage system is an intricate network of branching creeks, channels, intermittent rivers, and washes that are a compelling force in pediment dissection and active pedimentation. Similar to the southern Palm Canyon Wash, which may be a modifier for pediments' toes extending from the southern flank's mountain base, an eastern flank wash may have been a pediment modifier agent by segmentation. For instance, within the Capote Wash remnants

Figure 8. Pediment's Complicated Lithology. An exposed eastern pediment older strata reveals a complicated lithology and a fault trace (view photo from left to right). ***Source***: Stevens, Belinda. 2013. Personal photograph.

of a pediment line the wash's southern bank and possibility could have once been a part of a larger pediment that lies on the northern bank of the wash (32° 49′ 03.5076″ North, 115° 58′ 10.5739″ West). In the middle of Capote Wash lie fragments of a pediment and a few scattered knolls. This could imply the pediment extended from the southern bank to the northern bank of the wash approximately 1134 yards (1037 meters) at the widest point (See Figure 2).On the southern bank of the wash, one of the pediment remnants stands approximately 475 feet (144 meters) above sea level. One pediment lying within the wash stands at approximately 424 feet (129 meters) above sea level. The potential parent pediment on the northern bank stands approximately 500 feet (152 meters) above sea level. All three pediment segments have the same coverage from the non-marine Palm Spring Group, and older strata from the marine Imperial Group.

The Butaca Canyon, a major shifting intermittent river extending southeast and a tributary to Coyote Wash, is another pediment formational and/or modification agent that exhibits active pedimentation (32° 47′ 40.10″ North, 115° 58′ 58.18″ West) (See Figure 2). Categorized as PNS, a multiple level pediment shows a 3-tier landform terracing along the river's bank. The pediment coverage consists of alluvial deposits overlying a C-Type mud and sand older strata. It is possible the tri-leveled pediment is the result of historical climatic conditions that contributed to increased stream activities and alluviation (Tuan 1962, 63–64) (See Figure 9).

Figure 9. Multiple Pediments. These tri-level pediments are located near an eastern stream, which may have been a formation and/or modification agent as a result of historical climatic conditions. ***Source***: Stevens, Belinda. 2013. Personal photograph.

Summary

In summary, the eastern flank pediments reflected the concave, convex, and rectilinear pediment shapes and categorized in to pediments near faults (PNF), pediments near streams (PNS), or peripediments (PP) (no geographic association) according to their landscape position. Evidence of dissection was seen upon pediments' head, body, flanks, and toes. Pediments located within the interior of the ranges showed less dissection than pediments surrounding the base of the mountain, terracing along river banks, and spanning the length of washes. Prevailing evidence of formational and/or modifying agents includes faults, mountain tributaries, rills, gullies, washes, lateral erosion, and lateral planation. No structural benches were detected. The average eastern flank pediment slope falls in the range of 1–4 degrees.

GEOMORPHIC FEATURES OF THE NORTHERN FLANK

Introduction

The Coyote Mountains' northern flank, as defined in this study, extends northeast from 32° 50′ 05.80″ North, 116° 00′ 12.94″ West to northwest 32° 50′ 02.63″ North, 116° 09′ 19.54″ West. For identification purposes, the northern flank pediments are color coded a light lime green (See Figure 2). The northern flank is set apart from the southern and eastern flank by its fault scarps. The largest scarp stands approximately 600 feet high and has a marble strata (32° 50′ 28.83″ North, 116° 01′ 35.21″ West). The majority of the northern flank faults trend northwest and southwest. Fault traces are found along the mountain base, scarps, canyons, and within the open and closed basins. Most of the northern flank drainage system flows north towards Carrizo Creek Wash, which outflows east and west (See Figure 2). Similar to the southern and eastern flanks, the northern faults, streams, and washes are possibly formational and/or modifying agents of the northern flank pediments.

Pediment Distribution and Characteristics

The distribution of the northern flank pediments are found extending along washes, adjacent to faults, collectively in basins, and bordering the base of the mountain—similar to the southern and eastern flank pediments. Unlike the eastern flank, the northern flank has fewer washes and intermittent rivers. However, north of the eastern flank's Descanso Canyon and Wash, is a wash where pediments' marine and non-marine older strata lie side by side (32° 50′ 37.48″ North, 116° 00′ 15.77″ West) (See Figure 2). In most places, the non-marine Palm Spring Group C-Type clastics pale-yellow color aids in its identification around the ranges. However, on the northern flank the Group's C-Type clastics colors includes pale-orange to red claystones, and pale-orange silt and sandstones. Adjacent to the non-marine sediment is the marine Imperial Group C-Type clastics whose colors range from pale-yellow siltstones, pale-yellow to orange or brown sandstones, and gray claystones (Winker and Kidwell 1996, 300).

The northern flank has a major fault set that parallels along both the mountain ridge and a fault scarp (approximately 32° 50′ 23.52″ North, 116° 01′ 07.28″ West) (See Figure 2). The fault set trends northwest, and other faults parallel the southwest trending Elsinore Fault. The ridge, resembling the outer bend in a fishhook image, runs northeast of the range where several dip slope landforms and pediments extend off its steep cliffs. For example, one PNF pediment formed between the mountain's steep ridge and a dip slope landform has a regolith coverage and an offset non-marine C-Type clastic older strata due to faulting (32° 50′ 41.1684″ North, 116° 01′ 19.0530″ West). Both the coverage and older strata combined measures over 100 feet (30.48 meters) from the ground's base level to the top of the pediment's surface. West of the mountain ridge towers a large fault escarpment with pediments extending from its base. Their regolith coverage includes worm burrows and fossilized marble with oyster bore holes along the surfaces (Christensen 1957, 150 plate 1a–1b; Todd 2004, 7–8) (32° 50′ 28.83″ North, 116° 01′ 35.21″ West). The pediments' coverage of regolith overlies either a marine or non-marine L-Type or C-Type older strata.

The Coyote Mountains' only closed-in basin is centered between the southern and northern flanks. It measures approximately one mile long by a half mile wide, with the long axis running east to west (32° 48′ 48.60″ North, 116° 04′ 49.17″ West) (See Figure 2). Surrounded by mountains, the basin has become a catchment for pedimentation. Detlef A.Warnke refers to pediments within an embayment which "grow" towards each other from opposite sides of mountain ranges as Pediment Passes. The northern flank's closed-in basin consequent stream's drainage pattern and the pediments extend west, and is referred to as a Pediment Pass with symmetrical divides (Warnke 1969, 365). The closed-in basin pediments' older strata include both C-Type and L-Type clastics from the Imperial Group. Some of the pediments' coverage includes alluvium with surfaces enhanced by a variety of desert vegetation or cemented coquina shell layers (See Figure 10). Beyond the borders of the basin, pediments continue along the mountain base of the northern flank.

Similar to the southern and eastern flank pediments, the northern flank has pediments extending from the mountain's base. But, unlike the other two flanks the northern flank's extending pediments are surrounded by highly dissected ridges, large outcrops of sandstone, and mounds of cemented coquina shell landforms. Although the northern flank is absent of any major faults such as the southern flank's Elsinore Fault, there is evidence to suggest that faulting is a pediment's modifying agent. For example on the northwest end of the northern flank, three PNF pediments have a regolith coverage that overlies an older strata of non-marine C-Type clastics from the Palm Spring Group (area of 32° 50′ 09.57″ North, 116° 08′ 36.57″ West) (See Figure 2). All three of the pediment surfaces show evidence of fault traces, which could imply faulting was a formational and/or modifying agent for pediments along the northern flank (See Figure 11). It is possible the three pediments were once connected.

On the middle pediment, one fault has a left-lateral orientation of 388, 81 (Dr. Bykerk-Kauffman, March 2012 personal communication). A profile of the middle pediment displays a combination of a concave and convex shape, two fault traces, and a following pediment remnant (See Figure 12). The pediment profile numbers 5-110,113 identify the pediments from my research.

Figure 10. Pediment Pass. The closed-in basin pediments "grow" towards each other from opposite sides of the surrounding mountains. ***Source***: Stevens, Belinda. 2013. Personal photograph.

Figure 11. Pediment Surfaces with Fault Traces. The three largest pediments on the northwest flank show evidence of fault traces across the surfaces. ***Source***: Stevens, Belinda. 2013. Personal photograph.

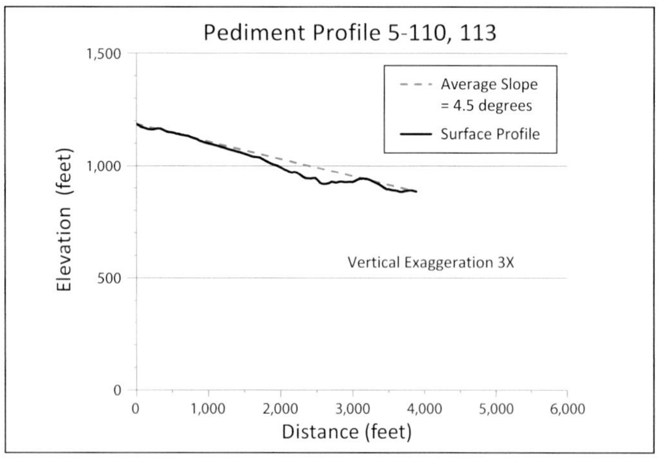

Figure 12. Pediment Profile with Faulted Surface.The northern flank pediment profile displays a combination of a concave and convex shape, and two fault traces.

An exposed flank unveils an angular unconformity at the end of one of the fault lines on the middle pediment, (32° 50′ 23.2980″ North, 116° 08′ 45.6079″ West). The exposed flank consists of easily erodible siltstones, sandstones, and mudstones from the non-marine Palm Spring Group, which suggest the exposure to be recent (See Figure 13).

Figure 13. Pediment's Angular Unconformity Older Strata. A northwestern pediment with fault traces across its surface has an exposed angular unconformity older strata of Palm Springs C-Type siltstones, sandstones, and mudstones. ***Source***: Stevens, Belinda. 2013. Personal photograph.

Summary

In summary, the northern flank pediment shapes are repetitive to the southern and eastern pediment shapes. Although the northern pediments extending from the mountain base, they are surrounded by highly dissected ridges, large outcrops of sandstone, and mounds of cemented coquina shell landforms. The pediments were able to be categorized into the three pediment groups pediments near faults (PNF), pediments near streams (PNS), and peripediments (PP) no geographic association. Evidence of dissection was seen upon pediments' surface, head, flanks, and toes. Prevailing evidence of formational and/or modifying agents includes faults, fault scarps, mountain tributaries, rills, gullies, washes, and lateral planation. The average northern flank pediment slope falls in the range of 4–5 degrees.

RESULTS OF PEDIMENT RESEARCH

My research on the distribution of pediments in the Coyote Mountains lead to a total of 520 mapped pediments, and associated fault traces, structural benches, and the identification of underlining lithology. Some highlighted results include the following:

1. All three types of pediment shapes: concave, convex, rectilinear, and combination shapes formed within all of the Coyote Mountains' rock units.
2. The average gradient for Coyote Mountains pediment groups PNF (7 degrees), PNS (4.5 degrees), and PP (6.7 degrees) fall into the southwestern United States pediment gradient measurements ranging from 0.5 to 11 degrees (Cooke 1970, 28; Cooke, Warren, and Goudie 1993, 192; Dohrenwend 1994, 327; Hadley 1967, 84,; Royse and Barsch 1971, 3177; Sharp 1940, 357).
3. Evidence supported the theory that pediments and a mountain's lithology could be different was validated (Cooke 1970, 33; Mammerickx 1964, 427–431). In fact, a single pediment can cross lithologic units. For example, the range's eastern flank lithology continues to be metasedimentary basement rock, but with increases of metaplutonic rock. Compared to an eastern flank PNS pediment with an exposed flank, which reveals a complicated older strata of: (1) marble, part of the basement complex, (2) sandstone, part of the Split Mountain Group, and (3) volcanic clastics, also part of the Split Mountain Group, all overlaid by regolith (See Figure 8).
4. The examination of some pediments implied there could have been more than one formational and/or modifying agent either simultaneously or at different stages of pedimentation (Hadley 1967, 85; Sharp 1940, 356). The southern flank has one example of two possible formational and/or modifying agents that fit this characteristic. First, the majority of the southern flank pediments adjoin the mountain where the active Elsinore fault line runs adjacent to some of the pediment heads. Second, Palm Canyon Wash parallels the southern flank, and may have obstructed the extension of pediment lengths. Both of these formational and/or modifying agents could have influenced pedimentation simultaneously or at different times.
5. Two areas show evidence of multi-level pediments (Tator 1952, 300–301). First example is on the southern flank at the mouth of the Cahuilla Canyon Wash. From the mouth of the wash active pedimentation can be seen as a four-tiered feature (32° 47′ 36.90″ North, 116° 03′ 15.67″ West). The lowest elevated tier lies along the wash's base level, and the highest tier is closer to the mountain base near the trace of the Elsinore Fault. Second example is on the eastern flank along the intermittent Butaca Canyon river. An exhibit of a three tier multi-level pediment lies along the bend of the river (See Figure 9).
6. A pediment's protective coverage proved to be a critical factor to the preservation of pediments (Bourne and Twidale 1998, 127; Miller 1950, 640; Twidale 1981, 427). Unlike the southern and eastern flank pediments, extending from the mountain base, the northern flank's extending pediments, from the mountain base, are surrounded by highly dissected ridges, large outcrops of sandstone, and mounds of cemented coquina shell landforms. Yet, there are three of the northern flank's largest pediments (32° 50′ 09.57″ North, 116° 08′ 36.57″ West). All have a regolith coverage that overlies an older stratum of easily erodible non-marine C-Type siltstones and sandstones from the Palm Spring Group. In addition, the pediments' protective surfaces show evidence of fault traces.

CONCLUSION OF PEDIMENT RESEARCH

The purpose of my pediment research in the Coyote Mountains of southern California was to provide a data base of pediment locations, surface configuration, and underlying lithology for future research relating to pediment development. The classification of pediments was associated to their geographic location and their landscape position. The distribution of pediments were found extending from the base of the ranges, at the mouths and heads of canyons, draped on tops of canyon walls, spanning the length of washes, present in adjoining rivers banks, collectively located in basins, and associated with paralleling and truncating faults. Suggestive formational and/or modifying agents involved in pedimentation include: faults, fault scarps, lateral stream planation, and the fluvial action of mountain tributaries, rills, gullies, and washes.

Several promising areas of future research were identified. First, locating and mapping pediments covered by lava flows should be conducted; this may aid in relative dating of some pediments. Second, examining pediments near faults and highly elevated areas may identify areas with more uplift activity, and aid in establishing rate, time of fault movement, and pediment morphogenesis. Third, comparison of crest drainages and pediments may suggest a formational and/or modifying relationship between drainage parameters and pediment lengths and gradients.

The mapped pediments of the Coyote Mountains in southern California can now be utilized as a foundation for continued work on the morphogenesis of the Coyote Mountains and the evolution of geomorphic landscapes.

GLOSSARY

Alluvial Fan: An outspread, gently sloping mass of alluvium deposited by a stream. It has a shape of an open fan, the apex being at the valley mouth.

Angular Unconformity: The younger sediments rest upon the eroded surface of tilted or folded older rocks.

Basal Sapping: The undercutting and retreat of a slope caused when weathering and/or erosion are concentrated at its base (Academic Dictionaries and Encyclopedias online, s.v. "Basal Sapping," http://geography_glossary.enacademic.com/95/basal_sapping [accessed April 2015]).

Base Level: The level of the ground surface.

Buttress Unconformity: An unconformity where onlapping strata is deposited on a steep, high relief scarp of older strata.

Dip-Slip Fault: A fault on which the movement is parallel to the dip of the fault.

Dip Slope Landform: A slope of the land surface that conforms approximately with the dip of the underlying rocks

Faults: A fracture or fracture zone in rock along which movement has occurred (Geology.com)

Fault Scarp: The cliff or escarpment formed by a fault that researches the earth's surface.

Geology: The study of the planet earth, the materials of which it is made, the structure of those materials, and the processes acting upon them.

Knoll: A small rounded hill.

Lateral Erosion: "The action of a meandering stream as it swings from side to side, impinging against and undercutting its banks" (*Dictionary of Geological Terms,* 3rd ed., s.v. "Lateral Erosion").

Lateral Corrasion: The process of wearing away earth surfaces laterally by running water.

Lateral Planation: The reduction of the land in an interstream area to a plain or nearly flat surface by lateral erosion.

Lithology: The description of rocks on the basis of characteristics as color, mineralogic, composition, and grain size.

Pediment: "A gently sloping erosional surface developed on bedrock or older unconsolidated deposits. The eroded surface may be subaerially exposed or covered by a continuous or discontinuous veneer of alluvial deposits, and the bedrock may include any lithologic type with any structural attitude" (Dohrenwend 1994, 323).

Pediment Sections

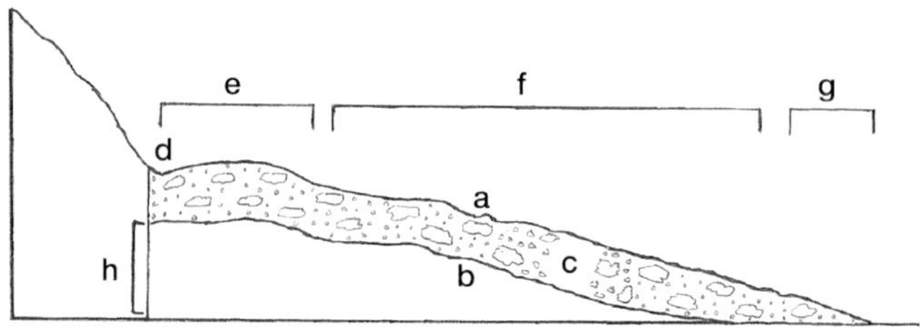

a) *Surface:* the upper top layer of a pediment.

b) *Bottom:* the basal bedrock surface.

c) *Coverage:* the pediment thickness consisting of various alluvial deposits, debris flow, colluviums, mass wasting deposits, and regolith.

d) *Piedmont Junction:* the slope break near the pediment head.

e) *Pediment Head:* the beginning slope following the piedmont junction.

f) *Pediment Body:* the pediment mass between the head and toes

g) *Pediment Toes:* the terminus point(s) of a pediment.

h) *Older Strata:* the underlying strata under a pediment coverage.

Pedimentation: "The range of processes bring about the formation of pediments" (Warnke 1969, 366)

Peripediment: Pediments that are not geographically located near faults or water type features.

Planation: The general lowering of the land.

PNF: Initialism for pediment near fault categorized group

PNS: Initialism for pediment near stream, wash, or water feature categorized group.

PP: Initialism for peripediments categorized group.

Strata: The plural of stratum (a layer of sedimentary rock).

Strike-Slip Fault: A fault that strikes parallel with the strike of the strata involved.

Structural Bench: A relatively planar surface breaking the continuity of a slope.

REFERENCES CITED

Alles, David L. ed. 2011. Geology of the Salton Trough. Western Washington University, Bellingham, Washington. http://fire.biol.wwu.edu/trent/alles/GeologySaltonTrough.pdf (Accessed Nov. 2013).

Bourne, J.A, and C.R. Twidale. 1998. Pediments and Alluvial Fans: Genesis and Relationships in the Western Piedmont of the Flinders Rangers, South Australia. Australian Journal of Earth Sciences 45:123–135.

Bykerk-Kauffman, Dr., February and March 2012 personal communication.

California Department of Parks and Recreation (CA Parks). Anza-Borrego Desert State Park. http://www.parks.ca.gov/pages/638/files/ABDSPmap.pdf (Accessed Oct. 2013).

Christensen, Andrew D. 1957. Part of the Geology of the Coyote Mountain Area, Imperial County, California. Master's thesis, University of California, Los Angeles.

Cooke, Ronald U. 1970. Morphometric Analysis of Pediments and Associated Landforms in the Western Mojave Desert, California. American Journal of Science, 269 [June]: 26–38.

Cooke, Ron, Andrew Warren, and Andrew Goudie. 1993. Pediments and Glacis. In Desert Geomorphology, 188-201. London: UCL Press.

Department of Geological Sciences (CSULB). n.d. Salton Trough Geologic Overview. California State University, Long Beach, California. http://geology.campus.ad.csulb.edu/VIRTUAL_FIELD/Salton_Sea/saltmain.htm (Accessed Nov. 2013).

Demere, Thomas A. 2006. The Imperial Sea: Marine Geology and Paleontology. In Fossil Treasures of the Anza-Borrego Desert, ed. George T. Jefferson and Lowell Lindsay, 29-41. California: Sunbelt Publications. https://www.sdnhm.org/archive/research/paleontology/FossilTreasuresABDCh2.pdf (Accessed Nov. 2013).

Denny, Charles S. 1967. Fans and Pediments. American Journal of Science 265:81–105.

Dohrenwend, John C. 1987. Basin and Range. In Geomorphic Systems of North America, ed. W.L. Graf, 303–342. Boulder, Colorado: The Geological Society of America, Centennial Special Volume 2.

_______1994. Pediments in arid environments. In Geomorphology of Desert Environments, ed Athol D. Abrahams and Anthony J. Parsons, 321–353. London: Chapman & Hall.

Dorsey, Rebecca. 2005. A Summary of Late Cenozoic Stratigraphy, Tectonics, and Basin Evolution in the Anza-Borrego Desert Region. Journal of Geology (Jul): 1–8.

_______, Rebecca J. et al. 2011. Stratigraphic Record of Basin Development within in San Andreas Fault System: Late Cenozoic Fish Creek-Vallecito Basin, Southern California. Geological Society of American Bulletin 123, iss. 5-6 (May): 771–793.

_______, Rebecca J. et al. 2012. Initiation of the Southern Elsinore Fault at ~1.2 Ma: Evidence from the Fish Creek-Vallecito Basin, southern California. Tectonics 31, TC2006: 1–21.

Fletcher, Kathryn E.K., Thomas K. Rockwell, and Warren D. Sharp. 2011. Late Quaternary Slip Rate of the southern Elsinore Fault, Southern California: Dating offset alluvial Fans via ^{230}Th/U on Pedogenic Carbonate. Journal of Geophysical Research 116, FO2006: 1–11.

GeoCommunity. GIS Data Depot: United States Geological Survey (USGS) Digital Elevation Models (DEM). MindSites Group. http://data.geocomm.com/dem/ (Accessed April 2012).

King, Hobart M., Geology.com Geoscience News and Information. http://geology.com/dictionary/glossary-f.shtml (Accessed Nov. 2014).

Google. 2013. Google Earth (version 7.1) [Computer Program]. Available at http://www.google-earth.gufile.com/free (Accessed 2012).

Hadley, Richard F. 1967. Pediments and Pediment-Forming Processes. Journal of Geological Education 15, no. 2 [April]: 83–89.

Mabbutt, J.A. 1955. Pediment Land Forms in Little Namaqualand, South Africa. The Geographical Journal 121, no. 1 [Mar.]: 77–83.

Mammerickx, Jacqueline. 1964. Quantitative Observations of Pediments in the Mojave and Sonoran Deserts (Southwestern United States). American Journal of Science 262 [April]: 417–435.

McGee, W.J. 1897. Sheetflood Erosion. Bulletin of the Geological Society of American 8 [Feb.]: 87–112.

Miller, Victor C. 1950. Pediments and Pediment-Forming Processes near House Rock, Arizona. The Journal of Geology 58, no. 6 [Nov.]: 634–645.

Oard, Michael J. 2004. Pediments formed by the Flood: Evidence for the Flood/post-flood boundary in the late Cenozoic. Journal of Creation, 18 iss. 2:15–27.

Oberlander, Theodore M. 1974. Landscape Inheritance and the Pediment Problem in the Mojave Desert of Southern California. American Journal of Science 274 [Oct]: 849–875.

Rich, John L. 1935. Origin and Evolutions of Rock Fans and Pediments. Bulletin of the Geological Society of America 46:999–1024.

Royse, Jr and Barsch. 1971. Terraces and Pediment-Terraces in the Southwest: An interpretation. Geological Society of America Bulletin, 82 [Nov.]: 3177–3182.

Rowlett, Russ. A Dictionary of Units of Measurement. Center for Mathematics and Science Education University of North Carolina at Chapel Hill. http://www.unc.edu/~rowlett/units/dictK.html (Accessed Aug. 2014).

Sharp, Robert P. 1940. Geomorphology of the Ruby-East Humboldt Range, Nevada. Bulletin of the Geological Society of America 51 [Mar]: 337–372.

Stevens, Belinda J. 2015. The Distribution of Pediment in the Coyote Ranges Southern, California. Master's thesis, California State University, Chico. http://csuchico-dspace.calstate.edu/handle/10211.3/145272

Tator, Ben A. 1952. Pediment Characteristics and Terminology. Annals of the Association of American Geographers 42, iss. 4 [Dec]: 295–317.

_______, B.A. 1953. Pediment Characteristics and Terminology (Part II). Annuals of the Association of American Geographers, 43: 47–53.

Todd, V.R. 2004. Preliminary geologic map of the El Cajon 30′ × 60′ quadrangle, Southern California. U.S. Geological Survey open file report 2004-1361 version 1.0 http://pubs.usgs.gov/of/2004/1361/ec1_pamphlet.pdf (Accessed July 2014).

Tuan, Yi-Fu. 1962. Structure, Climate, and Basin Land Forms, in Arizona and New Mexico. Annals of the Association of American Geographers 52 iss. 1:51–68.

Twidale, C.R. 1978. On the Origin of Pediments in different Structural Settings. American Journal of Science vol. 278 [Oct.]: 1138–1176

_______, C.R. 1981. Origins and Environments of Pediments. Journal of Geological Society of Australia 28:423–434.

U.S. Department of Agriculture (USDA) Farm Service Agency (FSA). Imagery Programs: National Agriculture Imagery Program. USDA Farm Service Agency. http://fsa.usda.gov/FSA/apfoapp?area=home&subject=prog-&topic=nai (Accessed April 2012).

U.S. Department of the Interior Bureau of Land Management (BLM). California Coyote Mountain Wilderness. http://www.blm.gov/ca/st/en/fo/elcentro/recreation/ecwilderness/coyote_mountains.print. html (Accessed Oct. 2013).

U.S. Geological Survey (USGS). Geologic Provinces of the United States: Basin and Range Province. http://geomaps.wr.usgs.gov/parks/province/basinrange.html (Accessed Nov. 2013).

U.S. Geological Survey (USGS) National Hydrography Dataset (NHD). Hydrography: National Hydrography Dataset and Watershed Boundary Dataset. United States Geological Survey. http://nhd.usgs.gov/ (Accessed April 2012).

Warnke, Detlef A. 1969. Pediment Evolution in the Halloran Hills, Central Mojave Desert, California. Annals of Geomorphology 13: 357–389.

The Weather Channel. Weather Underground (WU). http://www.wunderground.com/history/airport/KNJK/2012/2/13/MonthlyHistory.html?req_city=NA&req_state=NA&req_statename=NA (Accessed Sept. 2014).

Winker, Charles D. and Susan M. Kidwell. 1996. Stratigraphy of the Marine Rift Basin: Neogene of the Western Salton Trough, California. In Field Conference Guide. Pacific Section A.A.P.G. GB 73, Pacific Section S.E.P.M., edited by Patrick L. Abbott and John D. Cooper, 80:295–336.

III. HISTORICAL PAPERS

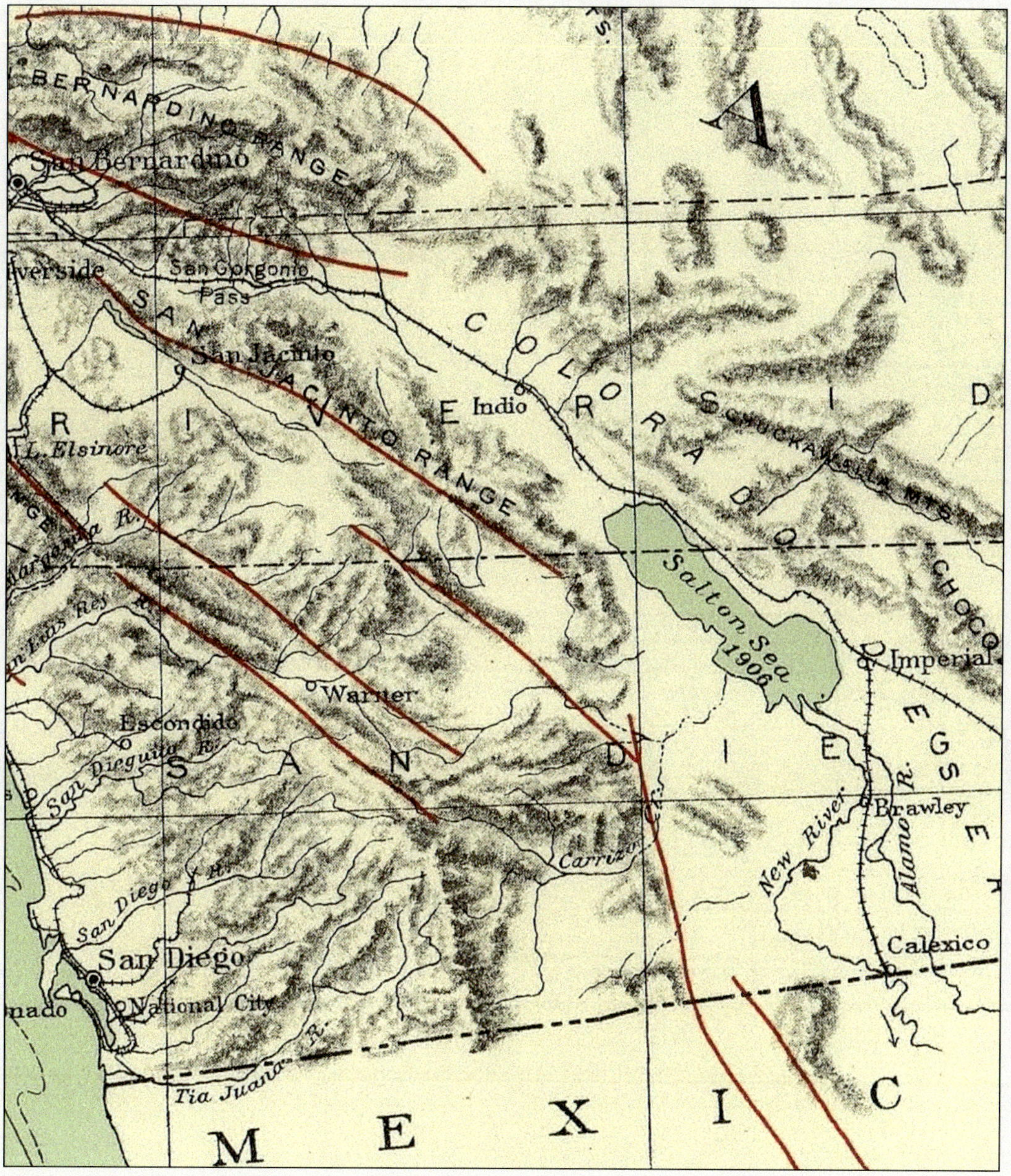

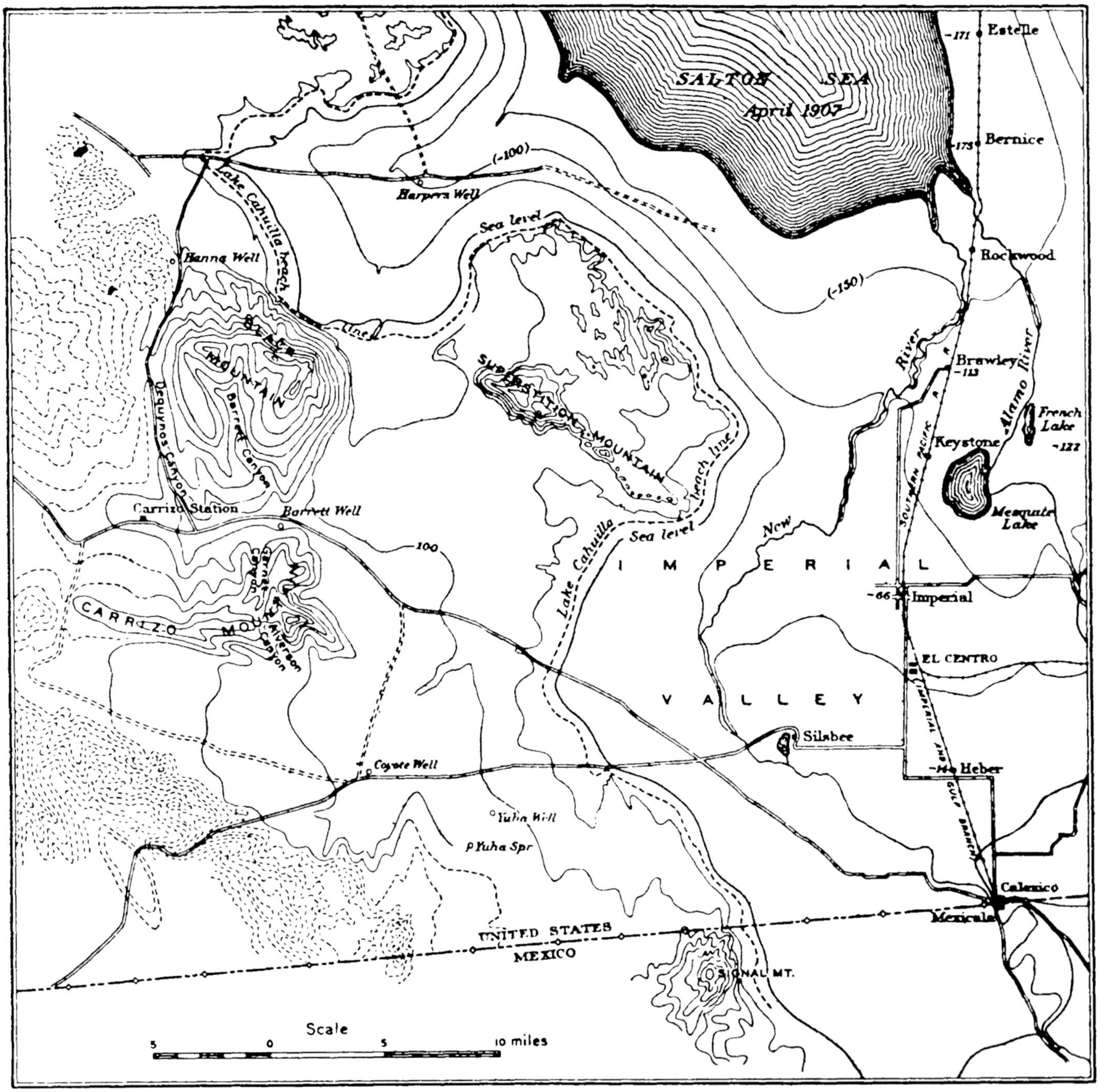

FIG. 1.—Sketch map of Carrizo Mountain and vicinity.

Previous Page: Portion of "Geomorphic map of California and Nevada with portions of Oregon and Idaho showing the diastrophic character of the relief, the steep descent from the sub-continental shelf to the floor of the Pacific, and the more important known faults," by the California State Earthquake Investigation Commission, scale 1:1,900,000, dated 1908.

NOTES ON THE GEOLOGY OF CARRIZO MOUNTAIN AND VICINITY, SAN DIEGO COUNTY, CAL.[1]

WALTER C. MENDENHALL
U. S. Geological Survey

INTRODUCTION[1]

Through the courtesy of Dr. J. C. Merriam, of the University of California, and Dr. Stephen Bowers, of Los Angeles, small collections of fossil corals and mollusks from the vicinity of Carrizo Mountain, San Diego County, Cal., were sent to Dr. T. Wayland Vaughan and Dr. Ralph Arnold,[2] of the U.S. Geological Survey, during the autumn of 1903. The material in these collections proved to be of exceptional interest to the paleontologists, and in order that a larger amount might be obtained for study and more definite information secured about the geology of the region, Dr. Bowers and the writer visited the field during the latter part of January, 1904.

A light camping equipment was secured at Imperial and the drive made to Carrizo Mountain from this point. One day was spent at the Yuha Oil Well, studying the stratigraphy and collecting, then camp was established at Coyote Well on the Jacumba Springs Road. From this point Alverson Canyon and the southern slopes of Carrizo Mountain were accessible. Two days were utilized in work in this vicinity. Camp was then re-established about two miles below Carrizo Station, on the old stage road from Yuma to San Diego via Julian. From this point Garnet Canyon and the north slopes of Carrizo Mountain and Barrett Canyon and the south slopes of Black Mountain could be reached readily. After representative collections, amounting in all to about 1,500 pounds of material, had been made from the several fossil localities visited, the return trip to Imperial was accomplished and the party disbanded. About ten days in all had been spent in the field.

[1] Published by permission of the Director, U.S. Geological Survey.

[2] A brief preliminary statement of the conclusions reached by Drs. Vaughan and Arnold after an examination of these collections appears in *Science*, N.S., XIX, No. 482 (1904), 503. At that time the fauna was regarded as lower Miocene; later conclusions are to the effect that it is upper Miocene.

Early Land Office maps, upon which neither roads nor relief are shown, were at that time the only official maps available, although since then, because of the rapid settlement of the Imperial Valley and the interest created in the region by the partial filling of the Salton Sink, the Geological Survey has issued a Reconnaissance Map which contains some details in the settled parts of the district but is very general in the vicinity of Carrizo and Black mountains, betause these masses lie to the west of the area actually surveyed.

In 1900 and 1901 a number of oil companies were organized to prospect along the west side of the desert, between the base of the Santa Rosa Mountains and the Mexican line, and a number of engineers entered the district in the employ of these companies. Mr. I. A. Hubon and Mr. C. S. Alverson, of San Diego, were among these, and they collected data which, when assembled in a sketch map at Dr. Bowers' request, served to guide us in our field-work. From all these sources and from some sketches made by the writer the accompanying generalized map (Fig. 1) has been prepared. It does not pretend to topographic accuracy but indicates merely general geographic relations and general topographic facts without detail.

GEOGRAPHY

Black and Carrizo mountains, known also as Fish Creek and Coyote mountains, are eastern outliers of the Peninsula Range that separates the depression occupied in part by the Gulf of California from the Pacific Ocean. They are in southeastern California near the western edge of the Colorado Desert, and from fifteen to thirty miles north of the international boundary. East of them the Colorado Desert, much of it below sea-level, extends to the Colorado River, while to the west low ridges extend to the base of the main Peninsula Range.

The two masses are separated by the valley of Carrizo Creek. This stream rises in Mexico, flows north for several

Source: *The Journal of Geology*, Vol. 18, No. 4 (May–Jun., 1910), pp. 336–355.
Published by: The University of Chicago Press. http://www.jstor.org/stable/30079345

miles, through a high valley in the Peninsular Mountains, then descends to the Desert level through a precipitous canyon. Nearly all of that part of its channel that lies within the desert is dry except during rare flood periods, when its waters join those of San Felipe Creek, north of Black Mountain, and eventually reach the Salton depression. At Carrizo Station, one of the relief stations of the old Butterfield stage line, a series of springs rise, and for one or two miles below this point flowing water is found in the creek bed, except during the hottest period of summer.

The desert floor at the eastern base of the peaks is generally from 100 to 200 feet above sea-level, but on the north side of Black Mountain the sea-level contour and the old beach of Lake Cahuilla,[3] 40 feet above sea-level, swing in against the mountain base. In the past the region has been rather difficult of access, because of its remoteness from settlements and its aridity. With the colonization of the Imperial Valley since 1900 and the building of the branch railroad from Old Beach to Calexico, however, this condition has been greatly modified. Now Carrizo Station or Coyote Well may be reached by one day's drive from Imperial or El Centro, and supplies are readily secured at many points in the valley. The old roads from the desert to San Diego, the one running north of Carrizo Mountain by way of Julian and the other south of the mountain by way of Jacumba and Campo, are still much used for direct communication between the Imperial Valley and the coast, although the Campo road below Mountain Springs is rough and after storms is nearly impassable.

EARLIER WORK

As yet there has been no detailed work done on the geological problems of the extreme southern part of California. Two important reconnaissances, however, have been carried out in that region and a number of other papers contain interesting notes.

Professor William P. Blake,[4] who accompanied one of the Pacific Railroad survey parties under Lieutenant Williamson through Southern California in 1853, wrote a comprehensive account of the region which he visited and made a number of contributions of permanent value. He explored the Colorado desert, studied the effects of wind erosion in it, examined the old water-line which is so conspicuous a feature on the west side of the valley opposite Coachella and Walters, and worked out correctly both the origin of the lake whose former presence is attested to by it and the cause of this lake's disappearance.

Professor Blake made three trips along the valley of Carrizo Creek with the expedition, and brought out a few fossils which he collected from one of the flat, sandstone-capped hills on the north side of the valley. These were examined by Mr. T. A. Conrad, who pronounced the species new but probably of Miocene age.

Dr. Harold W. Fairbanks[5] visited Carrizo Mountain in the early nineties for the California State Mining Bureau, and while the exigencies of publication were such that a complete expression of his observations and conclusions was not possible, his paper is none the less very definite and satisfactory. He made hurried trips to the slopes of Black Mountain, which lies north of Carrizo Creek, and to Carrizo Mountain itself, and made collections from each locality. He described the fossils and the general geology of the district, and mentioned the corals which occur at the base of the sedimentary section.

Dr. Stephen Bowers,[6] of Los Angeles, visited the west side of the Colorado Desert in the summer of 1901, for the California State Mining Bureau. At that time there were a number of companies in the district drilling in the sedimentary rocks for oil. It was in connection with this oil excitement that Dr. Bowers' visit was made. He secured some fossils from the Carrizo and Black mountain localities which were submitted to Drs. Merriam, Vaughan, and Arnold. The interest aroused by these small collections led to the planning of the trip described here.

Other writers on the Colorado Desert have devoted themselves to general observations or to the description of particular features outside the Carrizo Mountain region. Among them may be mentioned:

H. G. Hanks, "Mud Volcanoes and the Colorado Desert," 2d Ann. Report, State Mineralogist of California, 1880–82, 227–40.

Chas. R. Orcutt, "The Colorado Desert," 10th Ann. Report, State Mineralogist of California, 1890, 899–919.

Dr. Robt. E. C. Stearns, "The Fossil Freshwater Shells of the Colorado Desert, Their Distribution, Environment and Variation," Proc. U.S. Nat. Museum, XXIV (1902), 271–300.

3 *Nat. Geol. Mag.*, XVIII, No. 12, 830. In a note in this number of the *National Geographic Magazine*, Professor Blake proposes that the name Lake Cahuilla be applied to the vanished water body whose earlier existence is clearly proven by so many phenomena and whose history was first deciphered by Professor Blake himself. The name is most appropriate and the suggestion is most appropriately made by this distinguished worker.

4 *Pacific Railroad Reports*, "Geology" (1856), V.

5 Fairbanks, H. W., *11th Rept., State Mineralogist of California*, 1893, 88, 90.

6 Stephen Bowers, *Reconnaissance of the Colorado Desert Mining District*, California State Mining Bureau, 1901.

GEOLOGY

GENERAL

Carrizo and Black mountains are islands of granitic and metamorphic rocks, which rise through encircling terranes of later sediments and volcanics. These later beds are Miocene and younger, and the unconformity which exists between them and the older rocks upon which they lie is profound. The time interval represented by this unconformity is not known because the age of the altered rocks below it is a matter of uncertainty. Fairbanks[7] expresses the opinion that they are Carboniferous or older, the opinion being based presumably upon their general resemblance to upper Paleozoic rocks in other parts of California and upon the aspect of some shells found in a float piece of siliceous limestone. Accepting this determination as the best possible in the present state of our knowledge, we must conclude that the Triassic, Jurassic, and Cretaceous systems are without depositional representatives in this region. Either the Carrizo and Black mountain areas were land masses subject to erosion during this interval or the evidence of such periods of deposition as intervened was later removed by erosional processes.

The Miocene seems to have been inaugurated by volcanic activity. On the southern slopes of both Carrizo and Black mountains are bedded tuffs, volcanic conglomerates, and less extensive masses of dark lavas of andesitic aspect. On Black Mountain there are distinct sandstones interbedded with these and directly upon them lie the Miocene coral reefs. In Alverson Canyon, which drains south from Carrizo Mountain, red vesicular lavas are succeeded by green and lavender sandstones and conglomerates, whose constituent materials are volcanic, and these in turn grade into conglomerates with a diminishing proportion of volcanic pebbles. Above them are quartz conglomerates, tawny sandstones, and finally soft greenish-yellow clay shales.

An unconformity which is not especially conspicuous exists in the Miocene between the sandy shell-bearing beds, 100 feet or less in thickness, which immediately overlie the volcanics or the metamorphics, and the great mass of shales, greenish or yellowish at base, pink or pale red in general color-tone toward the top, which form the bad-land area (Fig. 2) that is especially well developed between Black and Carrizo mountains. Finally, across the planed edges of these shale beds, a sheet of river cobbles, well rounded, has been distributed unconformably throughout the Carrizo Valley. They are probably Pleistocene, but are earlier than the silts, sands, and gravels, which represent the offshore and beach deposits of the lake which until recently has occupied the Colorado Desert. The latest erosion has left these old stream deposits stranded upon the remnants of the earlier valley floor at heights of from 100 to 200 feet above the present bed of Carrizo Creek.

FIG. 2.—Lower Garnet Canyon and the adjacent bad lands cut in Miocene shales.

DESCRIPTIVE

Basal series.—The core of Carrizo Mountain (Fig. 9) is a series of metamorphic rocks in which a blue or gray crystalline limestone is predominant. Fairbanks reports that the limestones constitute the mass of the north face of the mountain. They are likewise present in great force on the divide between Alverson and Garnet canyons, where bands of graphitic schists are associated with them, but southwest along the first-named canyon, bands of dark biotite gneiss, which presumably represent early intrusives in the limestone, are abundant. Other fresh, dark, fine-grained intrusives, which may be related to the Miocene effusives, are found in narrow dikes.

The bedding of the marmorized limestones and the imperfect foliation in the gneisses are approximately parallel to each other and to the longer axis of Carrizo Mountain. They usually are nearly vertical, the dips in either direction being 70 or 80 degrees.

The basement of Black Mountain was observed at only a few points where the fundamental rock juts out into Barrett Canyon. Here it is a granitic plutonic, with little or no evidence of the action of metamorphic forces. Fairbanks, who examined it at a point somewhat farther east, also speaks of the rock as a granite.

Effusives.—West of Alverson Canyon (Fig. 3), along the north slope of Carrizo Mountain, is a conspicuous exposure of ashy, lavender-colored tuff, which appears to lie directly upon the metamorphic rocks. Within the canyon itself and about midway of its length, an exposure of the variegated lava, andesitic in character, has at its base a thin bed of sandstone and conglomerate, while near the mouth of the canyon the sedimentary beds rest directly upon 50 feet of tuffaceous beds, which in turn overlie the older rocks.

[7] Fairbanks, H. W., *11th Rept., California State Mineralogist*, 1893, 88, 90.

FIG. 3.—Alverson Canyon and the west slope of Carrizo Mountain.

No effusives were observed in the Garnet Canyon section on the north slope of Carrizo Mountain, but across Carrizo Creek, in the upper part of Barrett Canyon and especially in the ridge which separates Barrett Canyon from Deguyños Canyon, just west of it, is a heavy development of the lavas and tuffaceous beds.

Here, as on the south slope of Carrizo Mountain, there is some interbedding of sandstones with the flows, which evidently issued contemporaneously with the beginning of Miocene sedimentation. These interbedded sandstones are usually bright red or pink in hue, as though partly baked by the succeeding lava stream; hence they make conspicuous exposures among the more somber lavas. The uppermost lava flow, upon whose upper surface lies the coral reef (Fig. 6) at the head of Barrett Creek, is about 200 feet thick, and overlies a sandstone bed 20 to 50 feet thick. Below this more effusives extend below the bottom of the arroyo.

In this region the accumulated effusive materials are best displayed in maximum thickness in the ridge that has been mentioned between Barrett and Deguyños canyons. They must be more than 500 feet thick here. In Alverson Canyon their mass is much less.

FIG. 4.—View down Alverson Canyon from the south-slope of Carrizo Mountain.

Miocene conglomerates.—In the lower part of Alverson Canyon (Fig. 4) a heavy conglomerate bed 120 to 130 feet thick overlies a series of tuffaceous strata. This bed is composed of coarse material at the base but becomes finer toward the top. It is only moderately hard and along its upper margin is an abundantly fossiliferous horizon. Splendid coral heads are imbedded in these sandstones, and more delicate forms are found at the base of the superjacent sandy shales. These corals with the molluscan remains that accompany them, all of which await detailed examination and determination, prove the age of the inclosing rocks to be upper Miocene.

On the north slope of Carrizo Mountain, about the head of the easternmost arroyos which are tributary to Garnet Canyon, another series of fragments of a well-developed basal conglomerate are encountered. They extend well up the slopes of the older metamorphic rocks which form the axis of the mountain and dip away from it toward the north or northeast at the rate of 20 or 30 degrees. Being more resistant to weathering agencies than the soft overlying shales, these have been stripped from the sandstones at many points so that the old Miocene beach (Fig. 5), its sands indurated and its teeming life preserved only in fossil form, but yet exhibiting much the aspect and much the same relations which existed at the time of its deposition, is revealed for the modern student's inspection. These basal sands are not always found where their horizon is exposed. In many places the fine clays that were spread out over the sandstones were deposited directly upon the metamorphic rocks that form the core of the mountain and must at one time have formed the bottom and shores of the Miocene sea. The simplest interpretation of this relation is to suppose that before that change of conditions was complete which substituted muddy brackish water with oyster colonies for clear sea water and marine life, the sands of the earlier beach had been swept away, so that there is unconformity, without discordance, or at least without marked discordance in dips, and without a great time interval between the deposition of the sands and the deposition of the muds. The other hypothesis, namely, that these beach sands and the muds were deposited contemporaneously, because of differing local conditions, is made difficult to apply because the two physically different terranes overlie the basement rocks at very closely adjacent points, with no obvious explanation as to why such different conditions should have prevailed so near together.

The heavy sandstones occur at a number of places along the north slope of Carrizo Mountain, east of the head of Garnet Canyon. Many arroyos are incised in them, the stream channel in some cases being a mere notch but a few feet wide

FIG. 5.—Sandstones representing the Miocene beach on the north face of Carrizo Mountain.

and a hundred or more deep. Fossils, however, have not been reported in numbers except at the head of Garnet Canyon.

At the head of Barrett Canyon, which drains south from Black Mountain, the same general relations prevail that have been described in the area a dozen miles to the south on the slopes of Carrizo Mountain. But the fragments of the basal beds of the Miocene are even more widely scattered, and the sandstones and conglomerates are not so fully developed.

About 4½ miles above the mouth and one-half mile above the forks of the Arroyo, the basal beds of the Miocene flank the older rocks and extend across the valley from the west fork to the east fork. Dips here are 20 to 40 degrees to the south, i.e., away from the mountain. The beds are not so thick as on Carrizo Mountain but are succeeded, as is the case there, by soft yellow shales.

About a mile above this point, in a little cove at the head of a small western tributary of Barrett Creek, other outcrops of basal sandstone and conglomerate, not more than 10 feet thick, occur with the underlying igneous rocks all about them. Near this outcrop is a fossil coral reef lying directly upon the lavas and isolated from all the other sedimentaries (Fig. 6). A half-mile farther north a sheet of sandstone, folded into a basin and thus somewhat protected from erosion, still exists. Doubtless many other similar fragmental exposures would be revealed by more extended search.

Miocene shales.—Flanking Carrizo Mountain on nearly all sides and extending on the southeast practically to Signal Mountain, on the international boundary, are continuous exposures of the beds which overlie the basal conglomerate. They are well developed also on the southwest slopes of Black Mountain. The middle drainage basin of Carrizo Creek is a bewildering bad-land area (Fig. 7) formed by the sharp dissection of these soft clays. For many hundreds of feet above the conglomerate, the shale beds contain only

FIG. 6.—Fossil coral reef near the head of Barrett Creek.

FIG. 7.—Erosion in the Miocene shales of the Carrizo Valley.

occasional strata of thin brown nodular sandstone, hence they form smooth clay hills. They are entirely destitute of vegetation because of the aridity of the region, so the area in which they are found is desolate in the extreme. Fresh outcrops of the shales are to be seen only along the flood channels where there has been recent cutting. Ordinarily each shale hill is mantled by several feet of residual material, dust much the greater part of the time, soft adhesive mud during the occasional desert rains. This mantle is the result of the weathering of the soft shales. Exposed to the air, they disintegrate completely and rapidly. This action is probably aided by the efflorescence of certain of the alkali minerals which are abundant in the shales. Wherever a thin sandstone is interstratified with them, it partially preserves them from the rapid disintegration which ordinarily affects them and so usually caps a hill which stands above the general level of the unprotected shale. Such low structural monadnocks are numerou s in the neighborhood of Barrett's Well (Fig. 2).

Farther to the east and south, as in the vicinity of the Yuha Oil Well, at horizons which are presumably higher than those in the Carrizo Valley, although the conditions governing our brief reconnaissance were such that it was not possible to determine the relations with any certainty, the sandstone beds are more abundant. Many of them here weather in curious nodular forms of great variety. Such forms have been well described by Professor Blake.[8]

The Carrizo Valley is synclinal, the shales rising gently northward and southward from the axis of the valley toward the bases of Black and Carrizo mountains. Near the borders dips of 5° to 20° were measured, while in the center of the valley the beds are nearly horizontal or exhibit irregular attitudes. Faults of small displacement were noticed near the head of Garnet Canyon, others north of Carrizo Creek have been described and figured by Blake, and Fairbanks believes that the abrupt eastern face of Black Mountain overlooking the desert is a fault scarp.

In the vicinity of Yuha Oil Well more pronounced structures exist, at least one district anticline with a northeast-southwest axis having been observed half a mile south of the well.

No attempt was made to measure the thickness of these beds, but the buff clays in the middle Carrizo Valley must aggregate 1,000 feet or more, and the beds with a distinctly reddish tone, which are more prominent above Carrizo Station and southeast of Carrizo Mountain, overlie them. The oil well at Yuha starts in sandy strata which appear to be stratigraphically higher than the reddish clays, and at the time of our work in January, 1904, had penetrated over 700 feet of alternating sandstone, shale, gypsum, and shell beds.[9] The thickness of the basal conglomerate on either side of Carrizo Mountain is about 200 feet, as indicated in the generalized sections of Garnet Canyon (Fig. 10).

Superficial deposits.—The shale bluffs along Carrizo Creek, 50 to 100 feet high, are capped in many cases by a deposit of river cobbles three to ten feet in thickness. These cobbles are distributed over practically all of the lower shale hills within the valley whose tops are broad enough to retain the alluvium. Dissection has been so complete that the majority of these summits are reduced to mere points and lines. In these cases the arroyos and the slopes below the summits are often cumbered with the river wash which has slumped down as the hills have been reduced, but has not yet been removed. This material represents a variety of metamorphic and igneous rocks probably derived from the Peninsula Range and its outliers. Its deposition dates back to an earlier erosional cycle, when the present tops of the bad-land hills formed the bottom of the valley of the ancestor of Carrizo Creek.

Within Carrizo Valley proper and in the lowland south of Carrizo Mountain, there is the usual desert accumulation of washed material grading upward toward the mouth of each canyon into alluvial cones. West of Carrizo Station and at other favorable points in adjacent parts of the desert, there are many sand dunes, but these reach a greater development in the more open desert to the north and east.

Passing eastward beyond the valley of Carrizo Creek toward Imperial, the road at first traverses areas of well-reduced shales and sandstones, and then reaches a broad zone of beach gravels full of well-preserved shells of modern types.[10] This zone represents the beach line of the extinct lake that once occupied the Colorado Desert. The beach is not as distinct as a physical feature here as at many points farther to the north (Fig. 8) because the shore line itself was shelving and indefinite, but its general position is well marked by the molluscan remains.

Beyond and below the gravel zone is one of sand, and still farther east beyond this is the silt-covered bottom of the old lake Cahuilla. The deposits here are impalpably fine, laminated clays which, when stirred, as in a much-traveled road, become a tawny flour and when moistened are transformed into a smooth, adhesive mud.

8 "Explorations and Surveys for a Railroad Route from the Mississippi River to the Pacific Ocean" (War Dept., 1857), *Geological Report* by Wm. P. Blake, p. 102.

9 This well was afterward deepened to about 1,200 feet, but the full record is not available.

10 Robt. E. C. Stearns, Ph.D. "The Fossil Freshwater Shells of the Colorado Desert, Their Distribution, Environment, and Variation," *Proc. U.S. Nat. Museum*, XXIV 271-99.

Fig. 8.—Old water-line marking the shore of Lake Cahuilla west of Coachella.

RESUME OF GEOLOGIC HISTORY

The story of the development of this part of the country cannot be read with any approach to accuracy as yet for any period beyond the Miocene. The rocks which represent earlier time are marmorized limestones, schists, and gneisses as to whose age there is much doubt. The slight existing evidence points toward the Carboniferous as the period during which the limestones were deposited here. Whatever their age, their condition now indicates that their history previous to the Miocene was one involving deep burial and intense earth strain. They were upturned, intruded, and crystallized, uplifted and eroded into a mountainous topography, and at the beginning of the late Miocene formed islands in a sea teeming with life. Volcanic forces were active at this time and the flanks of the old land mass are partly buried under the effusive material which issued then, and the muds and the littoral whose fragments were supplied from volcanic sources are conspicuous at many points. But as the period advanced, vulcanism ceased and the present Carrizo and Black mountains were surrounded and perhaps for a part of the time were submerged beneath a clear sea in which the myriad forms of the life of the period swarmed. Still later in the Miocene the character of the sea changed. Instead of clear, salt water, some re-alignment of forces caused great quantities of muddy brackish water to spread about the old islands. Oysters of many forms, some of them of great size, some very tiny indeed, flourished. The heavy silts of these muddy waters accumulated to great depths as the land subsided. Finally the waters withdrew, presumably because of re-elevation, and the region was land again as it had been before, and the shells of many of the creatures which had lived in the clear and then in the muddy waters were preserved in the accumulated sediments. As the sea withdrew, the destructive forces of weathering and the erosive forces of wind and running water became active. The clays which had accumulated were now dry and were cut away again by these forces. The process was not long continued and the plain was not completed, those clay areas which were capped by protecting sandstones remaining as monadnocks above the wide valley floor. This valley, occupied by an earlier vigorous ancestor of Carrizo Creek, was strewn with rounded river cobbles brought from the higher mountains to the west. South and east of Carrizo Mountain large areas seem to have been reduced at this period well toward the condition of a peneplain. This plain lies perhaps two hundred feet above the later Pleistocene lake level with which it seems to have no connection. It is regarded as an earlier independent feature, perhaps Pliocene in age.

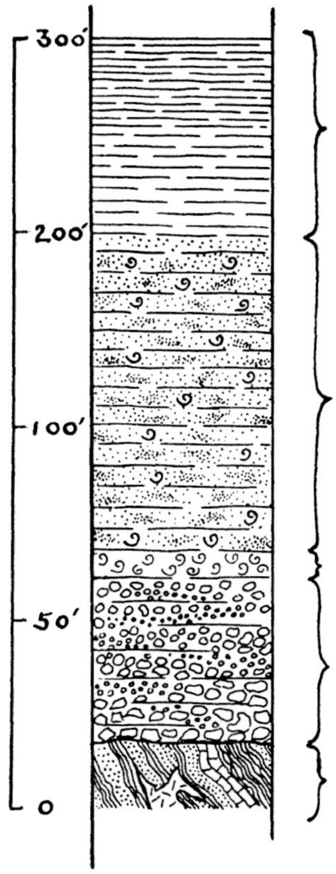

Fig. 10.—Columnar section of the rocks exposed in Garnet Canyon.

After the formation of this partially planed surface, over the soft rocks of Carrizo Valley, some change either in the relations of land and sea, or of climatic conditions, enabled the streams to dissect it again. The result of this dissection, which may well have been contemporaneous with the last occupancy of Colorado Desert by the Gulf of California, is seen in the Carrizo Creek bad lands of today.

The last important element in the development of the geography of this part of the desert was the formation

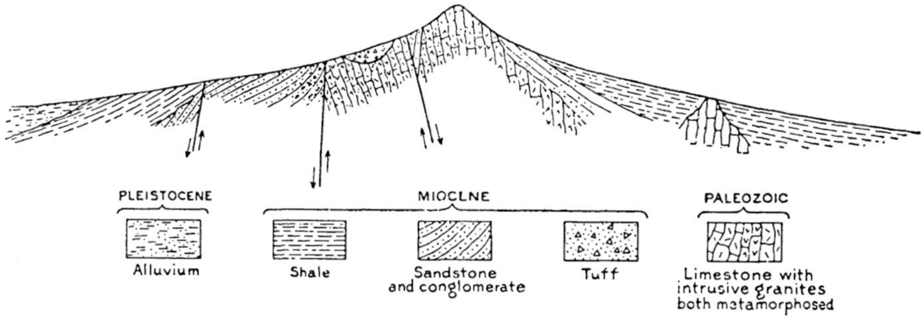

Fig. 9.—Diagrammatic section across Carrizo Mountain by way of Alverson and Garnet Canyons.

and the disappearance of the desert lake. So late is it, that the calcium carbonate incrustations which it left on its western shore (Fig. 8) show but little effect of erosive or solvent action since the waters left them, and the sandy beach, molded by the waves of the lake upon the alluvial fans which formed a large part of its shores, is still well enough preserved to be readily traced. Only the most modern gullies have cut it away. At one point a low sea cliff notched by the waves in steep alluvial-fan material still stands, as perfectly preserved as though the waters had just withdrawn.

CONCLUSION

In conclusion it is to be said that but the barest outlines of the history of this fascinating region are yet known. Its rocks contain a rich upper Miocene fauna, probably in large part new, and its exposures and structures are so clear that its geology will be an open book to the fortunate student to whom falls the pleasant task of deciphering it in detail.

FOSSIL LOCALITIES[11]

Shells are very abundant in the vicinity of Yuha Oil Well, but the variety is not as great here as at other localities where collections were made. A few hundred yards west of the well is an outcrop of an oyster bed which makes a conspicuous shell mound and other similar outcrops exist in the vicinity.

The horizon is the highest at which collections were made. It is probably a few thousand feet above the Carrizo and Black mountain horizons. Collections Nos. 160 and 161 were made near Yuha Well, and No. 162 was made about one and one-half miles southeast, but the horizons are not believed to be far apart.

Collection No. 165 is from a point about two miles east of the base of Carrizo Mountain. Shells are abundant here but species are limited, as is true of the vicinity of Yuha Well. These shells are probably stratigraphically lower than collections Nos. 160–62 but are higher than the others except possibly No. 168. Collection No. 168 was made on the county road, near Barrett's Oil Well.

The locality is nearly midway between Carrizo and Black mountains and must be from substantially the same horizon as Blake's original collection. It is also near the horizon of No. 165, but may be slightly lower.

No. 163 is from a small arroyo just east of Alverson Canyon, on the south side of Carrizo Mountain. The shells were taken from the yellow clays which immediately overlie the basal conglomerate. These clays contain a rich fauna which is not by any means fully represented in the collection. Stratigraphically these beds belong above the three collections yet to be mentioned. The length of the time interval between the two horizons depends upon the extent of the unconformity between the conglomerates and the clays. It is probable that this interval is slight, but there are no data at hand for estimating it.

Nos. 164 and 166 are from Alverson Canyon and the head of Garnet Canyon on the south and north slopes respectively of Carrizo Mountain. The horizons are identical, being in each case the sandstones which form the upper part of the arenaceous series at the base of the Miocene. These are the most conspicuous fossil localities in the region. The shells or their casts have weathered out and strew the slopes in great profusion. Corals, echinoids, ostrea, pectens, strombus, and malea are everywhere. The matrix, however, is coarse, and only large and robust types are well preserved. The locality has been noted by prospectors generally, because the occurrences are so conspicuous.

No. 167. This collection consists of corals almost entirely. The fossil reef is near the head of Barrett Canyon and lies directly upon the igneous rocks which served as a basement for Miocene sedimentation at this point. Whatever later beds may have originally covered it have been stripped away, so that the old reef is now isolated. There can be little doubt, however, that its position is at the base of the Miocene series and substantially equivalent to that of Nos. 164 and 166.

[11] These descriptive notes are introduced for the use of the paleontologists who may eventually study the collections now in the U.S. National Museum.

GEOLOGY AND MINERAL RESOURCES OF SAN DIEGO AND IMPERIAL COUNTIES

BY FREDERICK J. H. MERRILL, PH. D.

DECEMBER, 1914
California State Mining Bureau, California State Printing Office

CHAPTER II.

IMPERIAL COUNTY.

HISTORY.

This county formed part of San Diego County from the organization of the latter in 1851 to August, 1907, when it was separated and formed into a separate county by vote of the supervisors of San Diego County under an act of legislature giving them that power.

PHYSIOGRAPHY.

The most important feature of Imperial County is the broad and nearly level expanse of the Colorado River Delta which separates the Gulf of California from the Salton Basin and is generally known as the Imperial Valley. This basin lies in a synclinal valley bordering on the east the anticlinorium of the lofty granite ranges of San Diego and Riverside counties. Several small ranges of volcanics, metamorphic palæozic rocks and uplifted Tertiary beds, interrupt the general level of the delta plain and, on its eastern margin, is an extensive hilly area of volcanic rocks which continues eastward into Arizona. The great granite masses of San Diego County, barely appear east of the boundary line.

GEOLOGY.

In its general geology, Imperial County is closely linked with San Diego and, under that head, will be found a synopsis of the more important geological features. The chief item that is distinct is the history of the Salton Basin.

It appears that, in Tertiary time, this basin was the northern end of the Gulf of California, which extended some 160 miles farther north than at present, and extensive marine deposits were laid down, as shown by the marine Miocene fossils associated with the clays on its western margin. This condition of submergence probably continued after the close of the Tertiary and later, as the Pleistocene uplift of about 1,000 feet above present sea level which occurred on the Pacific coast of California probably involved this valley, a large part of the Tertiary Gulf of California became dry land. With the succeeding subsidence came another marine occupation of the valley. In the meantime, the delta deposits of the Colorado River, which debouched into this basin near Yuma, gradually increased in extent and altitude until they formed a natural dam of great width, isolating, from its connection with the gulf, that portion of the basin north of the present area of settlement. The Salton Basin was then occupied by a salt lake which, in that arid climate, was gradually desiccated and finally became dry, leaving behind large deposits of salt. From that time on, Salton Basin was alternately dry and flooded by the waters of the Colorado as, from time to time, the river found its outlet toward the gulf or toward the north.*

The local Indian tribes have various traditions of the presence of a great lake in the valley and its existence is proven by the presence of great numbers of fresh and brackish water shells, which gave to the northern part of the basin, in Mexican times, the name of the Conchilla Valley, barbarized by later inhabitants into Coachella. There is, also, substantial evidence of this lake in the old beach line which has been traced around the basin at an altitude of about 40 feet above sea level.[1] For this prehistoric body of water the name of Lake Cahuilla was proposed by its discoverer, Prof. W. P. Blake.†

The earliest inundation after the American occupation of the region occurred about 1849. The lake of that period also dried up and the Salton desert was without water until June, 1891, when another over flow of the Colorado River produced a lake 30 miles long and 10 miles in width.[2] This, in time, gradually evaporated and became much reduced in area, but in 1904, through the careless opening of the north dike or levee, in connection with the irrigation of the Imperial Valley, the river was again released and the Salton Basin was flooded and the submergence rapidly increased, doing great damage, until the Southern Pacific Railway Company, under presidential authority, closed the gap in February, 1907.

* See further U. S. G. S. Water Supply Paper No. 225, by W. C. Mendenhall, pp. 17–20. Also U.S.G.S. Water Supply No. 300, by H. D. McGlashan and H. J. Dean, pp. 29–33.

† *Nat. Geogr. Mag.*, Vol. 18. p. 830.

[1] Wm. P. Blake, *Pacific Railroad Reports*, Vol. V, 1853.

[2] E. B. Preston, R. XI, pp. 387–393.

MINERAL RESOURCES.

While of no great variety in its mineral resources, Imperial County is not deficient in natural wealth as, apart from its great agricultural possibilities, its deposits of crystalline limestone and clay promise a great supply of building material which should make it a center of production for a very large area. Its production of metals also, may be important in the near future.

In the following pages these resources are briefly outlined.

. . .

MARBLE.

BULLETIN 38, PAGE 107.

The chief deposits of this material in Imperial County are in Coyote Mountain, T. 15 S., R. 10 E. This mountain is largely formed of crystalline limestone but not all of it is fit for marble. Here valuable deposits, within 6 miles of the San Diego and Arizona Railroad, now under construction, are owned and controlled by the Golden State Mining and Marble Company, 309 Watts Building, San Diego, C. A. Walker, president; H. H. Sparks, secretary.

This company owns fourteen deposits, each deposit being of different grade and color. Several have the grade and texture of famous imported marbles.

The Creole group consists of four claims of 20 acres each, covering an outcrop of marble about 4,500 feet long and averaging 600 feet wide. All of the development has been done on Creole claim No. 1. Here are three distinct strata of marble. The first is deep blue-black with a hardness of 3½, free from quartz, chert, or other impurity. It is of uniform color, free from holes or blemishes and is impervious to moisture so that it can not be stained even by red ink. The quarry will produce blocks of large size even at the surface. This is suitable for borders, exteriors and pillars and is probably well suited for monuments.

The second stratum on the Creole group is cream pink in color, with occasional blotches of red and black lines which, in the translucent marble, look like the so-called moss in an agate. This marble somewhat resembles the Italian Paonazza. It has a hardness of 3½, is exceedingly tough and free from impurities, takes a high polish and has the desirable quality of translucency when lights are placed behind a slab. This stone also resists stains of ink or oil.

The third deposit of the Creole group is white with a bright green thread line. So far as known, this marble is in small blocks and suitable only for small borders or tiling.

About 2½ miles southerly from the above group is a deposit of marble varying in width from 30 to 400 feet and about 3,000 feet in length.

One end is mottled black and white marble, the color is distinctive. This deposit is solid at the surface. It takes a fine polish and does not show any strings or blemishes. It is suitable for corridors and for interior decoration.

Outcrops of Creole Marble, Coyote Mountain. F.J.H.M. Photo.

About 600 feet on the other end of this deposit, the marble is very fine grained, and is highly colored in red, yellow, white, purple and blue. In one opening the stone is banded like agate in blue, red and yellow. This marble is over 3½ in hardness, grained like wood and takes a high polish, showing the colors of the African Numidian, as well as lighter and darker tints not seen in that stone.

Near the above deposit the company owns one claim, containing 20 acres of boulders, which appear to be concretions remaining from some overlying formation which has disappeared. The stone is not quite as hard as the ordinary marble, but takes a smooth, glossy finish and shows a great variety of color, including pink, cream, red, yellow, orange and purple, with dashes of green and slender black lines, brecciating the entire mass. This marble is of the Sienna type, but more beautiful and though it can not be obtained in large pieces, it can be used in tiling, soda fountains, knobs and handles for plumbing work, fancy articles and curios.

Creole Marble, Bored by Marine Mollusks. F.J.H.M. Photo.

On the east end of the mountain and about one mile southeast of the above named quarries, this company owns a deposit of blue marble, closely resembling the Esperanza of Vermont. It is fine grained and hard, and is suitable for almost any work to which marble may be put, being especially adapted for monuments. Large blocks of uniform color are ready for shipment.

At the west end of the Coyote Mountain range is a deposit of marble, about one half of which is pure white and as translucent as opal. The other half is a peculiar marble, of lavender tint, without markings of any kind, and can be matched from any part of the quarry. Very little development work has been done here.

This company has erected buildings and ordered machinery for a plant at National City, on the line of the San Diego and Arizona Railroad for the exclusive cutting and finishing of this marble. One carload was shipped on the San Diego and Arizona Railroad last year via El Centro and Los Angeles, and 5 more carloads will be shipped during the coming year to the National City plant.

R. G. Fritz of San Diego also owns eight claims of marble in Coyote Mountain.

. . .

LIME AND CEMENT

On the south slope of the Coyote Mountain and at a point 3½ miles southwest of Coyote Wells are great deposits of limestone.

These deposits are from 3 to 5 miles distant from the survey of the San Diego and Arizona Railroad, and could be easily reached by spurs from that line. About a quarter of a mile from one of the largest deposits is a deposit of blue clay, averaging 70 feet in thickness and covering more than 300 acres of land.

The fact that this limestone carries little or no magnesia makes it ideal for cement manufacture as even a small content of magnesium is undesirable in high grade cement.

An analysis of a sample from Coyote Mountain by F. Salathé is here given:

Calcium carbonate	96.6
Magnesium carbonate	1.7
Silica	Trace
Alumina and iron sesquioxide	0.9
Calcium sulphate	0.5
Totals	99.7

The raw materials for Portland cement are very abundant on the west side of the Imperial Valley in great deposits of Tertiary clay and crystalline limestone, which are close together, and will soon be within reach of transportation. As the population of the valley increases, there will be a great demand for Portland cement for the lining of irrigation ditches and the local material will doubtless form the basis of a large manufacturing plant.

. . .

GYPSUM.

This material, the hydrous calcium sulphate or sulphate of lime, is of much importance in construction as the calcined product, plaster of Paris, is the base of nearly all wall plasters for interior work. It also forms the base of the material known as staff which is so extensively used in the construction of temporary buildings for exposition purposes. Gypsum is also important as a fertilizer in supplying lime and in correcting alkaline soils. In small quantities, it is used in the manufacture of Portland cement to retard its setting.

Miocene Tertiary Clays. East Side of Coyote Mountain. F.J.H.M. Photo.

The gypsum of Imperial County occurs mainly in an immense deposit some three miles in length from southeast to northwest, with an area of some 800 acres, which occupies a valley, among crystalline rocks, in the northwestern edge of the Fish Creek Mountains. This spur from the crystalline ranges of San Diego County, as shown by the geologic map, lies chiefly in townships 13 and 14 south, ranges 9 and 10 east. The gypsum outcrops are in Sections 19, 20, 28 and 29. T. 13 S., R. 9 E. It has not been thoroughly explored by boring to determine its depth but, from the exposure in ravines, its average thickness is estimated at 100 feet. This is claimed by the owners to be the largest deposit of pure gypsum on the Pacific Slope.

The following analysis made by J. O. Handy, of Pittsburg, Pa., is furnished by one of the owners:

Silica	0.92
Alumina	0.21
Iron protoxide	0.14
Lime	32.75
Magnesia	0.08
Sulphur trioxide	47.17
Water	18.72

The ownership of this deposit is distributed as follows:

Acres	Location	Owner
120	SE. ¼, SW. ½; S. ½, SE. ½; Sec. 19	Bert R. Chaplin, Suisun.
120	SE. ¼, NW. ¼; N. ½, SW. ¼; Sec. 20	S. L. Ward, San Diego.
160	SE. ¼, SW. ¼; SW. ¼, SE. ¼; Sec. 20 NE. ¼, NW. ¼; NW. ¼, NE. ¼; Sec. 29	Walter O. Hamilton, El Centro. L. E. Cooley, El Centro.
80	E. ½, NE. ¼; Sec. 29	W. H. Allen, San Diego.
40	NW. ¼, NW. ¼; Sec. 28	Sam C. Mack, Imperial.
40	SW. ¼, NW. ¼; Sec. 28	Ed H. Houck, Imperial.
80	NE ¼, NW. ¼, NW. ¼, NE. ¼; Sec. 33	David Chaplin, Suisun. W. L. Gillette, Holtville.
	The above are all patented under the timber and stone act.	
160	NE. ¼, SW. ¼; NW. ¼; SE. ¼; Sec. 28 S. ½, SE. ¼	C. H. Swallow, San Diego.

None of these are patented.

Gypsum Outcrop. Fishcreek Mountain.

Gypsum Outcrop. Fishcreek Mountain.

This deposit must necessarily become of great importance as soon as it is accessible to transportation. The new line of the San Diego and Arizona Railroad will pass within about 25 miles, and this will be rather too great a distance to haul a product of moderate value. Should the Atchison, Topeka and Santa Fe Railway build its contemplated line near the old highway from San Felipe through Carrizo Creek, the problem will be in a large measure solved.

Situated on the south slope of the Coyote Mountain range is another deposit of gypsum covering the better part of 10 acres. It is exposed over most of this area, has an average depth of 8 feet, and is said to show by measurements about 84,000 tons in sight, with a continuation under the overcap that can not be determined. The quality is said to be the same as that of the larger deposit to the north. This deposit lies about 3 miles northwest of Coyote wells and could be easily worked.

PETROLEUM.

This important fuel has been sought by boring on the west side of Imperial Valley, but without result.

While Miocene beds are present in this valley it would appear that the conditions under which they were deposited were not specially favorable to oil formation.

The principal well was the Yuha, Sec. 32, T. 16 S., R. 11 E., drilled by the Yuha Oil Company, Chas. Swallow, of San Diego, secretary-treasurer. It reached a depth of 1,363 feet, but without result. Another well was the Mesquite, about 850 feet deep, in Sec. 27, T. 12 S., R. 10 E. This brought a supply of flowing water.

The Barrett well drilled near Carrizo Creek, to a depth of 1,100 feet, was also unsuccessful.

SULPHUR.

There is a small exposure of this mineral on the east slope of Coyote Mountain. A claim has been located by Chas. Swallow of San Diego.

DISTRIBUTION AND AGE OF THE MARINE TERTIARY DEPOSITS OF THE COLORADO DESERT

BY W. P. WOODRING

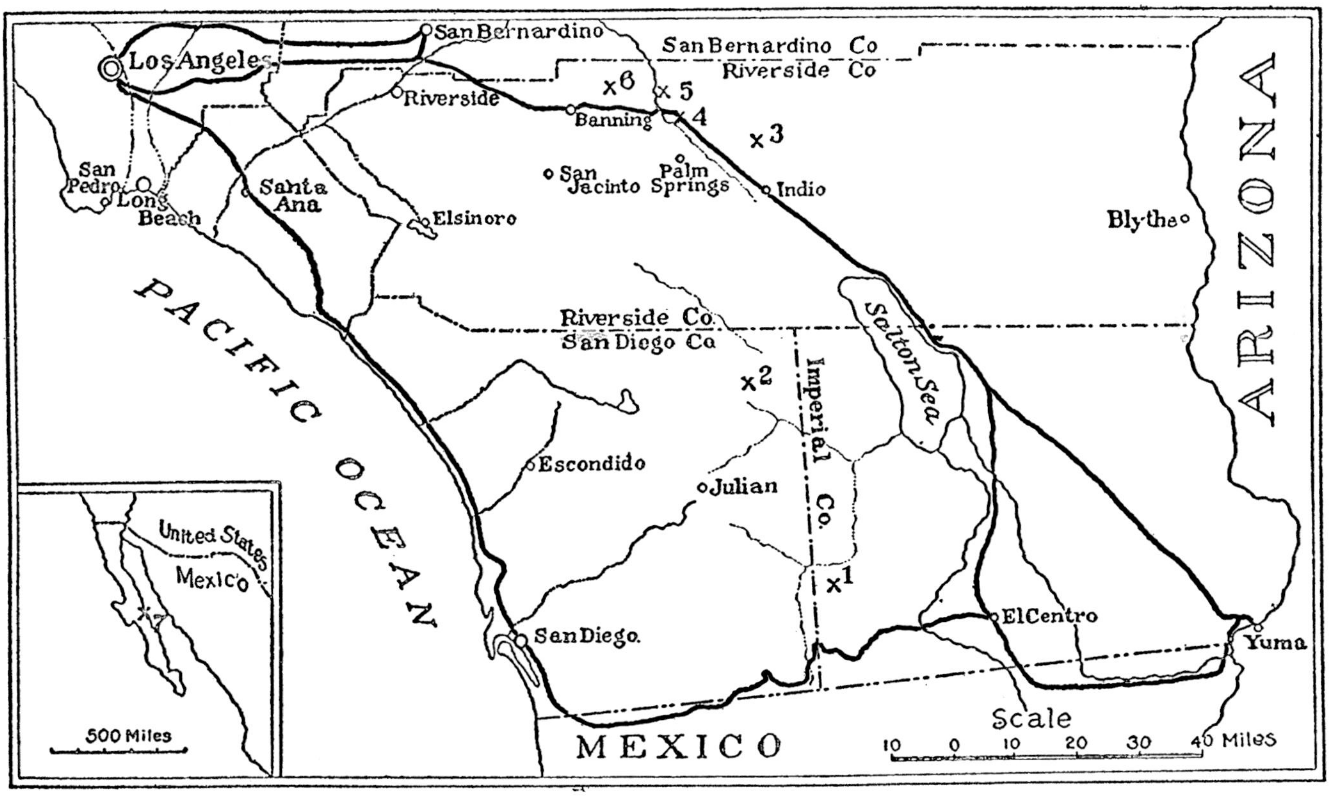

Fig. 1—Sketch maps of parts of Southern California and Mexico showing localities described.

INTRODUCTION

The marine Tertiary beds of the great desert trough in which the Salton Sea lies have attracted attention ever since they were first discovered by Blake in 1853. Blake saw them as he was traveling down the west side of the desert from San Gorgonio Pass with the party of engineers to which he was attached. They have been most extensively explored around Carrizo Mountain, a spur extending eastward into the desert from the Peninsular Range close to the Mexican boundary. Many geologists have examined the region around Carrizo Mountain and the fossils found there have been described in many reports, but the lack of an adequate map has retarded detailed mapping and precise stratigraphic work of the kind that is required to fully decipher the geologic history. Dr. J. C. Merriam long ago recognized the critical position that this region holds in any attempt to establish a synchronization of events in the geologic history of California and tropical America. During the past few years I have carried on work there as part of a project sponsored by the Carnegie Institution of Washington and the California Institute of Technology. Through the cooperation of Mr. H. S. Gale, aeroplane photographs covering the eastern part of Carrizo Mountain were available during the winter of 1929-30. With these photographs as a base, detailed mapping was carried

From: Contributions to Palaeontology: Papers Concerning the Palaeontology of California, Oregon and the Northern Great Basin Province. Carnegie Institution of Washington. Washington, D.C. Publication No. 418. July 1932. pp. 1–25.

on in April 1930. The mapping is not yet complete, but the results obtained seem to justify this preliminary report, which also sets forth the discovery of the marine beds in other parts of the desert trough and its borders. I hope to complete the mapping and to prepare an account of the geology and also a critical review of the fossils.

I have greatly profited from discussions with Mr. Gale and with Professor A. O. Woodford, of Pomona College, who kindly criticized the discussion of the Tertiary vulcanism of parts of California adjoining this area. Mr. G. H. Anderson examined a specimen of the lava from Carrizo Mountain and Mr. D. D. Hughes identified several collections of foraminifera.

The name Colorado Desert is used for this region as I think it should be revived as a general regional name comparable with Mohave Desert. Unfortunately it has fallen into disuse and no other name except "Salton Sea region" or "Salton Basin," which are less satisfactory, has taken its place. Coachella Valley and Imperial Valley cover only part of the area.

DISTRIBUTION

CARRIZO MOUNTAIN AND VICINITY

(LOCALITY 1, FIG. 1)

Blake[1] followed the old road along Carrizo Creek three times. I saw or surmised the essential geologic features of the region. Anyone who knows the country and reads his pages must admire his keen observations and lucid, vivid descriptions, not only in this area but along other parts of the routes he covered. His description of wind action in San Gorgonio Pass and of the little pedicles of feldspar protected from wind erosion by tips of garnet ("pointing like jewelled fingers in the direction of the prevailing wind") is one of the most beautiful and inspiring that I know of in geological literature. Mendenhall's[2] and Kew's[3] accounts of the geology around Carrizo Mountain are the most detailed now available, but others have been issued by Fairbanks,[4] who made the first extensive explorations, Bowers,[5] Orcutt[6] and Brown.[7] The sketch maps by Kew and Brown are the only attempts at a geologic map published.

The oldest rocks on Carrizo Mountain are phyllite, mica schist, quartzite, schistose limestone and marble that are intruded by granitic and gabbro-like rocks. These metamorphics are intensely deformed and at places lie in great recumbent folds. They certainly are not younger than Paleozoic- and in degree of metamorphism resemble Paleozoic rocks of the San Bernardino Range. Lying on the metamorphics and intrusives is a series of basalt flows, flow-breccias, tuffs, conglomerate and sandstone. Generally resting on the volcanics, but at places interbedded with them, are the marine deposits. I will return at greater length to the relation of volcanics and marine beds under the discussion of the age. Where the volcanics are absent the marine beds rest directly on the metamorphics.

The classification of the marine beds and overlying deposits here adopted and its relation to earlier arrangements are shown in the following table [next page].

The marine beds were first formally named by Kew,[8] who called them the "Carrizo formation" ("Carrizo Creek formation" is the form first used). Inasmuch as the name Carrizo sandstone was already in use for Eocene deposits in Texas,[9] as Vaughan[10] pointed out, Hanna[11] proposed Imperial formation as a substitute name. Unfortunately he also used the same name in a greatly restricted sense for the so-called coral reef in Alverson Canyon, and introduced additional names, as shown in the preceding table, for other parts of the section. The so-called coral reef is a lens in detrital deposits. For the part of these detrital deposits overlying the so-called reef, Hanna proposed the name "Latrania sands." This scheme would leave unnamed the beds below the so-called reef. Further objections to Hanna's nomenclature arise from the interbedding of the "Yuha reefs" at various horizons in the "Coyote Mountain clays." It is here proposed to simplify this nomenclature by using the name Imperial formation in Hanna's broad sense, but redefining it to embrace the entire series of marine deposits bordering Carrizo Mountain, and to exclude the overlying nonmarine beds. The Imperial formation is here redefined consists of a basal member and an overlying siltstone member. These subdivisions correspond to Kew's

[1] W. P. Blake, *Geological report, Williamson's Reconnaissance in California,* U. S. Pacific R. R. Expl. (U. S. 33d Cong., 2d sess., Senate Ex. Doc. 78 and House Ex. Doc. 91), vol. 5, pt. 2, pp. 80–105, 107–108, 120–122, 174–176, 1857.

[2] W. C. Mendenhall, *Notes on the Geology of Carrizo Mountain and Vicinity, San Diego County, California,* Jour. Geol., vol. 18. pp. 336–355, 1916.

[3] W. S. W. Kew, *Tertiary Echinoids of the Carrizo Creek Region in the Colorado Desert,* Univ. Calif. Pub. Bull. Dept. Geol., vol. 8, No. 5, pp. 39–60, 1914.

[4] H. W. Fairbanks, *Geology of San Diego County, also of Portions of Orange and San Bernardino Counties,* Calif. State Min. Bur., 11th Ann. Rpt. State Mineralogist, pp. 88–90, 1893.

[5] Stephen Bowers, *Reconnaissance of the Colorado Desert Mining District,* Calif. State Min. Bur., 19 pp., 1901.

[6] C. R. Orcutt, *The Colorado Desert,* West American Scientist, vol. 12, No. 1, pp. 2–14, 1901.

[7] J. S. Brown, *The Salton Sea Region, California,* U. S. Geol. Survey Water-Supply Paper 497, pp. 44–46, 1923.

[8] W. S. Kew, loc. cit., p. 40, 1914.

[9] J. Owen, Texas Geol. Survey First Ann. Rpt., pp. 70, 73, 1889.

[10] T. W. Vaughan, U. S. Geol. Survey Prof. Paper 98, p. 357, 1917.

[11] G. D. Hanna, Proc. Calif. Acad. Sci., 4th ser., vol. 14, No. 18, p. 434, 1926.

Classification of Tertiary marine beds and younger deposits of Carrizo Mountain and vicinity

<table>
<tr><th colspan="2">Kew 1914</th><th>Hanna 1926</th><th colspan="2">This report</th><th>Approximate thickness</th></tr>
<tr><td rowspan="3">Carrizo formation</td><td>?</td><td>?</td><td colspan="2">Palm Spring formation (non- marine): unconsolidated or poorly consolidated sands, light chocolate-brown and faint brick-red silts.</td><td>1000 feet (est.)</td></tr>
<tr><td>Upper Division</td><td>Yuha reefs.
Coyote Mountain clays</td><td rowspan="2">Imperial formation (marine)</td><td>Siltstone member: buff silt- stone with occasional hard beds of very fossiliferous limy sandstone.</td><td>1400 feet</td></tr>
<tr><td>Lower Division</td><td>Latrania sands
Imperial formation</td><td>Basal member: conglomerate, sandstone, coralliferous limestone, and at places basalt flows and flow-breccias</td><td>Few inches to 600 or 700 feet</td></tr>
</table>

"Lower division" and "Upper division," except that there is nothing in his account to indicate that he did not also include in his "Upper division" at least the basal part of the nonmarine beds here named the Palm Spring formation.

The thickness and composition of the basal member of the Imperial formation are very variable. It may consist of only a few inches of limy sandstone or it may be made up of hundreds of feet of conglomerate and sandstone. Where it rests on the metamorphics, as near the head of Alverson Canyon on the south slope of Carrizo Mountain, the conglomerate is a mass of angular rubble that can hardly be distinguished from the dissected fan material that may overlie it. If the basement consists mostly of limestone, as near the east end of the mountain, the conglomerate is made up of limestone cobbles and blocks embedded in a limy matrix. The most curious conglomerate is the one that at places overlies the volcanics. It consists of closely spaced cobbles and boulders, or angular blocks, of basalt with a thin filling of limy matrix between them. This limy matrix has the appearance of calcite veinlets and the rock could readily be mistaken for an altered flow-breccia were it not for the fossils, many of which are large enough to completely fill the space between boulders. Cracks filled with the fossiliferous matrix extend down into the underlying basalt and it is difficult to determine just where the basalt ends and the conglomerate begins.

The conglomerate at many places grades upward into sandstone, with which limestone or limy sandstone carrying scattered heads and pieces of corals of reef facies may be interbedded, as in Alverson Canyon. I have not yet seen anything, however, that could properly be called a coral reef. Apparently most of the corals and other fossils in the coraliferous limestone in Alverson Canyon represent material that is not in the place where it grew. The oyster shells, which are far more abundant than corals in this bed, are badly worn. Fossils are to be found almost anywhere in the basal member except in most of the coarse conglomerate, but they generally are in a miserable state of preservation. The few accessible localities where good specimens can be obtained are pretty well cleaned out and new material weathers out slowly. A new locality recently discovered high on the crest of one of the main ridges yielded many fine specimens, but it already is in the class of places that are worked out.

The buff siltstone member conformably overlies the basal member with apparently no discontinuity. The basal member is so thin that the siltstone at many places seems to overlap it and to rest directly on the older rocks, but even at such places clean exposures show that the basal member, perhaps only a few inches thick, intervenes. Wherever the siltstone rests directly against the older rocks it is found to be in fault contact. Throughout the siltstone member are discontinuous beds of hard limy sandstone that form caps or dip slopes. These sandstones carry great numbers of a small oyster (called *Osirea vespertina* Conrad since Conrad's day, but probably representing a different species), or of a small jingle shell (*Anornia subcostata* Conrad). A few small scallop shells (*"Pecten" deserti* Conrad) and an occasional large oyster (*Ostrea heermanni* Conrad) may also be round. Some beds are almost pure oyster reefs, others are virtually pure jingle shell reefs. The siltstone itself gives rise to intricately dissected areas and weathers to a dust-like powder that effectively mantles the rock. In fresh exposures it is seen to be a hard siltstone that readily disintegrates in water. Marine fossils have been found at different horizons in the siltstone from the base almost to the top, but they are not at all obvious and apparently are absent throughout hundreds of feet. Beds that seem to be barren are found to carry a few foraminifera and plant débris. Some of the siltstone fossils are the same as those found in the basal member; others are different, but the difference may be due principally to facies control. The thickest continuous section of siltstone in this region is one lying between Carrizo Creek and Fish Creek Mountain that

Mr. Gale and I examined. The total computed thickness of siltstone in this section is 1250 feet, and the top is not exposed on account of faulting.

Clean contacts between siltstone and the underlying beds have not yet been seen, but the change takes place within a short interval. This considerable thickness of silt has some particular significance in the geologic history of the region. It is improbable that the material was suddenly furnished by the land areas that had been supplying the underlying coarse detrital débris. If this source is eliminated, the only apparent source is the ancient Colorado River; perhaps they represent the first recorded invasion of the Colorado into the Colorado Desert. This flood of silt, whatever its source, exterminated the corals and other sea animals that tolerate only clear water. The presence of normal marine fossils at different horizons shows that at times the salinity of the water was not lower than during the deposition of the basal member. The oyster reefs give an impression of brackish water, but so far as I know scallops and jingle shells do not tolerate water of low salinity. Perhaps the absence of fossils throughout a great part of the section may be attributed to particularly rapid increments of silt that smothered colonies that had succeeded in establishing themselves.

Conformably overlying the bluff siltstone with gradational contact are poorly consolidated or unconsolidated sands and silts of light shades of chocolate-brown and brick-red, which are well exposed in the vicinity of the ruins of the old stage station on Carrizo Creek. For these beds which have an estimated thickness of 1000 feet, name Palm Spring formation is proposed. This name is derived from the name of a spring along the lower part of Vallecito Creek, which is designated as the type region. Vallecito Creek is a south-eastward-flowing tributary entering Carrizo Creek about a mile above the old stage station. These beds have the appearance of nonmarine deposits, but the lower part may have been laid down in brackish water. At all events Blake (p. 105) recorded the finding of a *"Gnathodon"* (a brackish-water matroid now known as *Rangia*, *R. lecontei* (Conrad)) in the "red strata" on Carrizo Creek where the road descends to the creek. Toward the mountains the Vallecito formation probably grades into fanglomerates.

WEST SIDE OF COLORADO DESERT BETWEEN FISH CREEK MOUNTAIN AND SANTA ROSA MOUNTAINS
(Locality 2, fig. 1)

Farther north along the west side of the desert is an extensive area for which only fragmentary information is available. I have not yet seen any part of it. Blake (p. 103) found specimens of *"Gnathodon" lecontei* along "Salt Creek" (San Felipe Creek?), and Kew (p. 44) found the same species north of Superstition Mountain. These records apparently indicate outcrops of the lower part of the Palm Spring formation. The marine beds proper may turn up anywhere in this area.

INDIO HILLS
(Locality 3, fig. 1)

Buwalda and Stanton[12] have recently recorded the presence of marine beds along the south edge of the Indio Hills, which lie along the east side of the desert. They were also discovered independently by H. S. Gale. The marine beds, consisting of buff siltstone like that in the siltstone member of the Imperial formation, are found in isolated patches both east and west of the entrance to Thousand Palms Canyon. They dip steeply into the hills and are overlain by coarse nonmarine beds, called the "Indio formation" by Buwalda and Stanton.[13] Even from the highway the buff color of the marine beds stands in marked contrast to the somber overlying land-laid deposits. Almost pure reefs of *Ostrea "vespertina"* lie in the upper part of the series. I have collected the following fossils from the marine beds without having exhausted the possibilities:

Preliminary list of fossils from Indio Hills
(Calif. Inst. Tech. Loc. Nos. 187-191)

Gastropods:	Lamellibranchs:
Turritella imperialis Hanna	*Senilia*
Vermetid?	*Ostrea "vespertina"* Conrad
"Cerithium"	*"Pecten" deserti* Conrad?
	Anomia subcostata Conrad?
	"Crassatellites"
	Macoma?
	Tagelus?
	Corbula

GARNET HILL
(Locality 4, fig. 1)

Garnet Hill is an isolated low hill lying along the westward extension of the Indio Hills, a few hundred yards east of Garnet, or Palm Spring Station as it formerly was called, on the Southern Pacific Railroad. I was guided to this locality by Mr. Richard Bramkamp, of Pomona College, who discovered marine fossils there while looking for minerals weathered out of schist boulders. We found no

12 J. P. Buwalda and W. L. Stanton, *Geological Events in the History of the Indio Hills and the Salton Basin, Southern California*, Science n. s., vol. 71, pp. 104–106, 1930.

13 Unfortunately this name also is preoccupied, and by a curious coincidence it is in use for Eocene beds in Texas that lie in the same section as the Carrizo sandstone, a name that antedates "Carrizo formation," which Buwalda and Stanton use for the marine beds in the Indio Hills. I am indebted to Dr. T. W. Stanton for pointing out this earlier use.

clean exposures on the entire hill. The surface is littered with the kind of rubble that weathers out of fanglomerates. The highway cut skirting the west end of the hill shows such débris embedded in caliche-soaked fine material. The following fossils were picked up on the south slope of the hill, either lying loose among the rubble or partly embedded in caliche.

Preliminary list of fossils from Garnet Hill
(Calif. Inst. Tech. Loc. No. 192)

Lamellibranchs:
- *Ostrea heermanni* Conrad
- *Pecten carrizoensis* Arnold
- *"Pecten" purpuratus mendenhalli* Arnold?
- *"Pecten"* cf. *sanctiludovici* Anderson and Martin
- *"Pecten"*
- *Spondylus*

Echinoid:
- *Clypeaster*

Though no outcrops can be seen, it is concluded that beds of the Imperial formation are in this hill and are masked by a veneer of fanglomerate, for some of the fossils are too perfect to have been transported as constituents of the fanglomerate and no boulders from which they might have weathered out were observed in the fanglomerate.

PAINTED HILL

(Locality 55, fig. 1)

At least two desert-rovers independently discovered marine Tertiary beds adjoining Painted Hill, a schist hill lying east of Whitewater River in the foothills north of the Colorado Desert. My attention was first drawn to this locality by Messrs. R. B. Peters and K. Garner, who showed me some fossils collected there by a friend from San Bernardino. Acting on this information Buwalda, Stanton and I found marine beds in the second canyon east of the river, where they are in contact with the schist, apparently along a high-angle thrust. Here the beds consist of buff micaceous siltstone and fine micaceous sandstone. They apparently grade upward into sandstone and fanglomerate, called Coachella fanglomerate by F. E. Vaughan.[14] Though Vaughan considered this fanglomerate of Quaternary age, it certainly is much older and occupies the same stratigraphic position as the "Indio formation." After visiting this region I found that Bramkamp had discovered other localities in the first canyon east of the river, where the marine beds consist of conglomerate, sandstone and siltstone. My collections from Painted Hill consist of the following species. Bramkamp, who has collected there repeatedly, has many others.

Preliminary list of fossils from Painted Hill
(Calif. Inst. Tech. Loc. Nos. 185, 186, 194, 195, 624)
Species marked with asterisk also represented in Miocene "clay shale," overlying sandstone referred to Vaqueros formation as exposed in Los Sauces Canyon, Ventura County.

- *Amphistegina* cf. *vulgaris* d'Orbigny (Hughes):
- *Amphistegina* cf. *vulgaris* d'Orbigny
- *Bolivina* cf. sp. from Los Sauces Canyon
- *Bolivina*
- *Cassidulina* cf. *lævigata* d'Orbigny
- **Cibicides americanus* (Cushman)
- *Cibicides* cf. *lobatus* (d'Orbigny)
- *Cibicides* cf. *floridanus* (Cushman)
- *Cibicides* cf. *conoideus* Galloway and Wissler
- *Discorbis*
- *Elphidium* cf. *crispum* (Linné)
- *Elphidium* cf. *granulosum* Galloway and Wissler
- **Eponides*
- *Globigerina bulloides* d'Orbigny
- *Globigerina*
- **Nonion* ?
- *Planulina ariminensis* d'Orbigny
- *Quinqueloculina* cf. *seminulum* (Linné)
- *Quinqueloculina*
- *Robulus americanus spinosus* (Cushman)
- *Textularia*
- **Uvigerina*

Gastropods:
- *Acmæa*
- *Calliostoma*
- *Naticoid*
- *Turritella imperialis* Hanna
- *Serpulorbis*
- *Nassarius?*
- *Conus*
- *Bulla*

Scaphopod:
- *Dentalium*

Lamellibranchs:
- *"Arca"*
- *Barbatia* or *Fossularca*
- *Glycymeris*
- *Altrina*
- *Ostrea*
- *"Pecten" purpuratus mendenhalli* Arnold?
- *"Pecten"* 2 spp.
- *Spondylus*
- *Lima?*
- *"Modiolus"?*
- *"Crassinella"*
- *Chama*
- *Trigoniocardia*
- *"Transennella"*
- *Antigona*
- *Solecurtus*
- *Corbula*

These collections are particularly interesting for the large number of foraminifera and for the small genera of mollusks that are not represented in any collection from Carrizo Mountain. All the foraminifera listed are from one locality (194).

SAN GORGONIO PASS

(Locality 6, fig. 1)

F. E. Vaughan[15] announced in 1918 the discovery of marine beds along the north side of eastern San Gorgonio Pass. The few fossils that were found showed that these beds are of the same age as those around Carrizo Mountain.

[14] F. E. Vaughan, *Geology of San Bernardino Mountains North of San Gorgonio Pass*, Univ. Calif. Pub. Bull. Dept. Geol. Sci., vol. 13, No. 9, pp. 386-387, 1922.

[15] F. E. Vaughan, *Evidence in San Gorgonio Pass, Riverside County, of a late Pliocene Extension of the Gulf of California* (abstract), Bull. Geol. Soc. Amer., vol. 20, pp. 164-165, 1918.

In his report on the geology of the eastern San Bernardino Mountains, Vaughan[16] named the fossiliferous bed the Lion sandstone. I visited with Bramkamp the branch of Lion Canyon described by Vaughan. We found no fossils in the sandstone that we took to be his Lion sandstone, but the following were collected in a siltstone cropping out on the left bank of the canyon about 100 feet below the outcrop of the sandstone. These beds dip steeply northward, and unless the section is overturned the siltstone underlies the sandstone.

Preliminary list of fossils from branch of Lion Canyon
(Calif. Inst. Tech. Loc. No. 193)

Foraminifera:	Lamellibranchs:
Amphistegina	*Anadara*
Quinqueloculina	*Ostrea*
Coral:	*"Pecten"*
Solenastrea	*"Callocardia"*
Gastropod:	*"Transennella"*
Turritella imperialis Hanna	*Tellina*
Anachis?	*Macoma*
Plant:	*Semele?*
Leaf imprint	

It should be noted that a piece of coral of reef facies and a leaf imprint were found in this siltstone.

No marine beds have been found farther west in San Gorgonio Pass, but the following fossils, collected by R. B. Moran in section 2, T. 3 S., R. 2 E., show that the fossiliferous siltstone crops out east of Lion Canyon:

Preliminary list of fossils from siltstone in sec. 2, T. 3 S., R. 2 E.
(Calif. Inst. Tech. Loc. No. 197)

Gastropods:	*"Callocardia"*
Callistoma	*"Transennella"*
Turritella imperialis Hanna	*Tellina*
Lamellibranchs:	*Semela?*
Anadara	*Tagelus*
Pecten	*Corbala*
"Pecten"	Echinoid:
"Protocardia"	*Clypeaster* cf. *deserti* Kew

SANTA ROSALIA, LOWER CALIFORNIA
(LOCALITY 7, FIG. 1)

Fossils resembling those from the Colorado Desert are recorded by Arnold[17] from Santa Rosalia, a mining district lying on the Gulf side of Lower California about halfway down the Peninsula. According to a recent account of the geology of the Santa Rosalia region, marine fossils are found in pyroclastics and other sediments overlying a lava described as an augite andesite.[18] The few fossils listed in this report (determinations by Grant) are not strongly suggestive of the Colorado Desert fossils and apparently are younger.

Three Carrizo Mountains Pectens (*Pecten carrizoensis* Arnold, *"Pecten" deserti* Conrad, and *"Pecten" mendenhalli* Arnold) are recorded from Santa Antonita Point, south of Santa Rosalia, from beds considered of upper Pliocene age.[19] These three species are closely allied to living ones.

AGE

The marine beds of the Colorado Desert have been called "doubtfully Cretaceous," Miocene and Pliocene. The earliest opinion by T. W. Vaughan[20] that the corals discovered by Fairbanks indicated a doubtful Cretaceous age was soon abandoned. Arnold[21] first called them lower Miocene, or simply Miocene, but later,[22] on considering that one of the fossils, *"Pecten" deserti*, was represented in the Etchegoin formation of the San Joaquin Valley, referred them to the upper Miocene. Inasmuch as the Etchegoin formation is now considered of Pliocene age, this was the first assignment to the Pliocene. According to Kew,[23] the echinoids indicate a "comparatively late age," later[24] called Pliocene. Vaughan's[25] final opinion on the corals was that "lower Pliocene appears to be the maximum age which may be assigned to the fauna." After examining the mollusks, Hanna[26] concurred in regarding the beds as Pliocene and even considered the greater part of the series as middle and upper Pliocene. At a late date, however Dickerson[27] again called attention to the Caribbean Miocene affinities of the fauna.

16 F. E. Vaughan, Univ. Calif. Pub. Bull. Dept. Geol. Sci., vol. 13, No. 9, pp. 375-376, 1922.

17 Ralph Arnold, *The Tertiary and Quaternary Pectens of California*, U. S. Geol. Survey Prof. Paper 47. pp. 84-85, 1906.

18 M. E. Touwaide, *Origin of the Boleo Copper Deposit, Lower California, Mexico*, Econ. Geol., vol. 25, No. 2. pp. 113-114, 1930.

19 G. D. Hanna and L. G. Hertlein, Proc. Calif. Acad. Sci., 4th ser., vol. 16, No. 6, pp. 144-145, 1927.

20 T. W. Vaughan, U. S. Geol. Survey Mon. 39, p. 19, 1900.

21 Ralph Arnold, Science n. s., vol. 19, p. 503, 1904; U. S. Geol. Survey Prof. Paper 47, p. 21, 1906.

22 Ralph Arnold, U. S. Geol. Survey Bull. 396, p. 44, 1909. Ralph Arnold and Robert Anderson, U. S. Geol. Survey Bull. 398, p. 139, 1910.

23 W. S. W. Kew, Univ. Calif. Pub. Bull. Dept. Geol., vol. 8, No. 5, p. 46, 1914.

24 W. S. W. Kew, Univ. Calif. Pub. Bull. Dept. Geol., vol. 12, No. 2, pp. 32, 33, 56, 59, 60, 61, 137, 1920.

25 T. W. Vaughan. U. S. Geol. Survey Prof. Paper 98. p. 368, 1917.

26 G. D. Hanna, *Paleontology of Coyote Mountain, Imperial County, California*, Proc. Calif. Acad. Sci., 4th ser., vol. 14, No. 18, p. 434, 1926.

27 R. E. Dickerson, *Mollusca of the Carrizo Creek beds and their Caribbean Affinities* (abstract), Bull. Geol. Soc. Amer., vol. 29, p. 148, 1918.

EVIDENCE FROM THE HISTORY OF VULCANISM

The volcanic rock of Carrizo Mountain is an olivine basalt, according to G. H. Anderson, of the California Institute of Technology, who kindly examined a specimen from Alverson Canyon. His description of the rock is as follows:

"The rock is dark reddish gray and carries phenocrysts of altered olivine in a fine-grained holocrystalline groundmass composed essentially of laths of plagioclase and abundant prisms of augite. Extinction angles that could be measured indicate that the plagioclase is labradorite. The augite occurs principally in very pale green to colorless prisms that have almost no pleochroism, and also as irregularly shaped grains. Numerous minute grains of magnetite or ilmenite are scattered through the specimen. A few grains of titanite and a little zircon were recognized. Biotite is absent or very rare.

"The phenocrysts originally were olivine, but are greatly altered. They are mostly in euhedral crystals of varying size. On the margins and along cleavage planes the original mineral has been changed to iron oxides. In the interior, alteration has produced a fibrous, almost colorless weakly pleochroic mineral of moderate birefringence, probably closely allied to chrysotilite. Hornblende and chlorite appear less frequently as alteration products.

"In contrast to the extremely altered condition of the phenocrysts, the groundmass presents a relatively fresh appearance. Within the groundmass the ferromagnesian minerals appear to be in somewhat greater amount than the feldspar, and they are in considerable excess, taking into account the phenocrysts."

Both Mendenhall and Kew have described the occurrence of sediments under the lava. At most localities examined these beds appear to be land-laid. They consist of reddish and greenish sandstone, conglomerate and tuff. On the southeast slope of Carrizo Mountain, particularly in the area drained by the first big canyon southwest of Painted Gorge, into which a tourist-frequented road runs, the basal marine beds are interbedded with the basalt. In this area three or four beds of sandstone carrying marine fossils lie between and interfinger with basalt. The upper surface of each sandstone bed is altered to a quartzite that is as hard and as ancient-looking as the quartzite in the basement complex. The best collection of fossils so far obtained came from the lower unaltered part of one of these beds. Some of the quartzites form great dip slopes, one of which is a conspicuous feature as Carrizo Mountain is passed on the San Diego highway. It is so dark and rough-looking that I had assumed that it was a lava flow. No evidence was seen indicating that the basalt flowed into the sea, though no extensive search was made to find such evidence. At other places nearby, the basal marine bed lies on an eroded surface of lava and is made up principally or solely of material derived from it. This relation and the interbedding of lava and marine beds indicate that the region was close to sea-level during the period of vulcanism, and the lava-covered sandstones may have been above sea-level when the lava flowed over them.

At Santa Rosalia marine beds, composed largely of tuffaceous material and probably of the same age as those at Carrizo Creek, overlie a lava described as an augite andesite.

The lava of Carrizo Mountain apparently represents the period of vulcanism that began in lower Miocene time, the effects of which are wide-spread in southern California and adjoining regions. If this vulcanism were limited to a short interval of geologic time it would afford a means in establishing synchronization more precise than a good many paleontologic zones, but unfortunately it is not. According to published accounts, the lavas in the Mohave Desert,[28] at the south end of the San Joaquin Valley,[29] in the Cuyama Valley,[30] and farther south in Southern California,[31] range in age from lower Miocene to Pliocene or even Quaternary. Evidence pointing to a much earlier period of vulcanism during Martinez (Paleocene) time in Southern California has recently been found in the Elsinore Valley by J. C. Sutherland, of the California Institute of Technology. The stratigraphic succession at Carrizo Mountain more closely parallels the sequence at the south end of the San Joaquin Valley than in any other area in Southern California where lavas and marine beds are found in the same section, as is shown in the following summary:

[28] C. L. Baker, *Notes on the Later Cenozoic History of the Mohave Desert Region in Southeastern California*, Univ. Calif. Pub. Bull. Dept. Geol., vol. 6, No. 15, pp. 333–383, 1911.

C. D. Hulin, *Geology and Ore Deposits of the Randsburg Quadrangle, California*, Calif. State Min. Bur. Bull. 95, pp. 42–60, 1925.

[29] R. W. Pack, *The Sunset-Midway Oil Field, California*, U. S. Geol. Survey Prof. Paper 116, pp. 258–265, 1930.

H. W. Hoots, *Geology and Oil Resources along the Southern Border of San Joaquin Valley, California*, U. S. Geol. Survey Bull. 812, pp. 258–265, 1930.

[30] W. A. English, *Geology and Oil Prospects of Cuyama Valley, California*, U. S. Geol. Survey Bull. 621, p. 205, 1916.

[31] W. S. W. Kew, *Geology and Oil Resources of a Part of Los Angeles and Ventura Counties, California*, U. S. Geol. Survey Bull. 753, pp. 91–93, 1924.

W. A. English, *Geology and Oil Resources of the Puente Hills Region, Southern California*, U. S. Geol. Survey Bull. 768, pp. 46–47, 1926.

A. O. Woodford, *The San Onofre Breccia, its Nature and Origin*, Univ. Calif. Pub. Bull. Dept. Geol. Sci., vol. 15, No. 7, pp. 178–179, 1925.

Comparison of stratigraphic succession at Carrizo Mountain and south end of San Joaquin Valley

Carrizo Mountain	*South end of San Joaquin Valley*
Marine beds of Imperial formation overlying or interbedded with basalt.	Marine beds referred to Vaqueros formation overlying or interbedded with basalt.
Basalt.	Andesite followed by basalt, or basalt only.
Land-laid conglomerate, sandstone, and tuff.	Land-laid Tecuja beds of Stock.[32]

The temptation to generalize from this parallelism is inviting, but the probability that the parallelism itself has any age significance is very remote. About the best that can be said for the lava is that on the assumption that it represents the most wide-spread period of vulcanism in Southern California it fixes the lower age limit for the Imperial formation as lower Miocene; it might represent any part of the Miocene, or Pliocene, or even Quaternary. The basaltic composition perhaps favors a time later than the very beginning of the period of vulcanism, as the more acid lavas of this Miocene-Quaternary group of volcanics generally are earlier than the basic ones.

DIASTROPHIC EVIDENCE

Even in a region so unstable as California some time significance may be attributed to extensive floodings by the sea in different regions. Aside from the partial flooding of the San Joaquin Valley, the Pliocene seas were limited to relatively small coastal basins. The lower Miocene Vaqueros sea represents a pronounced transgression following a long period of emergence. This feature is clearly brought out in the forthcoming study of the Vaqueros formation by Arnold, Loel and Corey, in which they are assembling in the most painstaking manner everything that is known about the formation and its fossils. At many places the Vaqueros lies on or begins with red beds, which represent material accumulated during the period of emergence or reworked from it. This long period of emergence accounts for the strikingly meager record of marine Oligocene deposits in California. Part of what has been called Oligocene is Eocene,[33] and the San Ramon formation of B. L. Clark with its modernized fauna looks suspiciously like Miocene.[34] This line of evidence lends further support to the possible age significance of the parallelism between the sections at Carrizo Mountain and the south end of the San Joaquin Valley, but is by no means conclusive.

The middle Miocene transgression was even more extensive than the Vaqueros one, but only a short interval of emergence intervened and at places there was none. Extensive land-laid beds at the base of the middle Miocene Temblor formation are rare, except at the localities where virtually all the underlying Vaqueros consists of nonmarine deposits. Likewise the upper Miocene sea covered a still greater area but land-laid beds are absent beneath the deposits laid down in it, except at the margin of a basin, as in the Cuyama Valley.

Diastrophic grounds favor a Miocene age rather than a Pliocene, and if the long period of emergence attested to by the absence of any earlier marine Tertiary beds and by the presence of land-laid deposits below the lavas corresponds to the longest general period of Tertiary emergence in the Coast Ranges, it points to a lower Miocene age.

PALEONTOLOGIC EVIDENCE

No matter what period of geologic time is represented by these deposits, the fauna that they carry is unlike any in the Coast Ranges, for it represents a different faunal region. This feature is responsible in large measure for the widely varying views that have been held as to the age, and indicates that there was no connection between the head of the embayment and the Coast Ranges. Nevertheless it should be possible to find some phyla that are represented both in the Colorado Desert embayment and in the Coast Ranges or on the outer side of Lower California, just as some modern phyla are found in both regions. If the marine beds of the desert are Pliocene, at least a few closely allied species should be found in one or more of the Pliocene faunas in southern California, and Lower California. No close affinities are apparent in the lower Pliocene fauna found in Elsmere Canyon near Newhall. Aside from the record of *Clypeaster deserti* Kew and *"Pecten" mendenhalli* Arnold at Cedros Island off Lower California,[35] the middle or upper Pliocene

32 Chester Stock, *An Early Tertiary Vertebrate Fauna from the Southern Coast Ranges of California*, Univ. Calif. Pub. Bull. Dept. Geol., vol. 12, No. 4, pp. 267–276, 1920.

33 W. P. Woodring, *Upper Eocene Orbitoid Foraminifera from the Western Santa Ynez Range, California, and their Stratigraphic Significance*, Trans. San Diego Soc. Nat. Hist., vol. 6, No. 4, pp. 145-170, 1930.

34 The modernization of the marine Miocene fauna is clearly brought out by Dr. A. Morley Davies in his recent presidential address before the Geologists' Association. (*Faunal Migrations since the Cretaceous Period*, Proc. Geol. Assoc., vol. 40 [1929], pp. 307–327, 1930.)

The Miocene affinities of the San Ramon fauna have been discussed by Stewart in a report that was issued after this paper was written. (R. B. Stewart, *Gabb's California Cretaceous and Tertiary Type Lamellibranchs*, Acad. Nat. Sci. Phila., Special Pub. No. 3, pp. 19–20, 1930.)

35 E. K. Jordan and L. G. Hertlein, Proc. Calif. Acad. Sci., 4th ser., vol. 15, No. 14, pp. 416, 419, 1926.

An undetermined coral of the genus *Solenastrea* is recorded from the Pliocene of Maria Madre Island (E. K. Jordan and L. G. Hertlein, Proc. Calif. Acad. Sci. 4th ser., vol. 15, No. 4, p. 211, 1926).

San Diego fauna, which is found as a distinct and readily recognized warm-water faunal zone from Lower California to the Ventura Basin,[36] also holds no close resemblances. Both Carrizo Mountain species recorded from Cedros Island closely resemble living ones and it is doubtful whether much weight can be attached to them unless no other close affinities can be found. It is hardly necessary to consider any younger deposits, for I think no one would seriously consider that the marine beds of the Colorado Desert are younger than the San Diego formation. Nor is it necessary to consider the Jacalitos or Etchegoin formations of the San Joaquin Valley. The evidence on which Arnold concluded that the Carrizo Mountain fossils bore some resemblance to those of the Etchegoin formation is too slender and has for the most part already been disposed of.

If no marked Pliocene affinities are apparent, the question arises as to whether any Coast Range Miocene affinities can be recognized. Arnold suggested a lead when he listed *"Turritella* (cf.) *hoffmanni* Gabb" from Carrizo Mountain.[37] It had never occurred to me that this *Turritella*, named *T. imperialis* by Hanna, was allied to *T. inczana* Conrad, an earlier name for the species Gabb called *T. hoffmanni*, until W. H. Corey, of the University of California, orally pointed out its striking resemblance to "frilled" forms of *T. inezana* that he is describing in collaboration with W. F. Loel in their monograph of the Vaqueros fauna. If this resemblance is due to actual genetic affinity instead of to homeomorphy, other Vaqueros resemblances should be apparent. Another can be seen in the resemblance of a large undescribed *"Pecten"* to *"Pecten" magnolia* Conrad, another characteristic Vaqueros fossil, though the resemblance is not so striking as in the case of the *Turritella*. Another is apparent in the resemblance on a small scale of *Ostrea loeli* Hertlein to *O. heermanni* Conrad. This is a short list, but it embraces two extinct phyla and it is very suggestive in the absence of any comparable Pliocene affinities.

Nothing is strongly suggestive of the middle Miocene fossils of the Coast Ranges. So long as the genus *Turritella* is represented, a middle Miocene Gulf of California fauna would be expected to include the phylum of *Turritella ocoyana*, which is found in Lower California and southward into the tropics on both the Pacific and Atlantic sides.[38] Though such negative evidence is untrustworthy, some Coast Range middle Miocene stocks should be represented.

Likewise nothing definitely suggests Coast Range upper Miocene. The Santa Margarita species *"Pecten" sanctiludovici* Anderson and Martin, recorded by Hanna from Carrizo Mountain, represents a wide-spread group of Pectens, the species of which may not be intimately allied. *"Pecten" mediacostatus* Hanna is described as resembling *"Pecten" estrellanus* Conrad, which has a considerable range through the Miocene and into the Pliocene. The Vaqueros and Temblor species *"Pecten" miguelensis* Arnold belongs to this group of Pectens.

I have always been impressed with the Miocene aspect of this fauna, but I have been unwilling to entirely trust my judgment for fear it might be somewhat distorted by long familiarity with Caribbean Miocene faunas. Caribbean Miocene affinities are unmistakable in the fossils from Carrizo Mountain, as Dickerson and others have pointed out, but these affinities alone are not conclusive evidence that the fauna is Miocene, for such affinities also are unmistakable in the present fauna of the Gulf of California and the coast of Mexico and Central America farther south. The Mazatlantic and Panamic regions harbor the end members of many Caribbean Miocene phyla. Though this feature must be kept in mind, certain phyla that are extinct along the Pacific Coast deserve special consideration. *Turritella imperialis* represents one of them. It closely resembles, as everyone has recognized, *T. altilira* Conrad, of the middle Miocene Gatun formation of Panama and Costa Rica, and the closely allied forms found in the Miocene of the Caribbean region, for which far too many names have been proposed. This phylum is wide-spread in Caribbean lower Miocene deposits (Anguilla, Haiti, Dominican Republic, Costa Rica, Panama, Colombia, Venezuela, Trinidad) and during middle Miocene time its distribution was even more extensive (Dominican Republic, Jamaica, Costa Rica, Panama, Colombia, Venezuela, Trinidad, Peru, Pacific coast of Panama). A species in the upper Miocene Cerros de Sal formation of the Dominican Republic is the last representative in the tropics, but a Pliocene species from Florida, *T. perattenuata* Heilprin, is regarded as the final member. If these Caribbean Turritellas and *T. inezama* are derived from the same stock, as Corey believes and as seems reasonable, it may be significant that of all the Caribbean representatives *T. crocus* Cooke,[39] from the

[36] W. P. Woodring, *Miocene Deposits North of Simi Valley, California*, Proc. Calif. Acad. Sci., 4th ser., vol. 19, No. 6, pp. 57–64, 1930.

[37] Ralph Arnold, U. S. Geol. Survey Prof. Paper, 47, p. 22, 1906.

[38] W. P. Woodring, *Miocene Mollusks from Bowden, Jamaica, Pt. 2: Gastropods and Discussion of Results*, Carnegie Inst. Wash. Pub. No. 385, pp. 96, 98, 1928. *Distribution in tropical America of Turritellas of the Phylum of Turritella ocoyana* (abstract), Bull. Geol. Soc. Amer., vol. 40, No. 1, pp. 256–257, 1929.

[39] C. W. Cooke, Carnegie Inst. Wash. Pub. No. 291, p. 121, pl. 4, fig. 8. 1919.

Anguilla formation of the island of Anguilla, is most like *T. inezana s. s.* If this evidence furnished by one phylum means anything, it is a clear indication as to what part of the lower Miocene is represented by the Vaqueros formation. The Anguilla formation embraces the oldest Miocene zone now recognized in the Caribbean region and is considered of Aquitanian age.

A Carrizo Mountain cone, called *Conus regularis* Sowerby by Hanna, closely resembles *C. proteus* Hwass, living West Indian species that is found as far back as the middle Miocene, and doubtfully in the lower Miocene. Like the Caribbean Miocene specimens those from Carrizo Mountain retain, with unusual perfection, the color pattern of many spiral rows of small, rectangular, brown, blotches.[40] Mr. E. G. Vanatta, of the Philadelphia Academy of Natural Sciences, kindly confirmed my impression that no cone closely resembling *C. proteus* is recorded as living along the Pacific Coast of America. This fossil species is more similar to *C. proteus* than it is to *C. regularis* Sowerby, *C. scalaris* Valenciennes, and *C. gradatus* Mawe, all of which are living Pacific Coast species. Mr. Vanatta reminded me that Dall[41] had specimens of *C. proteus* from the Pacific Coast of Panama, but that he considered the locality record erroneous. I remember those specimens in the West Coast collections of the United States National Museum labeled "Panama?." The possibility of *proteus* living in the Panamic region can not be lightly dismissed, for any living species that is found as far back as Miocene time is likely to be established now on both sides of Central America.

The most striking example of Caribbean affinities, involving the extinction on the Pacific Coast of a genus, is afforded by *Cassis*. This genus has a long history in West Indian and Floridian waters, extending from middle Oligocene time to the present. The Carrizo Mountain species, *C. subtuberosa* Hanna, is the only one, recent or fossil, from the entire Pacific Coast of America. It represents a clear ease of a migrant from the West Indies that gained a temporary foothold in the eastern Pacific.

In addition to these few species, which represent extinct, or apparently extinct, phyla, so far as Pacific waters are concerned, many other Carrizo Mountain fossils are similar to Miocene species from the Caribbean region, including several genera and groups of species that are extinct in the Caribbean Sea (*Malea, Solenosteira,* group of *Cancellaria obesa* Sowerby, group of *Conus fergusoni* Sowerby, group of *Ficus decussata* (Wood)). In fact, no species, so far known, in the Imperial formation would be out of place in a Caribbean Miocene fauna.

If the Caribbean Miocene affinities mean free faunal migration between the two regions—and they clearly do—when did the migration take place? It is well known from areal geology and from the distribution of sea and land animals, that the sea extended across parts of Central America during lower and middle Miocene time, but there is no clear evidence for later interoceanic communication. The extinct phyla discussed and many others represented in the Imperial formation are known to have been living in the Caribbean region when the seas were in communication. It does not necessarily follow, that the beds of the Colorado Desert are Miocene, for these animals might have migrated from the Caribbean Sea into the Pacific, or in the opposite direction, during Miocene time and still survive until Pliocene time in the Gulf of California. According to my experience, however, many marine migrants lose their foothold in the region to which they have migrated when communication with the area from which they have come is broken, not because a constant stream of newcomers is required to keep up the stock, but because the closing of such communications has a profound effect on oceanic circulation, which represents a delicate adjustment of an intricate series of interrelated factors. This disarrangement of oceanic circulation and the resulting upsetting of oceanographic equilibria seems adequate to account for the extinction of animals that failed to become adjusted to the change. Why some animals should succeed and others fail is a question that is shrouded in mystery. This clearly is the sequence of events in the wholesale extinction of genera among many kinds of marine invertebrates in the Caribbean Sea toward the close of Miocene time, an extinction for which I have suggested that one factor alone—temperature—may be responsible,[42] but that solution is inadequate and far too simple. If this line of reasoning is accepted, it is improbable that the extinct Caribbean migrants survived until Pliocene time. If the fauna is Miocene, the percentage of mollusks referred by Hanna, to species living in the Gulf of California (65 per cent), is much too high.[43] Perhaps this figure is not to be taken too seriously, for only a small part of the fauna then living is represented and much of

[40] W. P. Woodring, Carnegie Inst. Wash. Pub. No. 385, pp. 204–205, 1928.

[41] W. H. Dall, *Summary of the Shells of the Genus Conus from the Pacific Coast of America in the U. S. National Museum*, Proc. U. S. Nat. Mus., vol. 38, p. 227, 1910.

[42] W. P. Woodring, Carnegie Inst. Wash. Pub. No. 385, pp. 28-29, 1928.

Stewart has recently emphasized the inadequacy of the temperature factor. (R. B. Stewart, Acad. Nat. Sci. Phila., Special Pub. No. 3, p. 24, 1930.)

[43] For figures covering a number of Miocene faunas see Woodring, Carnegie Inst. Wash. Pub. No. 385, pp. 106-107, 1928.

the material studied is poorly preserved. The small genera from Painted Hill now being examined by Bramkamp may considerably change the percentage.

T. W. Vaughan has shown that the corals have even more unmistakable Caribbean affinities. All six genera represented in the Imperial formation are now living in the Caribbean Sea and Gulf of Mexico, but only one (*Porites*) is living in the eastern Pacific. On account of the resemblance to Pliocene and Recent Floridian and West Indian species and on account of the absence of West Indian Miocene genera now living in Indio-Pacific waters (*Stylophora, Pocillopora* and *Gonlopora*) Vaughan concluded that the corals could not very well be older than Pliocene and that they indicated a late Tertiary communication with the Gulf of Mexico after the extinction there of the Miocene Indo-Pacific genera. Perhaps the evidence can be analyzed further, particularly in view of later information derived from exploration of the rich Miocene coral reefs of Haiti and the Dominican Republic. These data are summarized in the table on page 23 [below].

It will be seen from this tabulation, that not only all the genera represented in the Imperial formation were living in the West Indies during lower or middle Miocene time but that at least five of the eight species with which the Imperial corals are closely allied were living then. These new records effectively break down the marked Pliocene affinities of the Imperial coral fauna. They do not in themselves demonstrate that the fauna is of Miocene age, though they support such a view. If the same line of reasoning is applied to the corals as to the mollusks, the probability is that, they are not younger than Miocene. If they are Miocene, it seems strange that *Stylophora, Pocillopora, Asterosmilia, Antillia* and *Gonipora* are not represented, though their absence is not conclusive evidence that the age is not Miocene.

According to Kew, four of the five species of echinoids closely resemble living species from the Gulf of California, whereas the fifth, *"Clypeaster" carrizoensis* Kew has a doubtful resemblance to *"Scutella" gabbli* (Rémond), from the lower part of the San Pablo series of B. L. Clark. Hanna has added a sixth doubtfully determined species, *Metalia spatagus* (Linné)?, which is living in the Gulf. If the Imperial echinoids show that the beds are Pliocene, as Kew maintained, the Gatun echinoids show that the Gatun formation is Pliocene, though a middle Miocene age for the Gatun formation is accepted by all the modern workers, both American and European. Aside from a *Schizaster*, the Gatun echinoids, representing the genera *Clypeaster* and

Corals from Imperial formation and most closely allied species

Species from Imperial formation	Most closely allied species, all from Florida or West Indies (Vaughan 1917,[44] 1919[45])	Earliest appearance in West Indies of most closely allied species or of a similar one[46]
Eusmilia carrizensis Vaughan	*E. fastigiata* (Pallas), Plioc., Rec.	*E.* sp., undetermined part of Yaque group (lower or middle Miocene), Dominican Republic.
Dichocœnia merriami (Vaughan) var. *crassisepta* (Vaughan)	*D.* sp. Plioc.; D. stokesi Milne Edwards and Haime, Rec.	*D. stokesi*, Las Cahobas formation (early middle Miocene), Haiti.
Solenastrea fairbanksi (Vaughan) var. *columnaris* (Vaughan) var. *normalis* Vaughan var. *minor* Vaughan	*S. hyades* (Dana) and *S. bournoni* Milne Edwards and Haime, Plioc., Rec.	*S. hyades*, Las Cahobas formation (early middle Miocene), Haiti. *S. bournoni*, Thomonde formation (late lower Miocene), Haiti.
Mæandra bowersi Vaughan	*M. labyrinthiformis* (Linné) Plioc., Rec.	*M.* aff. *labyrinthiformis*, "probably Oligocene" (lower Miocene?), Dominican Republic.
Siderastrea mendenhalli Vaughan	*S. dalli* Vaughan. Plioc., and *S. siderea* (Ellis and Solander), Rec.	*S. siderea*, Thomonde formation (late lower Miocene), Haiti.
Siderastrea californica Vaughan	*S. pliocenica* Vaughan, Plioc., and *S. radians* (Pallas), Rec.	*S. radians*, Artibonite group (probably early middle Miocene part), Haiti.
Porites carrizensis Vaughan	*P. astreoides* Lamarck, Plioc., Rec.	*P.* aff. *astreoides*, lower or middle Miocene, Trinidad; Artibonite group (probably early middle Miocene part), Haiti.

44 T. W. Vaughan, U. S. Geol. Survey Prof. Paper 98, pp. 355–395, 1917.

45 T. W. Vaughan, U. S. Nat. Mus. Bull. 103, pp. 223–224, 1919.

46 Data after Vaughan in following reports: T. W. Vaughan and W. P. Woodring, Tertiary and Quaternary Stratigraphic Paleontology, Dom. Rep. Geol. Survey Mem., vol. I, pp. 89–168, 1921; W. P. Woodring, J. S. Brown, and W. S. Burbank. Geology of the Republic of Haiti, Rep. Haiti Geol. Survey, pp. 178–218, 1924; T. W. Vaughan and J. E. Hoffmeister, Miocene Corals from Trinidad, Carnegie Inst. Wash. Pub. No. 344, 105–134, 1926.

Encope, are very much species from the Imperial formation, as Jackson has pointed out.[47] It should also be noted that "*Clypeaster* aff. *deserti* Kew" is recorded from the Miocene of Lower California.[48]

The foraminifera furnish additional paleontological evidence that heretofore has not been considered. In addition to identifying the material from Painted Hill already listed, Mr. D. D. Hughes has kindly examined a sample collected near the base of the siltstone member of the Imperial formation on the north slope of Carrizo Mountain. This collection consists of the following species:

Preliminary list of foraminifera collected near base of siltstone member of Imperial formation
(Field No. 7-7-1; identified by D. D. Hughes)
Species marked with asterisk also represented in the Miocene "clay shale," overlying sandstone referred to Vaqueros formation, as exposed in Los Sauces Canyon, Ventura Co.

Buliminella	**Lagena*
Cibicides cf. *floridanus* (Cushman)	**Nonion* cf. *incisum* (Cushman)
Guadryina	*Planulina* cf. *ariminensis* d'Orbigny
**Guttulina*	**Tritaxia* cf. *pyramidata* Reuss
Haplophragmoides?	*Textularia*
Lagena cf. *melo* (d'Orbigny)	**Valvulineria*

Another sample, from a horizon lying several hundred feet above the base of the siltstone member, carries only one species, *Elphidium* cf. *hughesi* Cushman and Grant.

According to Mr. Hughes, this fauna, like the one from Painted Hill, is different from any in the coastal region, but several of the species, or a number of similar ones, are found in the "clay shale" that overlies sandstone referred to the Vaqueros formation, as exposed in Los Sauces Canyon, Ventura County. The "clay shale" is considered of late lower or early middle Miocene age. The *crispum*-like and *granulosum*-like species of *Elphidium* give the fossils a late Pliocene or Pleistocene appearance in terms of the succession of foraminiferal faunas in the Los Angeles and Ventura Basins, but such species are found as far back as the lower Miocene in the West Indies and Florida. Likewise the *Elphidium*, similar to *E. hughesi*, may indicate brackish water, rather than the Pliocene age that is generally associated with it. Of the described Caribbean and Floridian fossil foraminifera, Mr. Hughes considers that those from the upper Miocene Choctawhatchee formation of Florida show the closest affinity to the fossils from the Colorado Desert. It should be pointed out, however, that the foraminifera of a great many West Indian Miocene deposits have not yet been described.

[47] R. T. Jackson, *Fossil Echini of the Panama Canal Zone and Costa Rica*, Proc. U. S. Nat. Mus., vol. 53, 489–501, 1917; U.S. Nat. Mus. Bull. 103. pp. 103–116, 1919.

[48] L. G. Hertlein and E. K. Jordan, *Paleontology of the Miocene of Lower California*, Proc. Calif. Acad. Sci., 4th Ser., vol. 16, No. 19, pp. 607, 611, 1927.

CONCLUSIONS AS TO AGE

The paleontological evidence points to the conclusion that the marine beds of the Colorado Desert are of Miocene age, and the other two lines of evidence passively support this conclusion. There is little on which to base a conclusion as to what part of the coastal Miocene section is represented, but the evidence that is now apparent indicates late Vaqueros age; that is, late lower Miocene. According to this view, if vertebrate remains are found in the nonmarine beds underlying the basalt, they would correspond in age to those from the Tecuja beds of Stock, at the south end of the San Joaquin Valley. If they are found in the Palm Spring formation, they would be of Temblor or Barstow age. The "Indio formation" and "Coachella fanglomerate" occupy the same stratigraphic position with reference to the marine beds as the Palm Spring formation, but as both are closer to the head of the trough, their base may lie at a lower horizon.

This region is an interesting one, not only on account of its fascinating marine fauna, but also because it offers the opportunity to establish a time scale at a place where marine beds and land-laid deposits are found in superposition. Such places are all too rare in California.

(Page numbers in parentheses)

Geologic map of the Coyote Mountains and surrounding area

Excerpted and modified from Figure 2 of Dorsey et al. (2011). EF—Elsinore Fault.

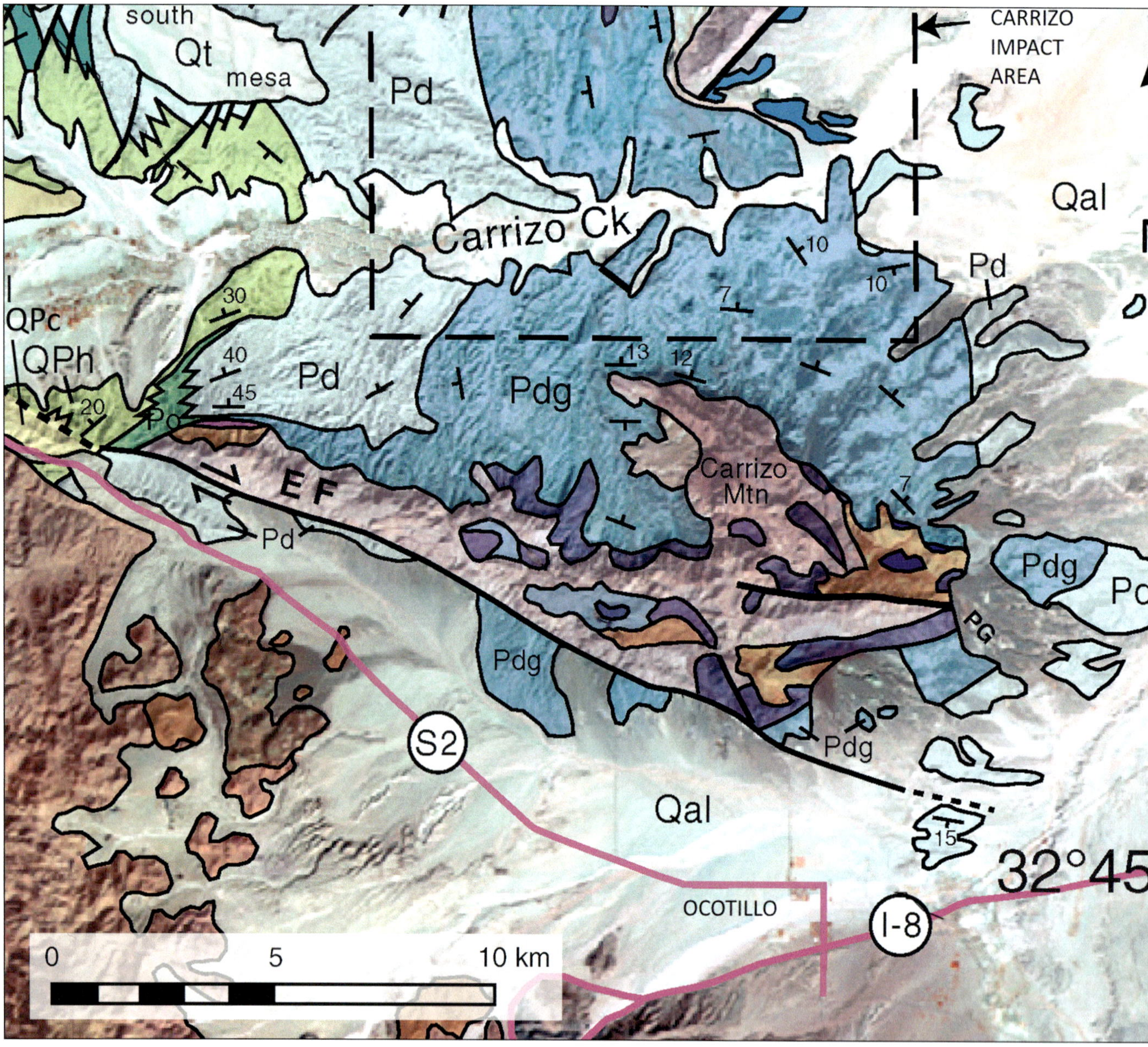

EXPLANATION

Pleistocene

- Qm - "Mesa" gravels

Plio-Pleistocene

- QPh - Hueso Formation
- Pt - Tapiado Formation
- QPc - Canebrake Congl.

Pliocene

- Po - Olla Formation
- Pd - A. Diablo Formation

PALM SPRING GROUP (non-marine): QPh, Pt, QPc, Po, Pd

- Pdg - Deguynos Fm
- PMl - Latrania Fm

IMPERIAL GROUP (marine): Pdg, PMl

Miocene

- Ms - SPLIT MT. GROUP

Cretaceous and pre-K crystalline rocks

- lower plate
- upper plate

REFERENCE:

Dorsey, R.J., Housen, B.A., Janecke, S.U., Fanning, C.M., and Spears, A.L.F., 2011, Stratigraphic rec of basin development within the San Andreas fault system: Late Cenozoic Fish Creek-Valle basin, southern California. Geological Society of America Bulletin published online 5 Janu 2011 as doi:10.1130/B30168.1

FROM ORIGINAL CAPTION:

Geologic map of the Fish Creek–Vallecito basin and surrounding area, modified from Winker (19 Winker and Kidwell (1996), Dibblee (1996), Axen and Fletcher (1998), Kairouz (2005), G. A (2008, personal commun.), and Steely (2006).

Geologic map of the Coyote Mountains

Explanation of mapped units (see inside front cover).
Excerpted from two Thomas W. Dibblee, Jr., maps (2008).

CENOZOIC — QUATERNARY (Holocene, Pleistocene) ? ? — TERTIARY (Pliocene, Miocene)

Qa | Qt

SURFICIAL SEDIMENTS: *Unindurated, undissected alluvial sediments; age, Recent*
Qa: Alluvial sand, gravel, silt, and clay of valleys areas
Qt: Terrace deposits, includes Qt1, younger terraces of gravel and sand, locally undifferentiated from Qa, and Qt2, older terraces of boulder to pebble gravel and sand, locally folded and faulted

UNCONFORMITY

To

OCOTILLO CONGLOMERATE (of Dibblee, 1954) Terrestrial clastic deposits, apparently derived from the Peninsular Range terrane during an episode of renewed uplift and erosion; age, Pleistocene or late Pliocene.

Locally called Ocotillo member of Toro Conglomerate, massive gray boulder to cobble comglomerate, grades basinward into bedded pebble-conglomerate and sandstone with occasional pink sandstones and clays.

Tps

PALM SPRING FORMATION (of Woodring, 1931): Stream-laid sediments accumulated as deltaic deposits draining from rising Peninsular Range terrane; age, middle Pliocene to mid-Pleistocene (Dibblee, 1954, and Woodard, 1961).

Light gray, green-gray to tan bedded arkosic sandstones and interbedded light red clays, with many sandstone strata containing calcareous concretions of various shapes; in places contains fragments of dark gray silicified iron wood (mesquite) with grain well preserved.

Ti

IMPERIAL FORMATION (of Woodring, 1931): Shallow brackish marine clastic, total thickness about 3600 feet; age, late Miocene to middle Pliocene.

Claystone, light gray-tan to yellow, conchoidally fractured, weathers to yellowish gray clay soil; contains interbeds of sandstone, buff to gray, laminated, friable to hard; commonly contains hard dark brown oyster reef fauna.

Tav

ALVERSON CANYON VOLCANICS: Sequence of mostly andesitic flows and terrestrial sediments; age, probably late Miocene (Dibblee, 1954).

Andesite, dark brown, basic, locally vesicular or amygdaloidal; locally includes minor interbeds of brown andesite agglomerate, gray tuff breccia, and non-marine sandstone and conglomerate.

Tsm

SPLIT MOUNTAIN CONGLOMERATE (of Tarbet and Holman, 1944): Non-marine sedimentary rocks, exposed in northwest Fish Creek Mountains and southeast foothills of Jacumba Mountains; rests on erosional surface cut into Paleozoic and Cretacous rocks; age, middle Miocene (Durham, 1954).

Granite- and gneiss-breccia, gray to brown, massive to bedded, boulder- to pebble-conglomerate and sandstone; contains andesite agglomerate and some basic andesite, and local minor red beds.

UNCONFORMITY

BASEMENT ROCKS

PALEOZOIC or older

gn-ms | ml gn | ms

METAMORPHIC ROCKS
Thick (up to 10,000 feet) complex of metamorphic rocks in Carrizo Peak and Fish Creek Mountain areas; age, Paleozoic or older.

gn/gn-ms: Gneiss or gneissoid granite, gray, medium grained, holocrystalline, with thin layers undulating or contorted.

ml: Gray to white limestone recrystallied to marble.

PERTINENT REFERENCES

Dibblee, T.W. Jr., 1954, Geology of the Imperial Valley Region, California, in: Jahns, R.H.,ed., Geology of southern California: California Division of Mines Bulletin 170, chpt. 2, cont. 2, p. 21–28, pl. 2, map scale 1:375,000.

Durham, J.W., 1954, The marine Cenozoic of southern California, in: Jahns, R.H., ed., Geology of southern California: California Division of Mines Bulletin 170, ch. 3, pt. 4, p. 23–31.

Tarbet, L.A., and Holman, W.H., 1944, Stratigraphy and micropaleontology of the west side of Imperial Valley, California: American Association of Petroleum Geologists Bulletin, v. 28, p. 1781–1782.

Woodard, G.D., 1961, Stratigraphic succession of the west Colorado Desert, San Diego and Imperial counties, southern California (abst.): Geological Society of America Bulletin, v. 72, no. 12, pt. 2 (Program, Cordilleran Section Meeting, p. 73–74).

Woodring, W.P., 1931, Distribution and age of the Tertiary deposits of the Colorado Desert: Carnegie Institute of Washington, Publication 148, p. 1–25.

Geologic map of the Coyote Mountains area

Explanation of mapped units (see back cover).

Excerpted and **modified from "Geologic map of Imperial County," Plate 1 of Morton (1977).**

Symbol	Unit	Description	Epoch	Period	Era
Qal	**ALLUVIUM**	Unconsolidated clay, silt, sand, and gravel occurring primarily as valley fill and streamwash deposits.	Holocene	QUATERNARY	CENOZOIC
Qc	**OLDER ALLUVIUM**	Partly dissected largely unconsolidated, poorly sorted silt, and gravel of alluvial fans, desert pavement areas, margins of larger canyons, and terraces.	Pleistocene	QUATERNARY	CENOZOIC
Pcc	**CANEBRAKE CONGLOMERATE**	Composed largely of granitic debris. Represents marginal facies of Palm Spring and Imperial Formations.	Pliocene	TERTIARY	CENOZOIC
Pcp	**PALM SPRING FORMATION**	Interbedded nonmarine, light gray, arkosic sandstone and reddish clay; grades into Canebrake Conglomerate towards margins of Salton basin.	Pliocene	TERTIARY	CENOZOIC
Pi	**IMPERIAL FORMATION**	Marine sequence of light yellow-gray clay, interbedded arkosic sandstone with oyster-shell reefs, and fossiliferous calcareous sandstone.	Pliocene	TERTIARY	CENOZOIC
Mvp / Mva	**ALVERSON ANDESITE**	Varicolored andesite flows and interbedded tuff, breccia, and nonmarine sediments. Mvp = Jacumba pyroclastic rocks	Miocene	TERTIARY	CENOZOIC
Mcsu / Mcs / Mcsl	**SPLIT MOUNTAIN FORMATION**	Mcsu = Upper member; gray conglomerate and brown arkosic sandstone Mcsl = Lower member; reddish-brown arkosic sandstone, conglomerate, and sedimentary breccia.	Miocene	TERTIARY	CENOZOIC
ℙmls / ℙm	**METAMORPHIC ROCKS**	Predominantly biotite schist, quartzite, gneiss, and marble of Paleozoic(?) age. ℙmls = marble			UNDIVIDED PRE-TERTIARY

SYMBOLS:

——— – – – – · · · · – ? – **CONTACT** (dashed where approximately located; dotted where concealed; queried where inferred)

——— – – – – · · · · – ? – **FAULT** (dashed where approximately located; dotted where concealed; queried where inferred)